THE OPTIMUM SHAPE
Automated Structural Design

General Motors Research Laboratories Symposia Series

1985 J. A. Bennett, M. E. Botkin, eds., *The optimum shape: Automated structural design*, Plenum Press, New York, 1986.
1984 L. Evans, R. C. Schwing, eds., *Human behavior and traffic safety*, Plenum Press, New York, 1985.
1983 M. S. Pickett, J. W. Boyse, eds., *Solid modeling by computers: From theory to applications*, Plenum Press, New York, 1984.
1981 R. Hickling, M. M. Kamal, eds., *Engine noise: Excitation, vibration and radiation*, Plenum Press, New York, 1982.
1980 G. T. Wolff, R. L. Klimisch, eds., *Particulate carbon: Atmospheric life cycle*, Plenum Press, New York, 1982.
1980 D. C. Siegla, G. W. Smith, eds., *Particulate carbon: Formation during combustion*, Plenum Press, New York, 1981.
1979 R. C. Schwing, W. A. Albers, Jr., eds., *Societal risk assessment: How safe is safe enough?* Plenum Press, New York, 1980.
1978 J. N. Mattavi, C. A. Amann, eds., *Combustion modeling in reciprocating engines*, Plenum Press, New York, 1980.
1978 G. G. Dodd, L. Rossol, eds., *Computer vision and sensor-based robots*, Plenum Press, New York, 1979.
1977 D. P. Koistinen, N.-M. Wang, eds., *Mechanics of sheet metal forming: Material behavior and deformation analysis*, Plenum Press, New York, 1978.
1976 G. Sovran, T. A. Morel, W. T. Mason, eds., *Aerodynamic drag mechanisms of bluff bodies and road vehicles*, Plenum Press, New York, 1978.
1975 J. M. Colucci, N. E. Gallopoulos, eds., *Future automotive fuels: Prospects, performance, perspective*, Plenum Press, New York, 1977.
1974 R. L. Klimisch, J. G. Larson, eds., *The catalytic chemistry of nitrogen oxides*, Plenum Press, New York, 1975.
1973 D. F. Hays, A. L. Browne, eds., *The physics of tire traction*, Plenum Press, New York, 1974.
1972 W. F. King, H. J. Mertz, eds., *Human impact response*, Plenum Press, New York, 1973.
1971 W. Cornelius, W. G. Agnew, eds., *Emissions from continuous combustion systems*, Plenum Press, New York, 1972.
1970 W. A. Albers, ed., *The physics of opto-electronic materials*, Plenum Press, New York, 1971.
1969 C. S. Tuesday, ed., *Chemical reactions in urban atmospheres*, American Elsevier, New York, 1971.
1968 E. L. Jacks, ed., *Associative information techniques*, American Elsevier, New York, 1971.
1967 P. Weiss, G. D. Cheever, eds., *Interface conversion for polymer coatings*, American Elsevier, New York, 1968.
1966 E. F. Weller, ed., *Ferroelectricity*, Elsevier, New York, 1967.
1965 G. Sovran, ed., *Fluid mechanics of internal flow*, Elsevier, New York, 1967.
1964 H. L. Garabedian, ed., *Approximation of functions*, Elsevier, New York, 1965.
1963 T. J. Hughel, ed., *Liquids: Structure, properties, solid interactions*, Elsevier, New York, 1965.
1962 R. Davies, ed., *Cavitation in real liquids*, Elsevier, New York, 1964.
1961 P. Weiss, ed., *Adhesion and cohesion*, Elsevier, New York, 1962.
1960 J. B. Bidwell, ed., *Rolling contact phenomena*, Elsevier, New York, 1962.
1959 R. C. Herman, ed., *Theory of traffic flow*, Elsevier, New York, 1961.
1958 G. M. Rassweiler, W. L. Grube, eds., *Internal stresses and fatigue in metal*, Elsevier, New York, 1959.
1957 R. Davies, ed., *Friction and wear*, Elsevier, New York, 1959.

THE OPTIMUM SHAPE
Automated Structural Design

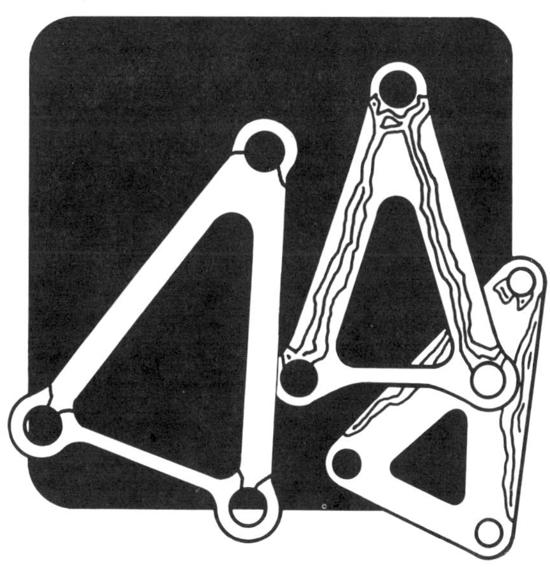

Edited by

J. A. BENNETT and M. E. BOTKIN

General Motors Research Laboratories

PLENUM PRESS • NEW YORK — LONDON • 1986

Library of Congress Cataloging in Publication Data

General Motors Symposium on the Optimum Shape: Automated Structural Design (1985: General Motors Research Laboratories)
The optimum shape.

(General Motors Research Laboratories symposia series)
Includes bibliographies and indexes.
1. Structural design–Data processing–Congresses. 2. Engineering design–Mathemtical models–Congresses. 3. Mathematical optimization–Congresses. I. Bennett, James A., 1942– . II. Botkin, Mark E. III. General Motors Corporation. Research Laboratories. IV. Title. V. Series.
TA658.G45 1985 620′.00425 86-21234
ISBN 0-306-42419-3

Proceedings of a General Motors Symposium on The Optimum Shape: Automated Structural Design, held September 30–October 1, 1985, at the General Motors Research Laboratories, Warren, Michigan

A Division of Plenum Publishing Corporation
233 Spring Street, New York, N.Y. 10013

Printed in the United States of America

PREFACE

This book contains the papers presented at the International Symposium, "The Optimum Shape: Automated Structural Design," held at the General Motors Research Laboratories on September 30—October 1, 1985. This was the 30th symposium in a series which the Research Laboratories began sponsoring in 1957. Each symposium has focused on a topic that is both under active study at the Research Laboratories and is also of interest to the larger technical community.

While attempts to produce a structure which performs a certain task with the minimum amount of resources probably predates recorded civilization, the idea of coupling formal optimization techniques with computer-based structural analysis techniques was first proposed in the early 1960s. Although it was recognized at this time that the most fundamental description of the problem would be in terms of the shape or contours of the structure, much of the early work described the problem in terms of structural sizing parameters instead of geometrical descriptions. Within the past few years, several research groups have started to explore this more fundamental area of shape design. Initial research has raised many new questions about appropriate selection of design variables, methods of calculating derivatives, and generation of the underlying analysis problem.

By 1985, it was apparent that sufficient progress had been made that a symposium devoted to assessing the state of the art and identifying new directions was appropriate. It was also clear that this symposium should include not just people who had worked in the traditional areas of structural optimization, but should also include workers in such diverse fields as geometric modeling, error analysis, adaptive analysis and finite element mesh generation.

The symposium was divided into four sessions: Derivatives and Algorithms, Analysis and Modeling for Shape Optimization, Applications, and New Frontiers in Shape Optimization. Following the formal presentation of each paper there was a discussion period, which was recorded and included in this book. At the end of the fourth session, Professor Lucien A. Schmit presented a summary of the topics covered in the symposium. This summary is also included in the book.

Many people played significant roles in planning and implementing this symposium. Our organizing committee, composed of Dean Richard H. Gallagher, Professor

Edward J. Haug, Professor Lucien A. Schmit, Professor Garret N. Vanderplaats and Professor Oleg C. Zienkiewicz, assisted us in identifying the key topics to be covered and the speakers to be included in the symposium. Professor Raphael T. Haftka, Professor Barna A. Szabo, Dean Richard H. Gallagher, and Dr. Jaroslaw Sobieski chaired the sessions and moderated the discussions, which were such a significant part of the symposium. The local arrangements were ably provided by Shirley Worth. Dolly Kenney, the symposium's secretary, was invaluable in handling not only the secretarial duties but also coordinating the many details associated with both the symposium and this book.

J. A. Bennett

M. E. Botkin

Publication of the book also required the able assistance of many people. Technical editing of both the discussions and the papers was handled by Dr. Martin Barone, Dr. Ji Oh Song, Dr. Dennis Vasilopoulos, and Dr. Ren-Jye Yang. Joan Kmenta edited the manuscripts and coordinated production, and Wendy Evans compiled the index. David Havelock and his group at the Research Laboratories were responsible for the artwork. We deeply appreciate the assistance of all these people in publishing this book.

James A. Bennett
Mark E. Botkin

CONTENTS

SESSION I
DERIVATIVES AND ALGORITHMS

Session Chairman

R. T. HAFTKA

Virginia Polytechnic Institute and State University
Blacksburg, Virginia

ADAPTIVE ANALYSIS REFINEMENT AND SHAPE OPTIMIZATION—SOME NEW POSSIBILITIES

O. C. ZIENKIEWICZ, A. W. CRAIG, and J. Z. ZHU

University College of Swansea
Swansea, United Kingdom

R. H. GALLAGHER

Worcester Polytechnic Institute
Worcester, Massachusetts

Abstract

Engineers have turned to shape optimization of structures to assure the efficient use of finite element analysis in producing safe and economical designs. Constraints on stresses and displacements should however be imposed with an accuracy commensurate with the degree of precision attainable in the analysis. A progressive refinement strategy can be used to increase the accuracy as the optimal design is approached and constraints are most critical. For this reason a simple and efficient error estimation capacity and an adaptive refinement strategy must be incorporated into the design program. This chapter will describe a new and efficient error estimation method based on mixed formulation concepts which can be incorporated into any existing program framework. In addition, a relatively simple refinement strategy will be shown which for a given problem can be designed to yield a specified accuracy of stress computation. Finally, a review of the methods used in shape optimization indicates the need for efficient mesh generation capabilities. If these can be combined with the indicators of error, then the objectives outlined above can be achieved.

INTRODUCTION

Sophisticated and elaborate finite element analysis is only justified in practice if it provides assurance on the performance and safety of engineering designs and if it provides more economical designs. For this reason many designers have turned to the subject of optimization and in particular to the optimization of the shape of engineering structures and components.

As the constraints and/or the objective functions of the problems include such products of the analysis as stresses and displacements, it is important that their accuracy be commensurate with the degree of precision specified by the criteria of design. Further, because the accuracy requirements are less stringent when the design is far from the optimal point, considerable computational savings can be achieved by a progressive refinement strategy. In this strategy, the accuracy requirements are progressively increased as the design approaches its optimum.

Such ideas require a simple and efficient error estimating capacity to be incorporated in the program and an adaptive refinement strategy. To date such concepts have not been widely applied, although their use has recently been suggested [1, 2]. Their effectiveness is clearly dependent on the efficiency of available error estimating capabilities that can be incorporated into any design program and on the optimization strategy to be used.

In this chapter we shall discuss these issues separately.

In the first three sections, a new and efficient method of error estimation will be presented. This is based on mixed formulation concepts and can be incorporated into any existing program framework. In addition we shall show how a relatively simple strategy of refinement can be designed to yield a specified accuracy of stress computation in a given problem.

In the last section we shall review the methods used in shape optimization [2, 3] and indicate the need for efficient mesh generation capabilities. If these can be combined with the indications of the accuracy requirements specified in preceding sections, then the objectives outlined in the introduction can be simply achieved.

GENERAL REMARKS ON ERROR NORMS

The finite element method presents a powerful tool for analysis of many engineering problems and it is extensively used in stress analysis. However the finite element method is an approximation to the true solution of the mathematical problem posed, and it is important to know not only the convergence characteristics but the actual magnitude of errors involved in any stage of the subdivision used. It is not surprising that much recent research has been devoted to this subject with the aim of devising a posteriori methods of estimating of such errors (i.e., methods which can be applied easily after the computation is completed). Pioneering work of Babuška [4, 5] and others [6–8] has led to procedures which, with some additional computational cost, can produce reasonably reliable estimates of error and can guide the mesh refinement (h convergence) or the increase of polynomial order in approximation (p convergence) in an efficient manner to produce answers with a desired accuracy. A survey of currently used procedures is given in reference [8] but to date few (if any) commercially available programs give this desired feature—

perhaps due to the difficulty of incorporating the needed additional computations into the program structure. We shall outline here a procedure which shows much promise and which is simple to incorporate into the existing structure of many codes.

However, before proceeding further it is important to make some remarks about the error and various measures of it.

In the context of the present paper we shall be concerned with standard stress analysis problems, but at this stage it is convenient to be more general and simply address the problem of solving

$$\mathbf{L}\mathbf{u} - \mathbf{f} = 0 \tag{1}$$

in a domain Ω with suitable boundary conditions. If $\bar{\mathbf{u}}$ is an approximate solution to the above, the local error is simply

$$\mathbf{e} = \mathbf{u} - \bar{\mathbf{u}}. \tag{2}$$

This is, however, an inconvenient form necessitating the definition of the whole field, and it is usual to measure the error as a norm which can give the required information in terms of a scalar quantity.

The energy norm most frequently used is defined as

$$\| \mathbf{e} \| = \left(\int_{\Omega} \mathbf{e}^T \mathbf{L} \mathbf{e} \, d\Omega \right)^{\frac{1}{2}} = \left(\int_{\Omega} (\mathbf{u} - \bar{\mathbf{u}}) \mathbf{L} (\mathbf{u} - \bar{\mathbf{u}}) \, d\Omega \right)^{\frac{1}{2}}. \tag{3}$$

If the operator $\mathbf{L}$ is self-adjoint, the above can be rewritten as

$$\| \mathbf{e} \|^2 \equiv \| \mathbf{u} \|^2 - \| \bar{\mathbf{u}} \|^2 \equiv a(\mathbf{e}, \mathbf{e}) \tag{4}$$

where the bilinear form $a(\cdot, \cdot)$ results from an integration by parts of equation (3).

Turning to the elastic stress analysis problem in which $\mathbf{u}$ is the displacement field and the operator $\mathbf{S}$ gives the strains $\boldsymbol{\varepsilon}$ as

$$\boldsymbol{\varepsilon} = \mathbf{S}\mathbf{u} \tag{5}$$

the bilinear form $a(\mathbf{e}, \mathbf{e})$ becomes simply the strain energy expression giving

$$\| \mathbf{e} \|^2 = \int_{\Omega} (\mathbf{S}\mathbf{e})^T \mathbf{D} (\mathbf{S}\mathbf{e}) \, d\Omega. \tag{6}$$

Also, $\mathbf{D}$ is the elasticity matrix defining stresses as

$$\boldsymbol{\sigma} = \mathbf{D}\boldsymbol{\varepsilon}. \tag{7}$$

For such problems the energy norm of the error $\| \mathbf{e} \|$ can be written in alternative ways as

$$\| \mathbf{e} \| = \left(\int_{\Omega} (\boldsymbol{\varepsilon} - \bar{\boldsymbol{\varepsilon}})^T \mathbf{D} (\boldsymbol{\varepsilon} - \bar{\boldsymbol{\varepsilon}}) \, d\Omega \right)^{\frac{1}{2}} \tag{8a}$$

References pp. 21–23

or, using (7),

$$\| \mathbf{e} \| = \left(\int_{\Omega} (\boldsymbol{\sigma} - \bar{\boldsymbol{\sigma}})^T \mathbf{D}^{-1} (\boldsymbol{\sigma} - \bar{\boldsymbol{\sigma}}) \, d\Omega \right)^{\frac{1}{2}} . \tag{8b}$$

The last form shows that this norm in fact gives a weighted root mean square (RMS) error of stress and as such is of considerable interest to the practitioner—particularly if local estimates of this can be found.

Other more direct norms can of course be used, but some of the simple features of equation (4) then disappear. For instance, a commonly used norm is the L_2 norm which, when applied to $\mathbf{u}$, is given as

$$\| \mathbf{u} \|_{L_2} = \left(\int_{\Omega} \mathbf{u}^T \mathbf{u} \, d\Omega \right)^{\frac{1}{2}} . \tag{9}$$

This focuses attention on the quantity $\mathbf{u}$ rather than on its derivatives and is of interest when the accuracy of $\mathbf{u}$ is to be determined.

In some of the examples given here we shall be interested in the L_2 norm of stress error defined as

$$\| \mathbf{e}_\sigma \|_{L_2} = \left(\int_{\Omega} (\boldsymbol{\sigma} - \bar{\boldsymbol{\sigma}})^T (\boldsymbol{\sigma} - \bar{\boldsymbol{\sigma}}) \, d\Omega \right)^{\frac{1}{2}} . \tag{10}$$

This is identical to the energy norm if the elastic matrix is taken as the identity matrix $\mathbf{I}$.

It is sometimes convenient to express the error in terms of average deviation of the quantity of interest over a certain area. Thus, for instance, using definition (10) we can express the local RMS stress error in an area of an element Ω_e as

$$\Delta \boldsymbol{\sigma} = \left\{ \frac{\| \mathbf{e}_\sigma \|_{L_2}^2}{\Omega_e} \right\}^{\frac{1}{2}} \tag{11}$$

where the norm is evaluated in the same area.

Assessments of error are often made by using successive finite element solutions to estimate the local error approximately at all points of the domain and then substituting this into the appropriate norm expressions. To save computational effort, much more accurate solutions are conveniently derived using the hierarchic element concept and performing the addition of variables one at a time [6–8]. In what follows, a simpler way of assessing the more accurate solution is developed.

MIXED ITERATIVE SOLUTION AND RESULTING ERROR ESTIMATES

For a given mesh size used in the displacement approximation

$$\mathbf{u} = \mathbf{N} \bar{\mathbf{u}} \tag{12}$$

the mixed representation with independent, continuous representations of strains (or stresses)

$$\boldsymbol{\varepsilon} = \mathbf{N}_\varepsilon \bar{\boldsymbol{\varepsilon}} \tag{13}$$

yields more accurate results (especially for strains/stresses) than the simple displacement forms [9], provided that the stresses and strains are reasonably smooth in the exact solution. Generally such mixed solutions are much more expensive than the displacement ones; however, recently introduced iterative procedures [10, 11, 12] allow a very considerable improvement of displacement type solutions to be obtained at very small cost.

The mixed form with independent interpolation of $\mathbf{u}$ and $\boldsymbol{\varepsilon}$ can be stated as a weak form of the strain displacement and equilibrium conditions as

$$\left(\int_\Omega \mathbf{N}_\varepsilon^T \mathbf{N}_\varepsilon \, d\Omega\right) \bar{\boldsymbol{\varepsilon}} - \left(\int \mathbf{N}_\varepsilon^T \mathbf{B} \, d\Omega\right) \bar{\mathbf{u}} = 0 \tag{14a}$$

$$\int \mathbf{B}^T \boldsymbol{\sigma} \, d\Omega = \mathbf{f} \tag{14b}$$

where, from (5) and (12),

$$\boldsymbol{\varepsilon} = \mathbf{S}\mathbf{N}\bar{\mathbf{u}} \equiv \mathbf{B}\bar{\mathbf{u}} \tag{15}$$

is the standard strain representation by discretized displacement.

If we further assume that the stresses can be interpolated from the nodal values by the same interpolation as strains, we can write

$$\boldsymbol{\sigma} = \mathbf{N}_\varepsilon \bar{\boldsymbol{\sigma}} \tag{16}$$

and

$$\bar{\boldsymbol{\sigma}} = \bar{\mathbf{D}} \bar{\boldsymbol{\varepsilon}} \tag{17}$$

and equations (14a) and (14b) become

$$\mathbf{A}\bar{\boldsymbol{\varepsilon}} - \mathbf{Q}\bar{\mathbf{u}} = 0 \tag{18a}$$

$$\mathbf{Q}^T \bar{\mathbf{D}} \bar{\boldsymbol{\varepsilon}} = \mathbf{f} \tag{18b}$$

where

$$\mathbf{A} = \int_\Omega \mathbf{N}_\varepsilon^T \mathbf{N}_\varepsilon \, d\Omega \tag{19a}$$

and

$$\mathbf{Q} = \int_\Omega \mathbf{N}_\varepsilon^T \mathbf{B} \, d\Omega. \tag{19b}$$

The iteration can proceed as follows:

$$\bar{\mathbf{u}}^{i+1} = \bar{\mathbf{u}}^i + \mathbf{K}^{-1}(\mathbf{f} - \mathbf{Q}^T \bar{\boldsymbol{\sigma}}^i) \quad i = 0, 1 \ldots \qquad \mathbf{u}^0 = 0 \qquad \boldsymbol{\varepsilon}^0 = 0 \tag{20a}$$

References pp. 21–23

$$\bar{\boldsymbol{\varepsilon}}^{i+1} = \mathbf{A}^{-1}\mathbf{Q}\bar{\mathbf{u}}^{i+1} \tag{20b}$$

$$\bar{\boldsymbol{\sigma}}^{i+1} = \bar{\mathbf{D}}\bar{\boldsymbol{\varepsilon}}^{i+1} \tag{20c}$$

where $\mathbf{K}$ is the standard displacement stiffness matrix and $\mathbf{u}^{(1)}$, the first step of the iteration, is simply the displacement type solution. The second part of the iteration directly represents the process of *smoothing* or variational recovery of continuous strain and stress. Equation (20a) can be viewed as successive correction of the equilibrium equation, adjusting the imbalance caused by stress smoothing [13].

The cost of each iteration is small if a diagonal form of the matrix $\mathbf{A}$ is assumed. This is most conveniently done by using nodal quadrature points when evaluating (19a), but other inexpensive possibilities exist [11].

The practical results of the iteration are generally excellent. Even a single iteration improves the stresses very considerably, and two or three iterations give a much improved displacement form. Clearly, even an $i = 0$ step carried out to smooth the stresses would allow these to be much better represented and could give a reasonable error estimate. We shall find that this indeed is the case if we evaluate the energy norm of the error, using equation (8), as

$$\|\mathbf{e}\| = \left(\int_{\Omega} (\mathbf{N}_{\varepsilon}\bar{\boldsymbol{\sigma}}^{i} - \mathbf{D}\mathbf{B}\mathbf{u}^{i})^{T}\,\mathbf{D}^{-1}(\mathbf{N}_{\varepsilon}\bar{\boldsymbol{\sigma}}^{i} - \mathbf{D}\mathbf{B}\mathbf{u}^{i})\,d\Omega\right)^{\frac{1}{2}}. \tag{21}$$

Indeed it is equally easy to use the expression giving the L_2 mean of the stress error in equation (10).

Figures 1 through 4 and the accompanying Tables 1 through 7 show how well the true errors are estimated for various examples in which the "exact" solution has been obtained by use of a very refined mesh. Figures 1 and 2 illustrate problems with bilinear elements, and the *effectivity indices* giving the ratio of estimated to actual error are approximately 0.8 for very coarse meshes with values tending uniformly to unity as the mesh is refined. For Figures 3 and 4, in which quadratic elements are used, coarse meshes produce an overestimate of error but again this tends to the correct values as the mesh is refined. The figures and tables show the estimated and actual values of the percent error defined as

$$\eta = \|\mathbf{e}\| / (\|\bar{\mathbf{u}}\|^2 + \|\mathbf{e}\|^2)^{\frac{1}{2}}. \tag{22}$$

For example, the estimates of the L_2 stress norm error as well as the average stress deviation are given in Figure 1.

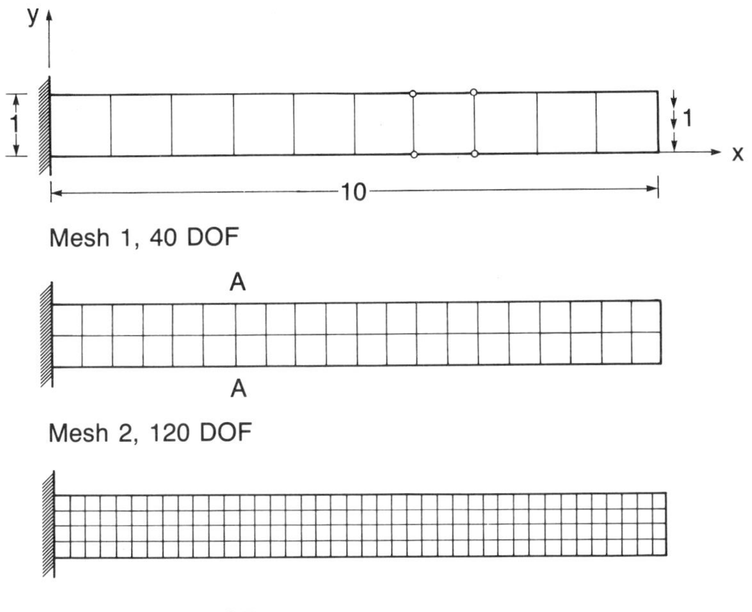

Figure 1. Cantilever beam – plane stress ($E = 10^5$, $\nu = 0.3$): analysis and error estimates for uniform subdivision of bilinear elements.

Table 1
Energy error norms, percentage errors and effectivity indices for cantilever beam: Example of Figure 1 (viz Equation 8(b))

DOF	$\|\| \bar{\mathbf{u}} \|\|^2$	${}^0\|\| \mathbf{e} \|\|$	${}^1\|\| \mathbf{e} \|\|$	${}^0\theta$	${}^1\theta$	$\eta(\%)$	${}^0\eta(\%)$	${}^1\eta(\%)$
40	2.72	0.802	0.878	0.70	0.765	57.1	43.8	47.0
120	3.58	0.591	0.626	0.881	0.934	33.4	29.8	31.4
400	3.90	0.331	0.338	0.917	0.937	17.9	16.5	16.9
Exact solution	$\|\| \mathbf{u} \|\|^2 = 4.0312$ (beam theory)							

${}^0\|\| \mathbf{e} \|\|$ = Error estimate by 0 iteration of mixed method

${}^1\|\| \mathbf{e} \|\|$ = Error estimate by 1 iteration of mixed method

${}^0\theta$, ${}^1\theta$ = Effectivity indices corresponding to above

η = Exact relative error

${}^0\eta, {}^1\eta$ = Computed relative error

$\Delta\sigma$ = $(\|\| \sigma \|\|^2_{L_2} / \Omega)^{\frac{1}{2}}$

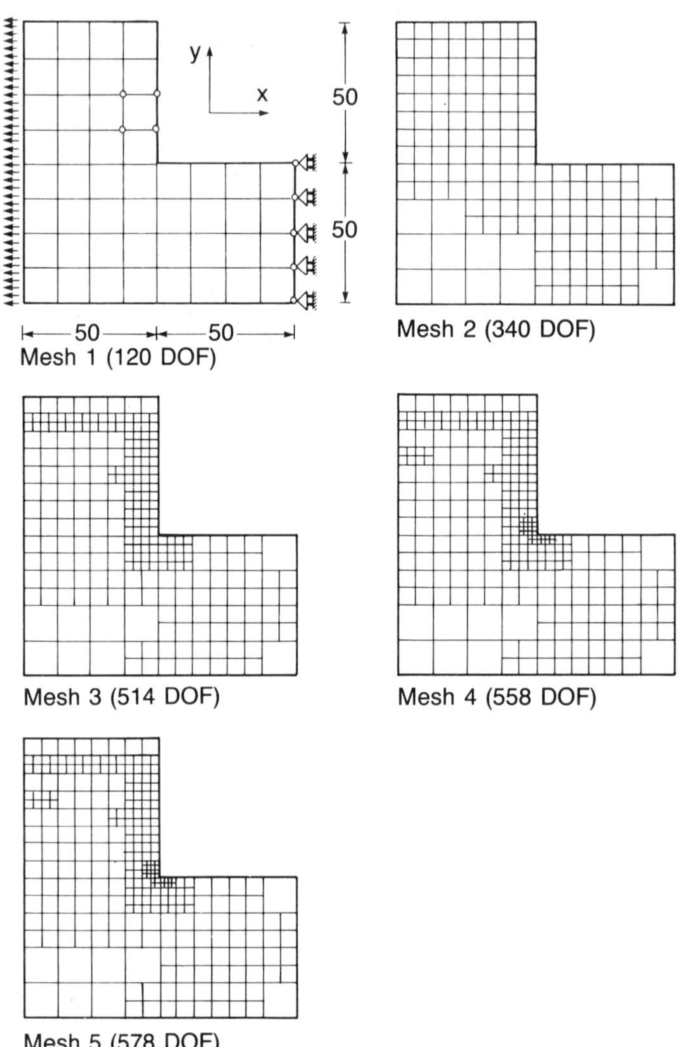

Figure 2. An L-shaped region in plane stress: sequences of mesh refinement of bilinear elements.

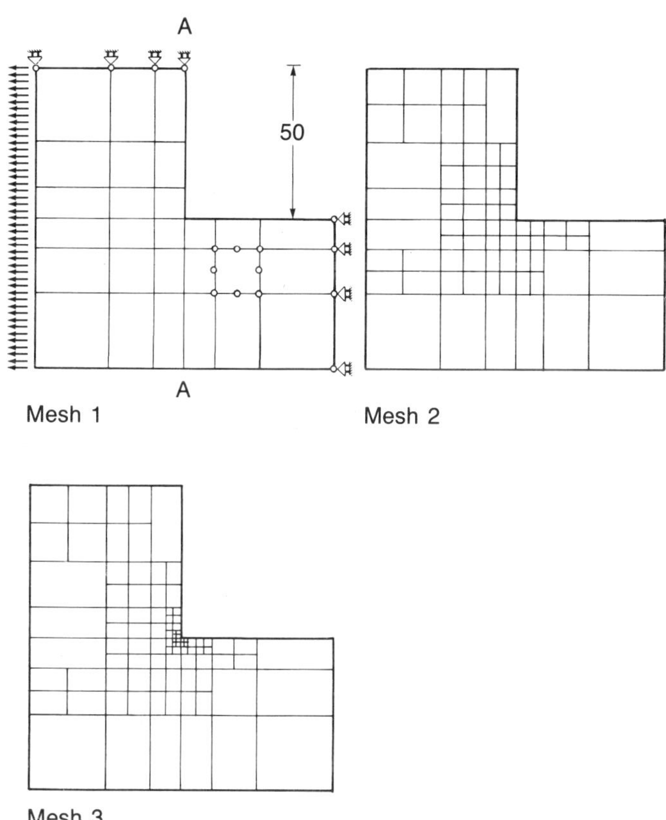

Figure 3. An L-shaped region in plane stress: sequences of mesh refinement (8 node quadratic element).

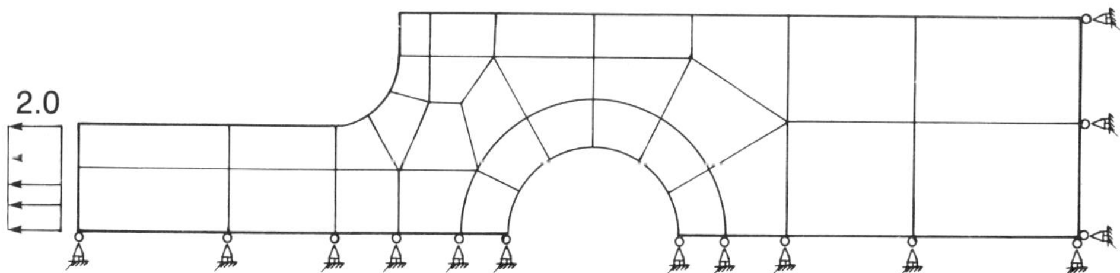

Figure 4. Perforated tension bar.

Table 2
L_2 error norms, percentage errors and effectivity indices for cantilever beam: Example of Figure 1 (viz Equation 10)

DOF	$\lVert \bar{\sigma} \rVert^2_{L_2}$	${}^0\lVert \mathbf{e} \rVert_{L_2}$	${}^1\lVert \mathbf{e} \rVert_{L_2}$	${}^0\theta$	${}^1\theta$	$\eta(\%)$	${}^0\eta(\%)$	${}^1\eta(\%)$	$\Delta\boldsymbol{\sigma}$
40	$0.266\text{x}10^6$	$0.158\text{x}10^3$	$0.199\text{x}10^3$	0.581	0.73	43.0	25.0	31.4	501.2
120	$0.135\text{x}10^6$	$0.135\text{x}10^3$	$0.146\text{x}10^3$	0.846	0.918	25.1	21.3	23.1	425.8
400	$0.390\text{x}10^6$	$0.764\text{x}10^2$	$0.787\text{x}10^2$	0.880	0.907	13.7	12.1	12.4	241.6
Exact solution	$\lVert \mathbf{u} \rVert^2_{L_2} = 80.4012x10^6$ (beam theory)								

$\Delta\boldsymbol{\sigma}$ global R.M.S. stress error

Table 3
L_2 error norms, percentage errors and effectivity indices for L-shaped domain: Example of Figure 2

DOF	$\lVert \bar{\mathbf{u}} \rVert^2$	${}^0\lVert \mathbf{e} \rVert$	${}^1\lVert \mathbf{e} \rVert$	${}^0\theta$	${}^1\theta$	$\eta(\%)$	${}^0\eta(\%)$	${}^1\eta(\%)$	$\Delta\boldsymbol{\sigma}$
120	0.299	0.0878	0.0925	0.795	0.838	19.8	15.9	16.7	0.321
340	0.306	0.0547	0.0568	0.810	0.842	12.1	9.83	10.2	0.20
514	0.308	0.0436	0.0450	0.839	0.867	9.30	7.82	8.07	0.159
558	0.309	0.0410	0.0422	0.879	0.904	8.36	7.35	7.57	0.150
578	0.309	0.0403	0.0414	0.904	0.929	8.0	7.24	7.43	0.147
Exact solution	0.3112536 (refined mesh and Richardson extrapolation)								

Table 4
Examples of Figure 3 with 8 node elements

DOF	$\|\| \bar{\mathbf{u}} \|\|^2$	$^0 \|\| \mathbf{e} \|\|$	$^0\theta$	$\eta(\%)$	$^0\eta(\%)$	$\Delta\sigma$
198	0.30943	0.04203	0.986	7.64	7.53	0.153
434	0.31041	0.02682	0.925	5.20	4.81	0.098
562	0.31096	0.01582	0.926	3.06	2.84	0.058
Exact solution	0.3112536 (refined mesh and Richardson extrapolation)					

Table 5
Example of Figure 3 with 9 node elements

DOF	$\|\| \bar{\mathbf{u}} \|\|^2$	$^0 \|\| \mathbf{e} \|\|$	$^0\theta$	$\eta(\%)$	$^0\eta(\%)$	$\Delta\sigma$
252	0.31000	0.04477	1.26	6.35	8.02	0.163
572	0.31067	0.02830	1.18	4.31	5.07	0.103
796	0.31104	0.01607	1.10	2.61	2.88	0.059
Exact solution	0.3112536 (refined mesh and Richardson extrapolation)					

Table 6
Example of Figure 4 with 8 node elements

DOF	$\|\| \bar{\mathbf{u}} \|\|^2$	$^0 \|\| \mathbf{e} \|\|$	$^0\theta$	$\eta(\%)$	$^0\eta(\%)$	$\Delta\sigma$
227	0.20089	0.0336	1.29	5.7	7.5	0.104
Exact solution	0.20156 (refined mesh and Richardson extrapolation)					

Table 7
Example of Figure 4 with 9 node elements

DOF	$\|\| \bar{\mathbf{u}} \|\|^2$	$^0 \|\| \mathbf{e} \|\|$	$^0\theta$	$\eta(\%)$	$^0\eta(\%)$	$\Delta\sigma$
289	0.20107	0.0339	1.53	4.9	7.5	0.099
Exact solution	0.20156 (refined mesh and Richardson extrapolation)					

In the computation presented, we have used linear strain (stress) distribution for linear elements (quadrilaterals and triangles). For quadratic 8 and 9 node elements we have however projected the stresses into a bilinear base rather than a quadratic one. Though this does not lead to a unique, mixed solution form, it does simplify computation and leads to equally reliable results as a quadratic projection.

We should remark here that simple stress smoothing, i.e., $i = 0$ in iteration of equation (20), leads throughout to reasonable estimates of error.

In Figures 5 and 6 we show the local effectivity indices for the examples of

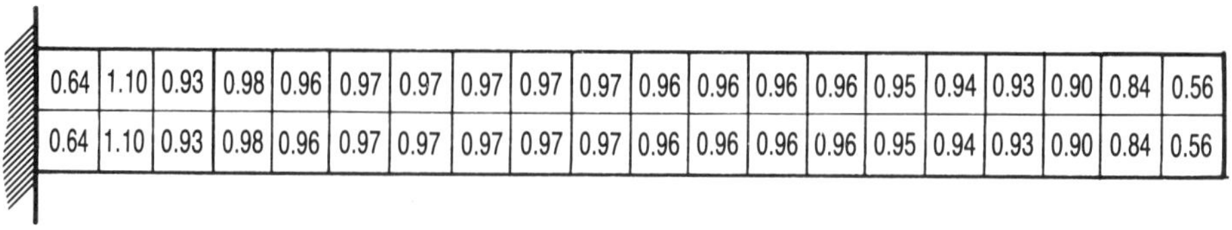

0.64	1.10	0.93	0.98	0.96	0.97	0.97	0.97	0.97	0.97	0.96	0.96	0.96	0.96	0.95	0.94	0.93	0.90	0.84	0.56
0.64	1.10	0.93	0.98	0.96	0.97	0.97	0.97	0.97	0.97	0.96	0.96	0.96	0.96	0.95	0.94	0.93	0.90	0.84	0.56

Figure 5. Local effectivity indices of Figure 1 (mesh 2).

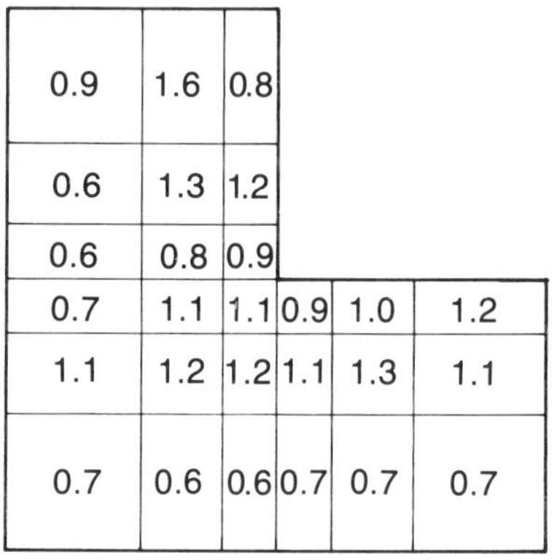

0.9	1.6	0.8			
0.6	1.3	1.2			
0.6	0.8	0.9			
0.7	1.1	1.1	0.9	1.0	1.2
1.1	1.2	1.2	1.1	1.3	1.1
0.7	0.6	0.6	0.7	0.7	0.7

Figure 6. Local effectivity indices of Figure 3 (mesh 1).

Figure 1 (mesh 2) and Figure 3 (mesh 1), indicating that the local energy norm error is well estimated by the procedures outlined. This is of some importance as frequently we are interested only in stresses at a given location and this allows such structures to be estimated with reasonable confidence. Of course this type of local estimate gives us little information about points where a singularity makes the stresses infinite, even though here it is valuable to have information about energy when stress intensity factors are being calculated. However, for problems where a finite stress enters an optimization calculation, we can reasonably assume the local average stress deviation to be well represented.

Finally we should remark that the stress smoothing process leads to more accurate stress representation than that directly resulting from the displacement method computation. Thus if such smoothed results are used (as indeed they always should be in optimization), the error estimate will be on the high side—obviously a conservative assumption. This is illustrated in Figures 7 and 8.

THE ADAPTIVE PROCESS

In the preceding section we have shown that the method presented gives a reliable global (and even local) estimate of errors. Now we can turn our attention to the adaptive strategy in which we shall strive to obtain an acceptable error range with the minimum number of steps.

A simple procedure can be explained in terms of the energy norm in which we

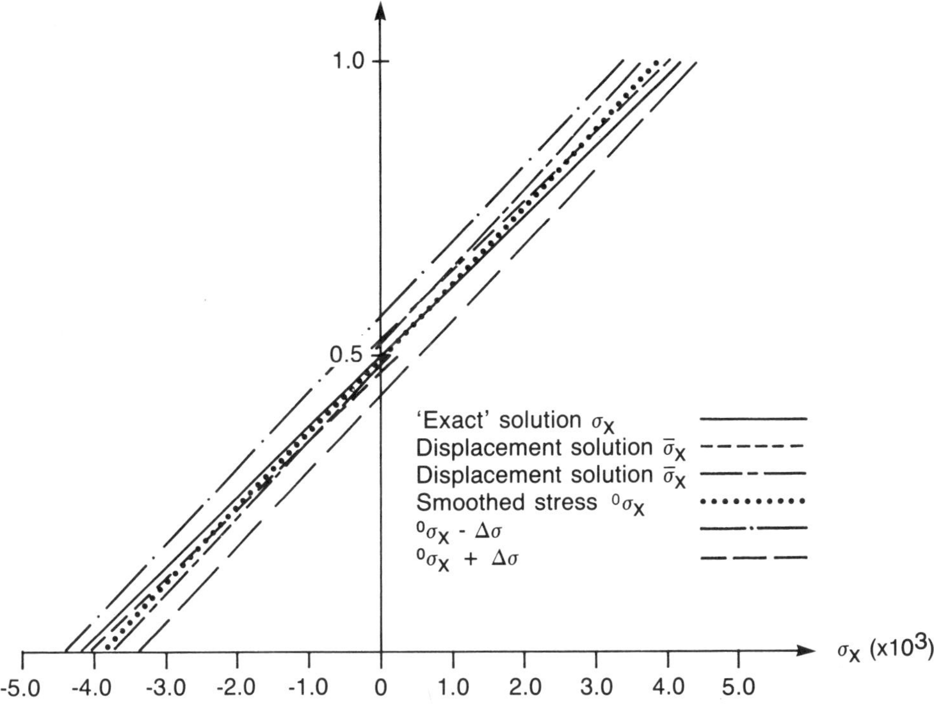

Figure 7. Plot of stress σ_x and distribution of $\Delta\sigma$ on section A-A of Figure 1 (mesh 2).

seek a given percentage accuracy for the whole region. We thus require that in the final subdivision the relative error for the whole domain (see equation (22)) falls below a certain value $\bar{\eta}$, i.e.,

$$\eta \leq \bar{\eta}. \tag{23}$$

If we choose a permissible relative error $\bar{\eta}$ and also assume that we want the energy error to be equally distributed amongst the elements, then for m elements the error in any element i, $\| \mathbf{e} \|_i$, should be bounded as

$$\| \mathbf{e} \|_i \leq \bar{\eta}(\| \bar{\mathbf{u}} \|^2 + \| \mathbf{e} \|^2)^{\frac{1}{2}} m^{-\frac{1}{2}}. \tag{24}$$

If we have element estimators ρ_i derived as in the last section such that

$$\begin{aligned} \| \mathbf{e} \|_i &\cong \rho_i \\ \| \mathbf{e} \|^2 &\cong \sum_{i=1}^{m} \rho_i^2 \end{aligned} \tag{25}$$

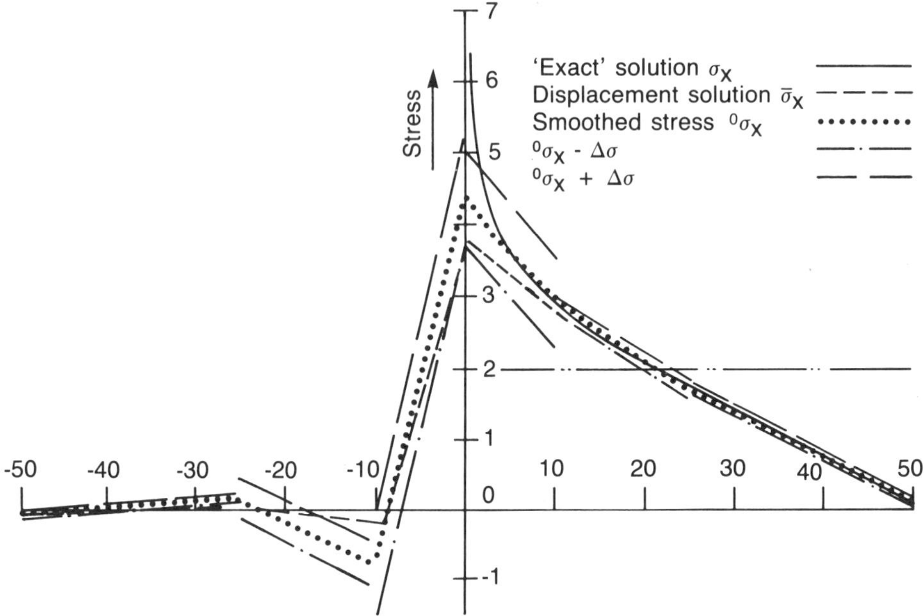

Figure 8. Plot of stress σ_x and distribution of $\triangle\sigma$ on section A-A of Figure 3 (mesh 1).

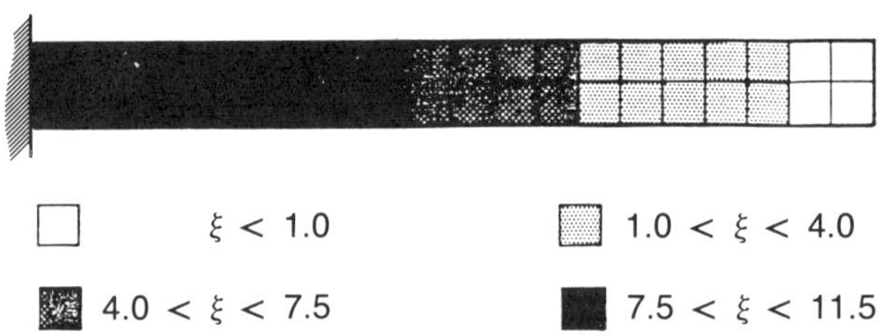

Figure 9. Distribution of ξ in the problem of Figure 1 (mesh 2).

then we may obtain a refinement criterion by substituting the estimators into inequality (24). That is, an element is acceptable if

$$\rho_i \leq \bar{\eta} \left(\| \bar{\mathbf{u}} \|^2 + \sum_{j=1}^{m} \rho_j^2 \right)^{\frac{1}{2}} m^{-\frac{1}{2}} (= \delta, \text{ say}). \tag{26}$$

The refinement strategy can now be guided by the element error ratio defined as

$$\xi_i = \frac{\rho_i}{\delta} = \frac{\| \mathbf{e} \|_i}{\text{permissible element error}} \tag{27}$$

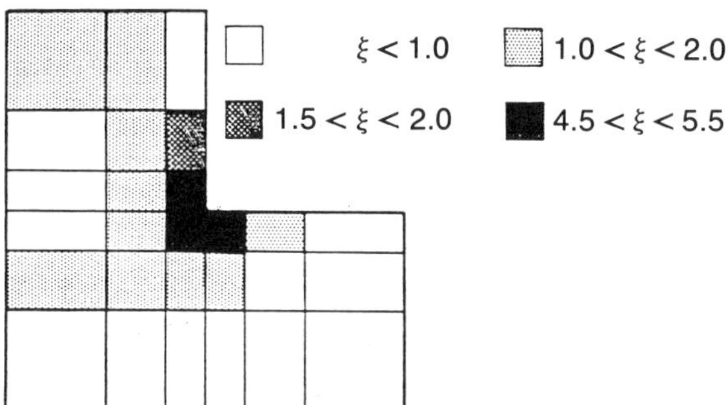

Figure 10. Distribution of ξ in the problem of Figure 3 (mesh 1).

with subdivision required when

$$\xi_i > 1. \tag{28}$$

In Figures 9 and 10 we show a plot of ξ_i on mesh 2 used in the example of Figure 1, and a plot of ξ_i on mesh 1 used in the example of Figure 3.

Obviously the magnitude of the number ξ_i indicates the degree of refinement needed and, if uniform refinement (of type h) is to be followed locally, we can predict approximately the local subdivision needed (or the number of new nodal points to be introduced) by assuming a certain rate of convergence. For instance if a single element is subdivided into K^2 new elements, the error will be reduced (in the absence of singularity) in relative terms to

$$\frac{\| \mathbf{e} \|_i}{K^p} \cong \frac{\rho_i}{K^p} \quad \text{or} \quad K \cong \xi^{1/p} \tag{29}$$

where p is the order of a complete polynomial in the trial functions used. For the bilinear element this gives us $p = 1$. Using this formula we get an indication of how many times we need to refine in order to reduce the error below our acceptable limit.

An approximate indication of subdivision requirements is obvious here, and in practice we have succeeded in obtaining the required accuracy with one or at most two mesh redesigns. Figures 2 and 3 demonstrate such adaptive refinement sequences in which, however, we have only used $K = 2$, thus converging more slowly to the requested solution.

Obviously other refinement criteria could be used. One of the more logical appears to be the requirement that the local stress deviation $\Delta\sigma$ as defined by equation (11) should be limited. Here we only need to evaluate this quantity in each element and refine where the permissible range is exceeded. In Figure 8 we show the distribution of $\Delta\sigma$ values which can serve as a guide to refinement.

In the next section we shall deal with the problem of shape optimization and show how essential it is to have a simple mesh generator which can regenerate the nodal coordinates as the structural shape is changed. A combination of the above mesh generation with progressively tightening criteria for the $\Delta\sigma$ error range can be used within the optimization process to achieve considerable economies.

LITERATURE REVIEW

In the 1981 review of structural optimization edited by Lev [14], it is observed that "shape optimization techniques are still in the stage of early development and their implementation in design practice is relatively rare." Even if this statement were associated with a review that was incomplete by virtue of neglecting certain references, it would remain an accurate appraisal when all possible references in shape optimization are measured against references in the overall field of optimization.

The literature on shape optimization is smaller still when we divide it with respect to classes of problems. Vanderplaats [3], in his 1981 review of shape optimization, distinguishes between *discrete structures* and *continuous structures*, where general shape parameters are the design variables. This paper deals only with the latter class of problem, and so this review is likewise limited.

As we address the topic of continuous structures, further subdivisions are appropriate. A significant portion of references to date treat the problem of shape optimization by mathematical programming techniques, with use of standard finite element formulations as the basis for constraint equations and sensitivities. Another segment of the literature employes optimality criteria or variational statements in the basic formulation of the optimization problem. Here, finite element representations and mathematical programming procedures are frequently used in the final computational process.

Some of the early work in the former category was introduced by Zienkiewicz and Campbell [16], who studied dam structures using isoparametric elements and sequential linear programming. The element node locations were the design variables. DeSilva [17], in a paper in the same volume and in work published elsewhere, optimized turbine disks as a shape optimization problem. Later, Quéau and Trompette [18, 29] used more advanced optimization procedures for a similar problem, as did Song and Lee [32].

Ramakrishnan and Francavilla [19] and Vitiello [20] used two-dimensional isoparametric elements wherein the design variables were the coefficients of the polynomials describing the element shape. Penalty function methods were applied in the minimization process. Ricketts and Zienkiewicz [1] employ sequential linear programming in similar work which, as already noted, raised the suggestion that is elaborated upon in the present work. Wang, Sun and Gallagher [21], in a fairly recent development, use the same ideas as in [1], [19] and [20] but establish equations for the constraint gradients, related to isoparametric elements, which are intended to produce efficiencies in the computational process.

Imam [22] solved shape optimization problems with three-dimensional isoparametric elements and "super-curves" whose parameters are optimized. He introduces a caution regarding the distortion of elements as a result of the revision of the shape of the structure in successive cycles. Botkin [23] and Bennett and Botkin [24] deal with the latter issue as well as others through a comprehensive scheme of optimization. This includes such features as the regionalization of the structure being optimized, the definition of a single constraint within the region, and remeshing based on an automated element triangularization scheme. A recognition of the role that adaptive mesh refinement can play in this process is the basis of Bennett and Botkin's paper [24]. It is also noteworthy that Botkin [23] gives the basis for shape optimization of plate and shell structures.

All of the above are concerned with the minimization of weight or mass. Before we leave this general area of shape optimization founded principally in algebraic procedures, we should cite some complementary studies which either have more practical orientation or which seek objectives other than weight or mass. For example, Middleton [25], in an optimization of pressure vessels, uses the geometric parameters of configurations such as the knuckle radius as design parameters to obtain a minimum shear stress. The selected optimization approach is the penalty function technique. A number of authors including Schnack [26], Kristensen and Madsen [27] and Francavilla et al. [30] address the issue of shape optimization to reduce stress concentration.

As noted above, a body of literature on shape optimization has emerged in conjunction with variational statements of the problem. Prager [35] is generally credited with motivating this line of endeavor, but it has advanced in 15 years from a principally analytical approach to one which fully exploits the finite element modeling and numerical minimization techniques. There are still other approaches founded in classical formulations of the optimization and these are reviewed in detail by Haug [15, 34].

Haug and coauthors [34] employ a variational statement of the optimization problem, a generalized steepest descent approach and finite element representations. They solve two problems numerically: the design of an elastic shaft under torsion and the design of the shape of a two-dimensional fillet in a tension bar. Tada and Seguchi [38], in a chapter in the same book, also use a variational formulation.

Na et al. [36] give a variational formulation of the optimal shape design of two-dimensional linearly elastic bodies. They introduce the term *structural remodeling* to denote the best form of modification within an available amount of resource. This might mean the addition of material, the subtraction of material or a combination of both for the same structure. Finite element discretization is used in the numerical computation.

Spillers and Singh [39] have outlined a shape optimization procedure, based on earlier work of Friedland [40], which is stated to be in the family of optimality criteria methods. Only a limited amount of detail is given in reference [39], however.

Clearly, shape optimization has come a long way in terms of the number of publications that have appeared and in the progress made in basic theory. Some general observations about that progress are pertinent.

References pp. 21–23

Firstly, we believe that the record shows that we have "shaken down" in terms of the broad approach to shape optimization. The chapters in this book will show that there are differences in the basic statement of the problem, either as a direct algebraic statement as in this chapter or as a variational formulation, but most share confidence in the use of finite elements as a tool of analysis and in the use of mathematical programming procedures as the ultimate algebraic minimization tool. Further, they grasp the importance of establishing means of coping with certain critical situations, such as the distortion of elements as the mesh is refined. There is a consciousness of the need to compress the problem to as few design variables as possible (a point emphasized by Zienkiewicz and Ricketts [1]) and to establish efficient and concise methods of representing the constraints. Although the use of element data as design parameters was appealing at the outset of work in structural optimization, this more efficient scheme of geometric design parameters has taken hold.

Secondly, we find it remarkable how rapidly this work is moving towards implementation of procedures in finite element analysis which are not widely adopted in analysis applications themselves and which, indeed, are not yet entirely crystallized. We refer here principally to adaptive mesh refinement, but as we have already seen, this is a topic that carries along with it such issues as convergence rates, error norms and mesh generation techniques. This chapter, and other publications that have already appeared, suggest the importance of increasing the refinement of the mesh in tandem with the process of optimization. It means that we are beginning to enjoy a greater collaboration of finite element analysts, structural optimizers and even computer graphics specialists than we have seen in the past. If this keeps up, the term computer aided design may very well take on its intended meaning.

Thirdly, it is noteworthy that although we have seen 40 years of parallel progress in mathematical programming, traditional algorithms seem to continue to hold their place. We refer to sequential linear programming. It is reminiscent of similar situations in finite element analysis—for example in solving equations where Gaussian elimination has remained attractive if not predominant despite the challenge from alternative approaches.

Fourthly, with increased national and international attention on manufacturing systems, shape optimization must properly account for manufacturing realities. Thus, the optimum shape must have a continuity from element to element. An optimization process that admits of a different size of each finite element model is in conflict with this requirement.

Finally, without trying to reflect upon the excellent applications reported in these proceedings, applications in practice, measured against their real potential, are very slim. One can cite classes of problems in the foregoing literature review that include the optimization of automotive components, turbine wheels, rotors, dams and little else. The intrinsic importance of shape optimization is quite evident, and an exciting future is in store as innumerable new problems are tackled using techniques described in this symposium.

REFERENCES

1. R. E. Ricketts and O. C. Zienkiewicz, Shape optimization of continuum structures, Ch. 6 in *New Directions in Optimum Structural Design* (Edited by E. Atrek, R. H. Gallagher, K. M. Ragsdell and O. C. Zienkiewicz). John Wiley & Sons (1984).

2. M. E. Botkin and J. A. Bennett, The application of adaptive mesh refinement to shape optimization of plate structures, Ch. 13 in *Adaptive Refinement on Error Estimates in Finite Element Analysis*. John Wiley & Sons (1986), to be published.

3. G. N. Vanderplaats, Numerical methods for shape optimization: An assessment of the state of the art, Ch. 4 in *New Directions in Optimum Structural Design* (Edited by E. Atrek, R. H. Gallagher, K. M. Ragsdell and O. C. Zienkiewicz). John Wiley & Sons (1984).

4. I. Babuška and W. C. Rheinboldt, Error estimates for adaptive finite element computations. *SIAM J. Numer. Anal.* **15** (4) (1978).

5. I. Babuška and M. Dorr, Error estimates for the combined h and p versions of the finite element method. *Numer. Math.* **25**, 257–277 (1981).

6a. D. W. Kelly, J. P. de S. R. Gago, O. C. Zienkiewicz and I. Babuška, A posteriori error analysis and adaptive processes in the finite element method–Part 1: Error analysis. *Int. J. Numer. Meth. Eng.* **19**, 1593–1619 (1983).

6b. J. P. de S. R. Gago, D. W. Kelly, O. C. Zienkiewicz and I. Babuška, A posteriori error analysis and adaptive processes in the finite element method–Part 2: Adaptive mesh refinement. *Int. J. Numer. Meth. Eng.* **19**, 1621–1656 (1983).

7. O. C. Zienkiewicz, J. P. de S. R. Gago and D. W. Kelly, The hierarchical concept in finite element analysis. *Comput. Struct.* **16**, 53–65 (1983).

8. O. C. Zienkiewicz and A. W. Craig, Adaptive mesh refinement and a posteriori error estimation for the p version of the finite element method, in *Adaptive Computational Methods for Partial Differential Equations* (Edited by I. Babuška, Chandra and Flaherty). SIAM (1983).

9. O. C. Zienkiewicz, *The Finite Element Method*, 3rd ed. McGraw-Hill, New York (1977).

10. O. C. Zienkiewicz, J-P. Vilotte, S. Toyoshima and S. Nakazawa, Iterative method for constrained and mixed approximation: An inexpensive improvement of FEM performance. *Comput. Meth. Appl. Mech. Eng.*, **51** 3–29 (1985).

11. O. C. Zienkiewicz, X-K Li and S. Nakazawa, Iterative solution of mixed problem and stress recovery procedures. *Commun. Appl. Numer. Meth.* **1** (3–9) (1985).

12. O. C. Zienkiewicz, X-K Li and S. Nakazawa, Dynamic transient analysis of a mixed, iterative method. *Int. J. Numer. Meth. Eng.* (1986), to appear.

13. G. Cantin, G. Loubignac and G. Touzot, An iterative algorithm to build continuous stress and displacement solutions. *Int. J. Numer. Meth. Eng.* **12**, 1493–1500 (1978).

14. O. Lev (Ed.), *Structural Optimization Recent Developments and Applications.* ASCE, New York (1981).

15. E. J. Haug, A review of distributed parameter structural optimization literature, in *Optimization of Distributed Parameter Structures* (Edited by E. Haug and J. Cea). Sijthoff-Noordhoff, The Netherlands (1981).

16. O. C. Zienkiewicz and J. S. Campbell, Shape optimization and sequential linear programming, in *Optimum Structural Design* (Edited by R. H. Gallagher and O. C. Zienkiewicz). John Wiley & Sons, New York (1973).

17. B. M. E. DeSilva, Feasible direction methods in structural optimization, in *Optimum Structural Design* (Edited by R. H. Gallagher and O. C. Zienkiewicz). John Wiley & Sons, New York (1973).

18. J. P. Quéau and Ph. Trompette, Two-dimensional shape optimal design by the finite element method. *Int. J. Numer. Meth. Eng.* **15**, 1603–1612 (1980).

19. C. V. Ramakrishnan and A. Francavilla, Structural shape optimization using penalty function. *J. Struct. Mech.* **4** (3), 403–422 (1974).

20. F. Vitiello, Shape optimization using mathematical techniques. *2nd Symp. on Structural Optimization.* AGARD CP-123 (Apr. 1973).

21. S. Y. Wang, Y. Sun and R. H. Gallagher, Sensitivity analysis in shape optimization of continuum structures. *Comput. Struct.* **20** (5), 855–867 (1985).

22. M. H. Imam, Three-dimensional shape optimization. *Int. J. Numer. Meth. Eng.* **18** (5) 635–673 (1982).

23. M. E. Botkin, Shape optimization of plate and shell structures. *AIAA J.* **20** (2) 268–273 (1981).

24. J. A. Bennett and M. E. Botkin, Structural shape optimization with geometric description and adaptive mesh refinement. *AIAA J.* **23** (3) 458–464 (1985).

25. J. Middleton, Optimal shape design to minimize stress concentration factors in pressure vessel components. *Proc. Int. Symp. on Optimum Structural Design.* Tucson, AZ (Oct. 1981).

26. E. Schnack, An optimization procedure for stress concentration by the finite element technique. *Int. J. Numer. Meth. Eng.* **14**, 115–124 (1979).

27. E. S. Kristensen and N. F. Madsen, On the optimum shape of fillets in plates subjected to multiple inplane loading cases. *Int. J. Numer. Meth. Eng.* **10**, 1006–1019 (1976).

28. K. Dems and Z. Mróz, Multiparameter shape optimization by the finite element method. *Int. J. Numer. Meth. Eng.* **13**, 247–263 (1978).

29. J. P. Quéau and Ph. Trompette, Optimal shape design of turbine and compressor blades. *Proc. Int. Symp. on Optimum Structural Design.* Tuscon, AZ (Oct. 1981).

30. A. Francavilla, C. V. Ramakrishnan and O. C. Zienkiewicz, Optimization of shape to minimize stress concentration. *J. Strain Anal.* **10** (2), 63–70 (1975).

31. R. A. Meric, Boundary element methods for optimization of distributed parameter systems, *Int. J. Numer. Meth. Eng.* **20** (10), 1291–1306 (1984).

32. J. O. Song, and R. E. Lee, Application of optimization to aircraft engine disk synthesis. *Proc. Int. Symp. on Optimum Structural Design.* Tucson, AZ (Oct. 1981).

33. Y. W. Chun and E. J. Haug, Two-dimensional shape optimum design. *Int. J. Numer. Meth. Eng.* **13**, 311–336 (1978).

34. E. J. Haug, K. K. Choi, J. W. Hou and Y. M. Yoo, A variational method for shape optimal design of elastic structures, in *New Directions in Optimum Structural Design.* (Edited by E. Atrek, R. H. Gallagher, K. M. Ragsdell and O. C. Zienkiewicz). John Wiley & Sons, New York (1984).

35. W. Prager, Conditions for optimality. *Comput. Struct.* **2**, 833–840 (1972).

36. M. S. Na, N. Kikuchi and J. E. Taylor, Optimal shape remodeling of linearly elastic plates using finite element methods. *Int. J. Numer. Meth. Eng.* **20** (10), 1823–1840 (1984).

37. M. S. Na, N. Kikuchi and J. E. Taylor, Optimal modification of shape for two-dimensional elastic bodies. *J. Struct. Mech.* **11**, 111–135 (1983).

38. Y. Tada and Y. Seguchi, Shape determination of structures based on the inverse variational principle/finite element approach, Ch. 8 in *New Directions in Optimal Structural Design.* (Edited by E. Atrek, R. H. Gallagher, K. M. Ragsdell and O. C. Zienkiewicz). John Wiley & Sons, New York (1984).

39. W. R. Spillers and S. Singh, Shape optimization: Finite element example. *Proc. ASCE, J. Struct. Div.*, Vol. 107, ST10, pp. 2015–2025 (Oct. 1981).

40. L. R. Friedland, Geometric structural behavior. Ph.D. Dissertation, Columbia University (1971).

DISCUSSION

D. Vasilopoulos *(General Motors Research Laboratories)*

In the beam example that you showed, you calculated the energy norm based on beam theory. Although the beam theory will give us very closely the displacement at the end, the energy of that solution may not be close to the solution of two-dimensional elasticity based on the fact that we have two singularities at the end. Since the energy norm contains derivatives, since the derivatives may be very different and since strains go to infinity, if we take the integral of the function as it goes to infinity, over a very small area that may be a significant contribution.

O. C. Zienkiewicz

Zienkiewicz

I agree with you in this particular case, if we just take the beam theory. In the L-shaped domain, however, we did not have a similar problem with singularity. I believe there was another example, not in this context but earlier on, when we did go through quite a bit of local refinement to see that difference. Barna Szabo has a good example of a very stubby beam, a square beam that is almost as long as it is broad. There is a very, very large difference with that beam. But in a beam of slender proportions, I believe we are very close to the exact solution using the beam theory. It was on that basis that the local efficiency (the microefficiency) was calculated. On the L-shaped domain, it was an exact solution which was used to find the total energy norm error which included singularities.

Vasilopoulos

My second question regards the theoretical basis for taking the difference between the smoothed and the direct value as an estimate of the local stresses. One might imagine that if our errors ranged from 40 to 60 percent, if we then smooth them out we would have an overall error of 50 percent; and if we take the difference we would have a difference of 10 percent, plus or minus. Is there a theoretical basis for taking the difference between smoothed and unsmoothed values as being the error?

Zienkiewicz

To be honest, we are still searching for a more precise theoretical basis. On error estimates for hierarchical and similar elements which were derived previously, we used a somewhat similar line of approach which involved integrating the residuals, etc., and also an estimate of the difference in the displacement situation. For these we have a sound theory. Here we proceed somewhat differently. We say that our next (mixed) solution is a bound of the overall error. The next speaker will possibly show an alternative derivation of bounding errors, the interpolation error. All I'm showing here is an intuitive interpretation of the same phenomenon. All of the examples done so far worked well. In fact we worked efficiency indices very comparable to those obtained by other methods—for instance, those used in earlier papers on hierarchic strategies.

It is well known that stress discontinuities provide a good idea of the accuracy of solution. It is precisely this that leads to the more quantitative basis used here.

Your question gives me a chance to say something I should have mentioned during the lecture. The problem with error estimation generally is the rather large cost of adding good error estimates into the program. With hierarchical projection, we have to do quite a bit of extra programming to obtain the local changes, and then integrals have to be carried out on an element-by-element basis, involving a special projection of residual integrals. This is liable to cost about as much as doing the original calculation. In the method given here, the calculation is extremely simple. It only takes one additional integration over each element. One projection suffices to get reasonable answers.

K. Izadpanah *(MacNeal-Schwendler Corporation)*

With regard to your L-shape domain in one of the sets of data that you presented, you have the worst effectivity index with a lot more degrees of freedom.

Am I misunderstanding your concept of the effectivity index? Do you think the effectivity index should converge?

Zienkiewicz

The local indices are not always precise, but the overall effectivity index actually did improve. Generally such improvement occurs also in the local one, although I didn't show it. Do you think that convergence of the effectivity index to unity should be guaranteed? This will probably occur, but a slight overestimate is acceptable.

Izadpanah

I would presume that with improving your mesh design and by having more degrees of freedom, your effectivity index would indicate the improvement, so that in your next iteration you would have guidance toward refinement or improvement of the model. However, I noticed in one of the diagrams that this was not the case.

Zienkiewicz

By and large the tables that I showed later do indicate an improvement of the effectivity index, but clearly you would get the guidance even if that were not true. Even if the effectivity index was oscillating a little, as long as the error was estimated within a certain bound, clearly you are improving on the solution. Theoretically, if the effectivity index has a very good foundation, it should go to unity in the limit. This one does, in most examples shown. For practical purposes, I would be happy to have effectivity indices of even 1.5 or thereabouts to get an indication of where I need to refine. I am referring to a very large reduction of error—sometimes you reduce the error locally five-fold.

Izadpanah

For a singular problem it has been shown that the results of the analysis will improve most when refining the mesh at the singularity. I was trying to figure out what exactly is the role of your effectivity index in an application. How could I use it if I have a mesh error situation?

Zienkiewicz

You would not use the effectivity index at all. Let me try to explain. The effectivity index is only known if the exact solution is known. Here it is used for the purpose of persuading the user that this error estimator is valid. In a real problem, I do not know the effectivity index. All I can do is assume that the estimate of error is reasonably correct. If the effectivity index is 1.5, it means the error estimate is incorrect by 50 percent; if it is .9, the error estimate is within 10 percent. I would use it here, in these dark-shaded areas of the diagrams, which show me that in certain areas we have approximately ten times more error than we are prepared to tolerate.

Izadpanah

What about those cases where other singularities are present in the problem?

Zienkiewicz

You start at uniform mesh. As you go on refining, you will come to a point where all areas are white, i.e., where the error is acceptable. This is what happened in the previous diagram where the refinement was specified.

L. A. Schmit *(University of California—Los Angeles)*

Would you propose that the refinement be guided somewhat by areas which seem to be critical in the design? In other words, if your error estimate was predicting significant errors in places that are not really design drivers, would you do the refinement anyway?

Zienkiewicz

No. Even the local error estimate is somewhat heuristic when the total error is outside permissible values. If I am only interested in the stress of some particular point of a structure, I would try to get the errors there down to some kind of sensible accuracy, without necessarily doing all the refinement near the singularity. All of us know that in the two theoretical problems I show, the stresses are infinite locally, so I cannot put a limit on them at the singularity. But I may be looking at some other stress and this may well decide the optimal design.

R. Haber *(University of Illinois—Champaign-Urbana)*

Is the stress-smoothing method that you are describing a variant of the Loubignac-Cantin iterative technique?

Zienkiewicz

It is similar but not identical. Loubignac uses simple averaging; we use a variational projection.

Haber

We found that there is a problem with this method for elements at the surface of a structure because they have no adjacent elements to average with. The improvement in accuracy at the surface is not as good as on the interior of the mesh. Have you experienced similar difficulties in your work?

Zienkiewicz

Yes. This takes one back to 1964-5 when simple triangles without mid-side nodes were used. Simple nodal averaging or smoothing gave results that were bad near the boundary. Similarly, if you look at the variational projection using the consistent matrix rather than the lumped form, the results are improved. The lumped form for triangular elements, however, corresponds to nothing else but averaging, and the results are generally worse.

MATERIAL DERIVATIVE METHODS FOR SHAPE DESIGN SENSITIVITY ANALYSIS

E. J. HAUG and K. K. CHOI

Center for Computer Aided Design
and
Department of Mechanical Engineering
The University of Iowa
Iowa City, Iowa

Abstract

Use of the material derivative concept of continuum mechanics to relate variations in structural shape to measures of structural performance is reviewed. Alternate formulations of design sensitivity analysis are obtained in the form of boundary integrals and domain integrals. Theoretical implications of the resulting formulations are developed to include the dependence of performance on only normal variation of the boundary and the need for evaluation of trace operators for projection of stress and strain related quantities onto the boundary. Computational implications of shape design sensitivity formulations are investigated, including compatibility of design sensitivity analysis with finite element and boundary element methods of structural analysis, parameterization of boundary shape and parameterization of design velocity fields. Implementation of shape design sensitivity analysis using finite element computer codes is discussed. Recent numerical results are used to illustrate the accuracy that is achievable using methods for material derivative shape design sensitivity analysis.

INTRODUCTION

A substantial literature has developed in the field of shape design sensitivity analysis and optimization of structural components [1, 2] over the past few years. Contributions to this field have been made using two fundamentally different approaches to structural modeling and analysis. The first approach uses a discretized structural model, based on finite element analysis, and proceeds to carry out shape design sensitivity analysis by controlling finite element node movement and differentiating the algebraic finite element equations [3-5]. While this approach permits direct application of classical design sensitivity analysis methods used earlier in structural optimization, it leads to algebraic complexity and problems of accuracy.

The second approach to shape design sensitivity analysis uses an elasticity model of the structure and the material derivative method of continuum mechanics to account for changes in shape of the structure [6-11]. Using this approach, expressions for design sensitivity in terms of domain shape change are derived in the continuous setting and evaluated using any available method of structural analysis; e.g., finite element analysis, boundary element analysis, photoelasticity, etc. While the theory underlying this approach is more complex than the discretization approach, it has produced better theoretical insights and more accurate numerical results [12].

The fundamental distinction between discrete and continuous approaches to shape design sensitivity analysis is not whether or not to discretize, but rather *when* to discretize. The discrete approach uses the finite element model to discretize the structure and carries out all calculations with that model. The continuous approach, in contrast, retains the elasticity formulation throughout the derivation of design sensitivity results and discretizes only to evaluate results.

The purpose of this paper is to present a summary of methods used in the development of the continuous approach for two- or three-dimensional elastic solids, and to draw a distinction between two representations of the resulting design sensitivities. Numerical considerations in implementing the alternate approaches are illustrated in two examples. Additional numerical experimentation that has been carried out recently is cited.

REPRESENTATION OF STRUCTURAL PERFORMANCE MEASURES

In rational design of structures, the cost function to be minimized and constraint functions to be bounded must be defined. Common quantitites associated with failure of a structural component are displacement, stress, strain energy density and other quantities that dictate failure if large values are achieved. In order to relate these quantities to the shape of the structural component, it is helpful to write them in the form of integrals over the structural component. For example, displacement $z^i(\hat{\mathbf{x}})$ at a specified point $\hat{\mathbf{x}}$ in a domain Ω of the structural component may be written, using the Dirac δ measure, as

$$\psi = \int\int_{\Omega} z^i(\mathbf{x})\delta(\mathbf{x} - \hat{\mathbf{x}})\, d\Omega. \tag{1}$$

Using this formulation, dependence of the displacement functional ψ on the shape of domain Ω arises in two forms. First, the explicit dependence of ψ on Ω in equation (1) is apparent. Second, the solution $\mathbf{z}(\mathbf{x})$ of the boundary value problem of elasticity depends on the shape of the domain and must be accounted for.

A more common form of structural performance measure involves stress in a structural component. Since stress is defined only in an integral mean square sense for general elasticity problems, it is natural to define stress performance measures as averaged over some small subdomain $\Omega_p \subset \Omega$. Such a measure may be written in the form

$$\psi = \int\int_{\Omega} g(\boldsymbol{\sigma}(\mathbf{z}))m_p(\mathbf{x})\, d\Omega \tag{2}$$

where $g(\boldsymbol{\sigma})$ is a stress measure, such as principal stress or von Mises stress, and $m_p(\mathbf{x})$ is an averaging or mollifying function that has a constant value on Ω_p and whose integral is one. Just as in the displacement measure of equation (1), the averaged stress measure depends on the shape of the domain Ω in two ways: first, it depends directly on the domain over which the integration is carried out, and second, on the displacement field $\mathbf{z}$ and associated stress $\boldsymbol{\sigma}$ that are solutions of boundary value problems that depend on the shape of the domain.

In order to account for the two forms of dependence of performance measures, such as those in equations (1) and (2), a variational formulation of the equations of elasticity is summarized in the next section, and its use with the material derivative concept of continuum mechanics is presented.

VARIATIONAL FORM OF ELASTICITY EQUATIONS

In order to be specific about the properties of shape design sensitivity analysis, it is helpful to formulate the variational equations for a typical problem. For this purpose, the three-dimensional linear elasticity problem for a body of arbitrary shape, shown in Figure 1, is considered. Results may then easily be reduced for planar models.

The strain tensor is defined as

$$\varepsilon^{ij}(\mathbf{z}) = \frac{1}{2}(z^i_j + z^j_i), \quad i,j = 1,2,3, \quad \mathbf{x} \in \Omega \tag{3}$$

where $\mathbf{z} = [z^1,\ z^2,\ z^3]^T$ is displacement. The stress-strain relation (generalized Hooke's law) is

$$\sigma^{ij}(\mathbf{z}) = \sum_{k,\ell=1}^{3} C^{ijk\ell}\varepsilon^{k\ell}(\mathbf{z}), \quad i,j,k,\ell = 1,2,3, \quad \mathbf{x} \in \Omega \tag{4}$$

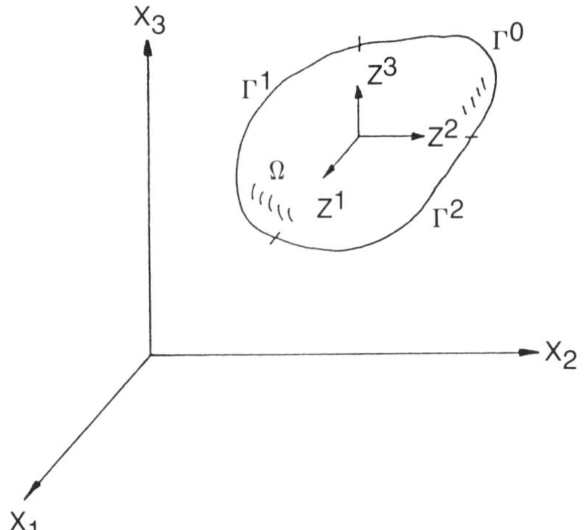

Figure 1. Three-dimensional elastic solid.

where **C** is the elastic modulus tensor, satisfying symmetry relations $C^{ijk\ell} = C^{jik\ell}$ and $C^{ijk\ell} = C^{ij\ell k}$, $i, j, k, \ell = 1, 2, 3$. The equilibrium equations [13] are

$$-\sum_{j=1}^{3} \sigma_j^{ij}(\mathbf{z}) = f^i, \quad i = 1, 2, 3, \qquad \mathbf{x} \in \Omega \tag{5}$$

with boundary conditions

$$z^i = 0, \quad i = 1, 2, 3, \qquad \mathbf{x} \in \Gamma^0 \tag{6}$$

$$T^{n_i}(\mathbf{z}) \equiv \sum_{j=1}^{3} \sigma^{ij}(\mathbf{z}) n_j = T^i, \qquad i = 1, 2, 3, \qquad \mathbf{x} \in \Gamma^2 \tag{7}$$

the boundary segment Γ^1 is traction free, where n_j is the jth component of the outward unit normal, $\mathbf{f} = [f^1, f^2, f^3]^T \in [C^1(\bar{\Omega})]^3$, and $\mathbf{T} = [T^1, T^2, T^3]^T \in [C^1(\Gamma)]^3$.

The formal operator equation of equation (5) may be reduced to a variational identity by multiplying both sides by an arbitrary virtual displacement vector $\bar{\mathbf{z}} = [\bar{z}^1, \bar{z}^2, \bar{z}^3]^T \in [H^1(\Omega)]^3$ and integrating by parts to obtain

$$\iiint_\Omega \left[\sum_{i,j=1}^{3} \sigma^{ij}(\mathbf{z}) \varepsilon^{ij}(\bar{\mathbf{z}}) \right] d\Omega - \iiint_\Omega \left[\sum_{i=1}^{3} f^i \bar{z}^i \right] d\Omega$$

$$= \iint_\Gamma \left[\sum_{i,j=1}^{3} \sigma^{ij}(\mathbf{z}) n_j \bar{z}^i \right] d\Gamma, \qquad \text{for all } \bar{\mathbf{z}} \in \left[H^1(\Omega) \right]^3. \tag{8}$$

If the boundary conditions of equations (6) and (7) are imposed, and with Γ^1 traction free, the variational equation is

$$a(\mathbf{z}, \bar{\mathbf{z}}) \equiv \int\int\int_{\Omega} \left[\sum_{i,j=1}^{3} \sigma^{ij}(\mathbf{z}) \varepsilon^{ij}(\bar{\mathbf{z}}) \right] d\Omega$$

$$= \int\int\int_{\Omega} \left[\sum_{i=1}^{3} f^i \bar{z}^i \right] d\Omega$$

$$+ \int\int_{\Gamma^2} \left[\sum_{i=1}^{3} T^i \bar{z}^i \right] d\Gamma \equiv \ell(\bar{\mathbf{z}}), \qquad \text{for all } \bar{\mathbf{z}} \in Z \tag{9}$$

where Z is the space of kinematically admissible displacements; i.e.,

$$Z = \{\mathbf{z} \in [H^1(\Omega)]^3 : z^i = 0, \qquad i = 1, 2, 3, \quad \mathbf{x} \in \Gamma^0\}. \tag{10}$$

For plane elasticity problems in which either all components of stress in the x_3 direction are zero or all components of strain in the x_3 direction are zero, equation (9) remains valid, with limits of summation running from 1 to 2 and an appropriate modification of the generalized Hooke's law of equation (4).

MATERIAL DERIVATIVE FOR SHAPE DESIGN SENSITIVITY ANALYSIS

The first step in shape design sensitivity analysis is development of relationships between a variation in shape of a structural component and the resulting variations in functionals that arise in the shape design problems described in the section on performance measures. Since the shape of domain that a structural component occupies is treated as the design variable, it is convenient to think of Ω as a continuous medium and to utilize the material derivative idea of continuum mechanics. In this section, the definition of material derivative is introduced and several material derivative formulas that will be used in later sections are derived.

Consider a domain Ω in one, two or three dimensions, shown schematically in Figure 2. Suppose that only one parameter τ defines the transformation $\mathbf{T}$, as shown in Figure 2. The mapping $\mathbf{T} : \mathbf{x} \to \mathbf{x}_\tau(\mathbf{x})$, $\mathbf{x} \in \Omega$, is given by

$$\left.\begin{aligned} \mathbf{x}_\tau &= \mathbf{T}(\mathbf{x}, \tau) \\ \Omega_\tau &\equiv \mathbf{T}(\Omega, \tau) \end{aligned}\right\} \tag{11}$$

The process of deforming Ω to Ω_τ by the mapping of equation (11) may be viewed as a dynamic process of deforming a continuum, with τ playing the role of time. At the initial time $\tau = 0$, the domain is Ω. Trajectories of points $\mathbf{x} \in \Omega$, begining at $\tau = 0$, can now be followed. The initial point moves to $\mathbf{x}_\tau = \mathbf{T}(\mathbf{x}, \tau)$. Thinking of τ as time, a design velocity can be defined as

$$\mathbf{V}(\mathbf{x}_\tau, \tau) \equiv \frac{d\mathbf{x}_\tau}{d\tau} = \frac{\partial \mathbf{T}(\mathbf{x}, \tau)}{\partial \tau}. \tag{12}$$

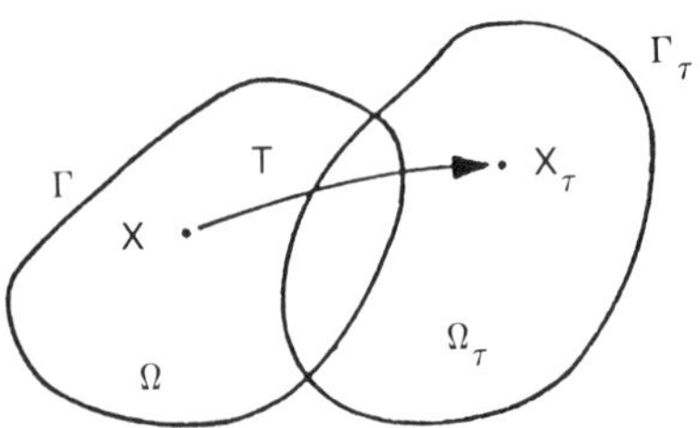

Figure 2. One parameter family of mappings.

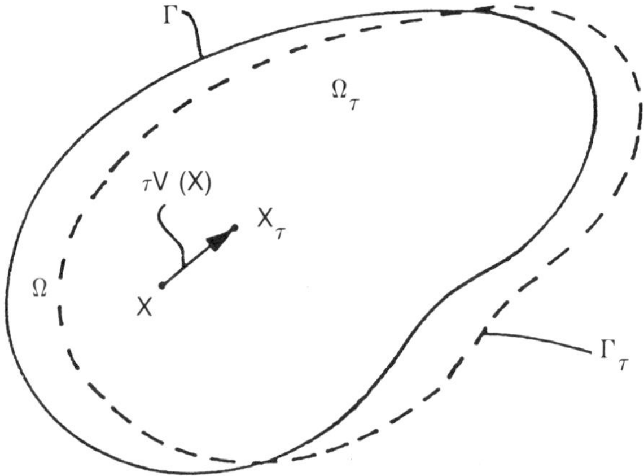

Figure 3. Variation of domain.

In a neighborhood of $\tau = 0$, under reasonable regularity hypotheses [2],

$$\begin{aligned}\mathbf{T}(\mathbf{x},\tau) &= \mathbf{T}(\mathbf{x},0) + \tau\frac{\partial\mathbf{T}}{\partial\tau}(\mathbf{x},0) + 0(\tau^2) \\ &= \mathbf{x} + \tau\mathbf{V}(\mathbf{x},0) + 0(\tau^2).\end{aligned}$$

Ignoring higher order terms,

$$\mathbf{T}(\mathbf{x},\tau) = \mathbf{x} + \tau\mathbf{V}(\mathbf{x}) \tag{13}$$

where $\mathbf{V}(\mathbf{x}) \equiv \mathbf{V}(\mathbf{x},0)$. In this paper, only the transformation $\mathbf{T}$ of equation (13) will be considered, the geometry of which is shown in Figure 3. Variations of the domain Ω by the design velocity field $\mathbf{V}(\mathbf{x})$ are denoted as $\Omega_\tau = \mathbf{T}(\Omega,\tau)$ and the boundary of Ω_τ is denoted as Γ_τ. Henceforth in the paper, the term *design velocity* will be referred to simply as *velocity*.

Let Ω be a C^k regular open set; i.e., its boundary Γ is closed and bounded and can be locally represented by a C^k function. Let $\mathbf{V}(\mathbf{x}) \in R^n$ in equation (13) be a vector defined on a neighborhood U of the closure $\bar{\Omega}$ of Ω, and let $\mathbf{V}(\mathbf{x})$ and its derivatives up to order $k \geq 1$ be continuous. With these hypotheses, it has been shown [14] that for small τ, $\mathbf{T}(\mathbf{x},\tau)$ is a homeomorphism (a one-to-one, continuous map with a continuous inverse) from U to $U_\tau \equiv \mathbf{T}(U,\tau)$, and that $\mathbf{T}(\mathbf{x},\tau)$ and its inverse mapping $\mathbf{T}^{-1}(\mathbf{x}_\tau,\tau)$ have C^k regularity and Ω_τ has C^k regularity.

Suppose $\mathbf{z}_\tau(\mathbf{x}_\tau)$ is a smooth solution of the elasticity equations. Then the mapping $\mathbf{z}_\tau(\mathbf{x}_\tau) \equiv \mathbf{z}_\tau(\mathbf{x}+\tau\mathbf{V}(\mathbf{x}))$ is defined on Ω, and $\mathbf{z}_\tau(\mathbf{x}_\tau)$ depends on τ in two ways. First, it is the solution of the boundary value problem on Ω_τ. Second, it is evaluated at a point $\mathbf{x}_\tau$ that moves with τ. The pointwise material derivative (which is shown to exist in reference [2]) at $\mathbf{x} \in \Omega$ is defined as

$$\dot{\mathbf{z}}(\mathbf{x}) = \frac{d}{d\tau}\mathbf{z}_\tau(\mathbf{x}+\tau\mathbf{V}(\mathbf{x}))\bigg|_{\tau=0} = \lim_{\tau\to 0}\frac{\mathbf{z}_\tau(\mathbf{x}+\tau\mathbf{V}(\mathbf{x}))-\mathbf{z}(\mathbf{x})}{\tau}. \tag{14}$$

If $\mathbf{z}_\tau$ has a regular extension to a neighborhood U_τ of $\bar{\Omega}_\tau$, then

$$\dot{\mathbf{z}}(\mathbf{x}) = \mathbf{z}'(\mathbf{x}) + \nabla\mathbf{z}^T\mathbf{V}(\mathbf{x}) \tag{15}$$

where

$$\mathbf{z}'(\mathbf{x}) \equiv \lim_{\tau\to 0}\frac{\mathbf{z}_\tau(\mathbf{x})-\mathbf{z}(\mathbf{x})}{\tau} \tag{16}$$

is the partial derivative of $\mathbf{z}$.

One attractive feature of the partial derivative is that, with reasonable smoothness assumptions, it commutes with the derivatives with respect to x_i[2]; i.e.,

$$(\frac{\partial \mathbf{z}}{\partial x_i})' = \frac{\partial}{\partial x_i}(\mathbf{z}'), \quad i = 1,2,3. \tag{17}$$

A pair of technical material derivative formulas that are used throughout the remainder of the paper are summarized in this section. Their proofs are presented in reference [2].

LEMMA 1: *Let ψ_1 be a domain functional, defined as an integral over Ω_τ,*

$$\psi_1 = \int\int_{\Omega_\tau} f_\tau(\mathbf{x}_\tau)d\Omega_\tau \tag{18}$$

where f_τ is a regular function defined on Ω_τ. If Ω has C^k regularity, then the material derivative of ψ_1 at Ω is

$$\psi_1' = \int\int_\Omega f'(\mathbf{x})d\Omega + \int_\Gamma f(\mathbf{x})(\mathbf{V}^T\mathbf{n})d\Gamma \tag{19}$$

or, equivalently,

$$\psi_1' = \int\int_\Omega [f'(\mathbf{x}) + \text{div }(f(\mathbf{x})\mathbf{V}(\mathbf{x}))]d\Omega. \tag{20}$$

References pp. 54–55

It is interesting and important to note that only the normal component $(\mathbf{V}^T\mathbf{n})$ of the boundary velocity appearing in equation (19) is needed to account for the effect of domain variation. In fact, it is shown by Theorem 3.5.2. of reference [2] that if a general domain functional ψ has a gradient at Ω and if Ω has C^{k+1} regularity, then only the normal component $(\mathbf{V}^T\mathbf{n})$ of the velocity field on the boundary is needed for derivative calculations.

In contrast to equation (19), use of the mathematically equivalent result given in equation (20) requires that the velocity field $\mathbf{V}(\mathbf{x})$ be defined throughout the domain Ω. Of course, it must be consistent with $(\mathbf{V}^T\mathbf{n})$ on Γ. Nevertheless, there are an infinite number of velocity fields that satisfy this condition, for each of which the result of equations (19) and (20) must be the same.

One of the principal points of this paper is that there are fundamentally different formulations of shape design sensitivity analysis that can be used to advantage with different numerical methods. The simple distinction between the second terms of equations (19) and (20), which follow from the divergence theorem, will be shown to lead to very different and complementary formulations of structural shape design sensitivity analysis.

Next, consider a functional defined as an integration over Γ_τ,

$$\psi_2 = \int_{\Gamma_\tau} g_\tau(\mathbf{x}_\tau)d\Gamma_\tau. \tag{21}$$

LEMMA 2: *Supppose* g_τ *in equation (21) is a regular function defined on* Γ_τ. *If* Ω *has a smooth boundary, the material derivative of* ψ_2 *is*

$$\psi_2' = \int_{\Gamma}[g'(\mathbf{x}) + (\nabla g^T\mathbf{n} + Hg(\mathbf{x}))(\mathbf{V}^T\mathbf{n})]d\Gamma \tag{22}$$

where H *is the curvature of* Γ *in* R^2 *and twice the mean curvature in* R^3.

ADJOINT VARIABLE FORMULATION OF SHAPE DESIGN SENSITIVITY ANALYSIS

As seen in the previous section, the static response of a structure depends on the shape of the domain. Existence of the material derivative $\dot{\mathbf{z}}$, which is proved in [2], and material derivative formulas presented in the previous section are used in this section to derive an adjoint variable method for design sensitivity analysis of general functionals. The derivation is presented here for an elastic solid without traction on the boundary, to demonstrate the method. The effect of tractions is accounted for in the more detailed development in the next section on stress shape design sensitivity analysis.

The variational equation of elasticity (9) on a deformed domain is of the form

$$\begin{aligned} a_\tau(\mathbf{z}_\tau, \bar{\mathbf{z}}_\tau) &\equiv \iint_{\Omega_\tau} c(\mathbf{z}_\tau, \bar{\mathbf{z}}_\tau)d\Omega_\tau \\ &= \iint_{\Omega_\tau} \mathbf{f}\bar{\mathbf{z}}_\tau d\Omega_\tau \equiv \ell_\tau(\bar{\mathbf{z}}_\tau), \qquad \text{for all } \bar{\mathbf{z}}_\tau \in Z_\tau \end{aligned} \tag{23}$$

where $\mathbf{z}_\tau$ is the space of kinematically admissible displacements on Ω_τ and $c(\cdot,\cdot)$ is a bilinear mapping that is defined by the integrand of equation (9).

Taking the material derivative of both sides of equation (23), using equation (19) and noting that the partial derivatives with respect to τ and $\mathbf{x}$ commute,

$$[a(\mathbf{z},\bar{\mathbf{z}})]' \equiv a'(\mathbf{z},\bar{\mathbf{z}}) + a(\dot{\mathbf{z}},\bar{\mathbf{z}}) = \ell'(\dot{\bar{\mathbf{z}}}), \qquad \text{for all } \bar{\mathbf{z}} \in Z \tag{24}$$

where, using equation (15),

$$\begin{aligned}[a(\mathbf{z},\bar{\mathbf{z}})]' &= \int\!\!\int_\Omega [c(\mathbf{z},\bar{\mathbf{z}}') + c(\mathbf{z}',\bar{\mathbf{z}})]d\Omega + \int_\Gamma c(\mathbf{z},\bar{\mathbf{z}})(\mathbf{V}^T\mathbf{n})d\Gamma \\ &= \int\!\!\int_\Omega [c(\mathbf{z},\dot{\bar{\mathbf{z}}} - \nabla\bar{\mathbf{z}}^T\mathbf{V}) + c(\dot{\mathbf{z}} - \nabla\mathbf{z}^T\mathbf{V},\bar{\mathbf{z}})]d\Omega \\ &\quad + \int_\Gamma c(\mathbf{z},\bar{\mathbf{z}})(\mathbf{V}^T\mathbf{n})d\Gamma \end{aligned} \tag{25}$$

and

$$\begin{aligned}\ell'(\bar{\mathbf{z}}) &= \int\!\!\int_\Omega \mathbf{f}\bar{\mathbf{z}}'d\Omega + \int_\Gamma \mathbf{f}\bar{\mathbf{z}}(\mathbf{V}^T\mathbf{n})d\Gamma \\ &= \int\!\!\int_\Omega \mathbf{f}(\dot{\bar{\mathbf{z}}} - \nabla\bar{\mathbf{z}}^T\mathbf{V})d\Omega + \int_\Gamma \mathbf{f}\bar{\mathbf{z}}(\mathbf{V}^T\mathbf{n})d\Gamma. \end{aligned} \tag{26}$$

The domain form of the last term from equation (20) could be used in equations (25) and (26). Use of this alternative is postponed until later.

For $\bar{\mathbf{z}}_\tau$, select $\bar{\mathbf{z}}_\tau(\mathbf{x} + \tau\mathbf{V}(\mathbf{x})) = \bar{\mathbf{z}}(\mathbf{x})$; i.e., choose $\bar{\mathbf{z}}$ as constant on the line $\mathbf{x}_\tau = \mathbf{x} + \tau\mathbf{V}(\mathbf{x})$. Then, since $[H^1(\Omega)]^3$ is preserved by $\mathbf{T}(\mathbf{x},\tau)$ (homeomorphism property noted in the previous section), if $\bar{\mathbf{z}}$ is an arbitrary element of $H^m(\Omega)$ that satisfies kinematic boundary conditions on $\Gamma, \bar{\mathbf{z}}_\tau$ is an arbitrary element of $H^m(\Omega_\tau)$ that satisfies kinematic boundary conditions on Γ_τ. In this case, using equation (15),

$$\dot{\bar{\mathbf{z}}} = \bar{\mathbf{z}}' + \nabla\bar{\mathbf{z}}^T\mathbf{V} = 0. \tag{27}$$

From equations (24), (25) and (26), using equation (27),

$$\begin{aligned}a'(\mathbf{z},\bar{\mathbf{z}}) &= -\int\!\!\int_\Omega [c(\mathbf{z},\nabla\bar{\mathbf{z}}^T\mathbf{V}) + c(\nabla\bar{\mathbf{z}}^T\mathbf{V},\bar{\mathbf{z}})]d\Omega \\ &\quad + \int_\Gamma c(\mathbf{z},\bar{\mathbf{z}})(\mathbf{V}^T\mathbf{n})d\Gamma \end{aligned} \tag{28}$$

and

$$\ell'(\bar{\mathbf{z}}) = -\int\!\!\int_\Omega \mathbf{f}(\nabla\bar{\mathbf{z}}^T\mathbf{V})d\Omega + \int_\Gamma \mathbf{f}\bar{\mathbf{z}}(\mathbf{V}^T\mathbf{n})\,d\Gamma. \tag{29}$$

Then, equation (24) can be rewritten to provide the result

$$\begin{aligned} a(\dot{\mathbf{z}}, \bar{\mathbf{z}}) &= \ell'(\bar{\mathbf{z}}) - a'(\mathbf{z}, \bar{\mathbf{z}}) \\ &= \iint_{\Omega} [c(\mathbf{z}, \nabla \bar{\mathbf{z}}^T \mathbf{V}) + c(\nabla \mathbf{z}^T \mathbf{V}, \bar{\mathbf{z}}) - \mathbf{f}(\nabla \bar{\mathbf{z}}^T \mathbf{V})] d\Omega \\ &\quad + \int_{\Gamma} [\mathbf{f}\bar{\mathbf{z}} - c(\mathbf{z}, \bar{\mathbf{z}})] \quad (\mathbf{V}^T \mathbf{n}) d\Gamma, \qquad \text{for all } \bar{\mathbf{z}} \in Z. \end{aligned} \tag{30}$$

Consider a general functional that may be written in integral form as

$$\psi = \iint_{\Omega_\tau} g(\mathbf{z}_\tau, \nabla \mathbf{z}_\tau) d\Omega_\tau \tag{31}$$

where $\mathbf{z} \in [H^1(\Omega)]^3$, and the function g is continuously differentiable with respect to its arguments. Taking the variations of the functional of equation (31), using the material derivative formulas of equations (17) and (19),

$$\psi' = \iint_{\Omega} [g_{\mathbf{z}} \mathbf{z}' + g_{\nabla \mathbf{z}} \nabla \mathbf{z}'] d\Omega + \int_{\Gamma} g(\mathbf{V}^T \mathbf{n}) d\Gamma. \tag{32}$$

Using equation (15), equation (32) can be rewritten as

$$\begin{aligned} \psi' &= \iint_{\Omega} [g_{\mathbf{z}} \dot{\mathbf{z}} + g_{\nabla \mathbf{z}} \nabla \dot{\mathbf{z}} - g_{\mathbf{z}} (\nabla \mathbf{z}^T \mathbf{V}) - g_{\nabla \mathbf{z}} \nabla (\nabla \mathbf{z}^T \mathbf{V})] d\Omega \\ &\quad + \int_{\Gamma} g(\mathbf{V}^T \mathbf{n}) d\Gamma. \end{aligned} \tag{33}$$

Equation (20) could be used to write this result as a domain integral. Note that $\dot{\mathbf{z}}$ and $\nabla \dot{\mathbf{z}}$ depend on the velocity field $\mathbf{V}$.

The objective now is to obtain an explicit expression for ψ' in terms of the velocity field $\mathbf{V}$, which requires rewriting the first two terms of the first integral on the right of equation (33) explicitly in terms of $\mathbf{V}$; i.e., eliminating $\dot{\mathbf{z}}$. An adjoint equation is introduced by replacing $\dot{\mathbf{z}} \in Z$ in equation (33) by a virtual displacement $\bar{\boldsymbol{\lambda}} \in Z$, and equating terms involving $\bar{\boldsymbol{\lambda}}$ to the energy bilinear form, yielding the adjoint equation for the adjoint variable $\boldsymbol{\lambda}$,

$$a(\boldsymbol{\lambda}, \bar{\boldsymbol{\lambda}}) = \iint_{\Omega} [g_{\mathbf{z}} \bar{\boldsymbol{\lambda}} + g_{\nabla \mathbf{z}} \nabla \bar{\boldsymbol{\lambda}}] d\Omega, \qquad \text{for all } \bar{\boldsymbol{\lambda}} \in Z. \tag{34}$$

To take advantage of the adjoint equation, evaluate equation (34) at $\bar{\boldsymbol{\lambda}} = \dot{\mathbf{z}}$, since $\dot{\mathbf{z}} \in Z$, to obtain the expression

$$a(\boldsymbol{\lambda}, \dot{\mathbf{z}}) = \iint_{\Omega} [g_{\mathbf{z}} \dot{\mathbf{z}} + g_{\nabla \mathbf{z}} \nabla \dot{\mathbf{z}}] d\Omega. \tag{35}$$

Similarly, evaluate the identity of equation (30) at $\bar{\mathbf{z}} = \boldsymbol{\lambda}$, since both are in Z, to obtain

$$a(\dot{\mathbf{z}}, \boldsymbol{\lambda}) = \ell'(\boldsymbol{\lambda}) - a'(\mathbf{z}, \boldsymbol{\lambda}). \tag{36}$$

Recalling that the energy bilinear form a(·,·) is symmetric in its arguments, the left sides of equations (35) and (36) are equal, so

$$\int\int_{\Omega} [g_{\mathbf{z}}\dot{\mathbf{z}} + g_{\nabla\mathbf{z}}\nabla\dot{\mathbf{z}}]d\Omega = \ell'(\boldsymbol{\lambda}) - a'(\mathbf{z}, \boldsymbol{\lambda}). \tag{37}$$

Using equations (30) and (37), equation (33) yields

$$\begin{aligned}\psi' = \int\int_{\Omega} [&c(\mathbf{z}, \nabla\boldsymbol{\lambda}^T\mathbf{V}) - \mathbf{f}(\nabla\boldsymbol{\lambda}^T\mathbf{V}) + c(\nabla\mathbf{z}^T\mathbf{V}, \boldsymbol{\lambda}) - g_{\mathbf{z}}(\nabla\mathbf{z}^T\mathbf{V}) \\ &- g_{\nabla\mathbf{z}}\nabla(\nabla\mathbf{z}^T\mathbf{V})]d\Omega + \int_{\Gamma} [g + \mathbf{f}\boldsymbol{\lambda} - c(\mathbf{z}, \boldsymbol{\lambda})](\mathbf{V}^T\mathbf{n})d\Gamma. \end{aligned} \tag{38}$$

The integrals in equation (38) can be transformed to either boundary or domain forms by using the variational identities (given in the previous section on variational forms of elasticity equations) for each structural component and associated boundary conditions. The fact that the design sensitivity ψ' can be expressed as a boundary or domain integral gives significant advantages in numerical calculations, depending on the method of analysis used.

Note that evaluation of the design sensitivity formula of equation (38) requires solution of equation (23) for $\mathbf{z}$. Similarly, equation (34) must be solved for the adjoint variable $\boldsymbol{\lambda}$. This is an efficient calculation, using finite element analysis, if the boundary value problem for $\mathbf{z}$ has already been solved; it then requires only evaluation of the solution of the same set of finite element equations with a different right side, called an adjoint load.

STRESS SHAPE DESIGN SENSITIVITY ANALYSIS

Shape design sensitivity analysis of locally averaged stress in linear elasticity is carried out here, using the adjoint variable method. For plane stress or plane strain problems, the formulas (3) through (10) remain valid, with limits of summation running from 1 to 2 and with an appropriate modification of generalized Hooke's law. The first result yields a boundary integral representation and the second yields a domain integral representation.

Boundary Representation—Consider the three dimensional elasticity problem with a mean stress constraint over a fixed test volume Ω_p, such that $\bar{\Omega}_p \subset \Omega$,

$$\psi = \int\int\int_{\Omega} g(\boldsymbol{\sigma}(\mathbf{z}))m_p d\Omega \tag{39}$$

where $\boldsymbol{\sigma}$ denotes the stress tensor, Ω_p is an open set, and m_p is a characteristic function that is constant on Ω_p (zero outside of Ω_p) whose integral is 1. Here, g is

References pp. 54–55

assumed to be continuously differentiable with respect to its arguments. Note that $g(\boldsymbol{\sigma}(\mathbf{z}))$ might involve principal stresses, von Mises failure criterion or some other material failure criteria. While the integrand in equation (39) could be written explicitly in terms of the gradient of $\mathbf{z}$, it will be more effective to continue with the present notation.

For boundary variation in the elasticity problem, it is supposed that the boundary $\Gamma = \Gamma^0 \cup \Gamma^1 \cup \Gamma^2$ is varied, except that the curve $\partial\Gamma^2$ that bounds the loaded surface Γ^2 is fixed, so the velocity field $\mathbf{V}$ at $\partial\Gamma^2$ is zero. For the case in which $\partial\Gamma^2$ is not fixed, variation of the traction term in equation (9) (given as an integral over Γ^2) gives an additional term that was not discussed. For this case, the interested reader is referred to reference [15]. A conservative load is applied that depends on position but not shape of the boundary. Taking the variation of equation (9), using equations (17), (19) and (22) and the fact that $f^{i'} = T^{i'} = 0$,

$$\begin{aligned}
&\iiint_\Omega \sum_{i,j=1}^{3} \left[\sigma^{ij}(\mathbf{z}')\varepsilon^{ij}(\bar{\mathbf{z}}) + \sigma^{ij}(\mathbf{z})\varepsilon^{ij}(\bar{\mathbf{z}}')\right] d\Omega + \iint_\Gamma \left[\sum_{i,j=1}^{3} \sigma^{ij}(z)\varepsilon^{ij}(\bar{\mathbf{z}})\right](\mathbf{V}^T\mathbf{n})d\Gamma \\
&\quad = \iiint_\Omega \left[\sum_{i=1}^{3} f^i \bar{z}^{i'}\right] d\Omega + \iint_{\Gamma^1\cup\Gamma^2} \left[\sum_{i=1}^{3} f^i \bar{z}^i\right](\mathbf{V}^T\mathbf{n})d\Gamma \\
&\quad + \iint_{\Gamma^2} \left[\sum_{i=1}^{3} T^i \bar{z}^{i'}\right] d\Gamma + \iint_{\Gamma^2} \sum_{i=1}^{3} \left[\nabla(T^i\bar{z}^i)^T\mathbf{n} + H(T^i\bar{z}^i)\right](\mathbf{V}^T\mathbf{n})\, d\Gamma, \\
&\quad \text{for all } \bar{\mathbf{z}} \in Z.
\end{aligned} \tag{40}$$

Using equations (15) and (27), equation (40) can be rewritten as

$$\begin{aligned}
a(\dot{\mathbf{z}}, \bar{\mathbf{z}}) &\equiv \iiint_\Omega \left[\sum_{i,j=1}^{3} \sigma^{ij}(\dot{\mathbf{z}})\varepsilon^{ij}(\bar{\mathbf{z}})\right] d\Omega \\
&= \iiint_\Omega \sum_{i,j=1}^{3} \left[\sigma^{ij}(\mathbf{z})\varepsilon^{ij}(\nabla\bar{\mathbf{z}}^T\mathbf{V}) + \sigma^{ij}(\nabla\mathbf{z}^T\mathbf{V})\varepsilon^{ij}(\bar{\mathbf{z}})\right] d\Omega \\
&\quad - \iiint_\Omega \left[\sum_{i=1}^{3} f^i(\nabla\bar{z}^{i^T}\mathbf{V})\right] d\Omega - \iint_\Gamma \left[\sum_{i,j=1}^{3} \sigma^{ij}(\mathbf{z})\varepsilon^{ij}(\bar{\mathbf{z}})\right](\mathbf{V}^T\mathbf{n})d\Gamma \\
&\quad + \iint_{\Gamma^1\cup\Gamma^2} \left[\sum_{i=1}^{3} f^i \bar{z}^i\right](\mathbf{V}^T\mathbf{n})d\Gamma \\
&\quad + \iint_{\Gamma^2} \sum_{i=1}^{3} \{-T^i(\nabla\bar{z}^{i^T}\mathbf{V}) + [\nabla(T^i\bar{z}^i)^T\mathbf{n} + H(T^i\bar{z}^i)](\mathbf{V}^T\mathbf{n})\}\, d\Gamma, \\
&\quad \text{for all } \bar{\mathbf{z}} \in Z.
\end{aligned} \tag{41}$$

Taking the variation of the functional of equation (39), and using material derivative formulas of equations (15) and (19) and $m'_p = 0$,

$$\psi' = \iiint_\Omega \left[\sum_{i,j=1}^{3} g_{\sigma^{ij}}(\mathbf{z})\sigma^{ij}(\mathbf{z}') \right] m_p d\Omega + \iint_\Gamma g(\boldsymbol{\sigma}(\mathbf{z})) m_p (\mathbf{V}^T\mathbf{n}) d\Gamma$$
$$= \iiint_\Omega \sum_{i,j=1}^{3} g_{\sigma^{ij}}(\mathbf{z})[\sigma^{ij}(\dot{\mathbf{z}}) - \sigma^{ij}(\nabla\mathbf{z}^T\mathbf{V})] m_p \, d\Omega \tag{42}$$

because $m_p = 0$ on Γ.

As in the general derivation of equation (34), the material derivative of state $\dot{\mathbf{z}} \in Z$ may be replaced by a virtual displacement $\bar{\boldsymbol{\lambda}}$ in the first term on the right side of equation (42) to define a load functional for the adjoint equation, obtaining, as in equation (34),

$$a(\boldsymbol{\lambda}, \bar{\boldsymbol{\lambda}}) = \iiint_\Omega \left[\sum_{i,j=1}^{3} g_{\sigma^{ij}}(\mathbf{z})\sigma^{ij}(\bar{\boldsymbol{\lambda}}) \right] m_p \, d\Omega, \qquad \text{for all } \bar{\boldsymbol{\lambda}} \in Z. \tag{43}$$

With smoothness assumptions, equation (43) is equivalent to the formal operator equation

$$-\sum_{j=1}^{3} \sigma_j^{ij}(\boldsymbol{\lambda}) = -\sum_{j=1}^{3} \left(\sum_{k,\ell=1}^{3} g_{\sigma k\ell}(\mathbf{z}) C^{k\ell ij} m_p \right)_j, \quad i = 1,2,3, \qquad \mathbf{x} \in \Omega \tag{44}$$

with boundary conditions

$$\lambda^i = 0, \quad i = 1,2,3, \qquad \mathbf{x} \in \Gamma^0 \tag{45}$$

$$\sum_{j=1}^{3} \sigma^{ij}(\lambda) n_j = 0, \quad i = 1,2,3, \qquad \mathbf{x} \in \Gamma^1 \cup \Gamma^2. \tag{46}$$

The derivative on the right of equation (44) is in the sense of the theory of distributions. The distributional derivatives m_{p_j}, $j = 1,2,3$, depend on the equations that represent the boundary of Ω_p[16].

Multiplying equation (44) by $\bar{\lambda} \in [H^1(\Omega)]^3$ and integrating by parts,

$$\iiint_\Omega \left[\sum_{i,j=1}^{3} \sigma^{ij}(\boldsymbol{\lambda})\bar{\lambda}_j^i \right] d\Omega - \iint_\Gamma \left[\sum_{i,j=1}^{3} \sigma^{ij}(\boldsymbol{\lambda}) n_j \bar{\lambda}^i \right] d\Gamma$$
$$= \iiint_\Omega \sum_{i,j=1}^{3} \left[\sum_{k,\ell=1}^{3} g_{\sigma^{k\ell}}(\mathbf{z}) C^{k\ell ij} m_p \right] \bar{\lambda}_j^i \, d\Omega$$
$$- \iint_\Gamma \sum_{i,j=1}^{3} \left[\sum_{k,\ell=1}^{3} g_{\sigma^{k\ell}}(\mathbf{z}) C^{k\ell ij} m_p \right] n_j \bar{\lambda}^i d\Gamma.$$

References pp. 54–55

Since $\sigma^{ij}(\boldsymbol{\lambda}) = \sigma^{ji}(\boldsymbol{\lambda})$ and $C^{k\ell ij} = C^{k\ell ji}$, using equations (3) and (4) the above equation becomes the variational identity

$$\begin{aligned}
&\iiint_\Omega \left[\sum_{i,j=1}^{3} \sigma^{ij}(\boldsymbol{\lambda})\varepsilon^{ij}(\bar{\boldsymbol{\lambda}})\right] d\Omega - \iiint_\Omega \left[\sum_{i,j=1}^{3} g_{\sigma^{ij}}(\mathbf{z})\sigma^{ij}(\bar{\boldsymbol{\lambda}})\right] m_p d\Omega \\
&= \iint_\Gamma \left[\sum_{i,j=1}^{3} \sigma^{ij}(\boldsymbol{\lambda}) n_j \bar{\lambda}^i\right] d\Gamma \\
&\quad - \iint_\Gamma \sum_{i,j=1}^{3} \left[\sum_{k,\ell=1}^{3} g_{\sigma^{k\ell}}(\mathbf{z}) C^{k\ell ij} m_p\right] \bar{\lambda}^i n_j d\Gamma, \quad \text{for all } \bar{\boldsymbol{\lambda}} \in [H^1(\Omega)]^3. \qquad (47)
\end{aligned}$$

Since $\dot{\mathbf{z}} \in Z$, equation (43) may be evaluated at $\bar{\boldsymbol{\lambda}} = \dot{\mathbf{z}}$ to obtain

$$a(\boldsymbol{\lambda}, \dot{\mathbf{z}}) = \iiint_\Omega \left[\sum_{i,j=1}^{3} g_{\sigma^{ij}}(\mathbf{z})\sigma^{ij}(\dot{\mathbf{z}})\right] m_p d\Omega. \qquad (48)$$

Similarly, since $\bar{\mathbf{z}} \in Z$ and $\boldsymbol{\lambda} \in Z$, equation (41) may be evaluated at $\bar{\mathbf{z}} = \boldsymbol{\lambda}$ to obtain

$$\begin{aligned}
a(\dot{\mathbf{z}}, \boldsymbol{\lambda}) &= \iiint_\Omega \sum_{i,j=1}^{3} [\sigma^{ij}(\mathbf{z})\varepsilon^{ij}(\nabla\boldsymbol{\lambda}^T\mathbf{V}) + \sigma^{ij}(\nabla\mathbf{z}^T\mathbf{V})\varepsilon^{ij}(\boldsymbol{\lambda})] d\Omega \\
&\quad - \iiint_\Omega \left[\sum_{i=1}^{3} f^i(\nabla\lambda^{i^T}\mathbf{V})\right] d\Omega - \iint_\Gamma \left[\sum_{i,j=1}^{3} \sigma^{ij}(\mathbf{z})\varepsilon^{ij}(\boldsymbol{\lambda})\right] (\mathbf{V}^T\mathbf{n}) d\Gamma \\
&\quad + \iint_{\Gamma^1\cup\Gamma^2} \left[\sum_{i=1}^{3} f^i\lambda^i\right] (\mathbf{V}^T\mathbf{n}) d\Gamma \\
&\quad + \iint_{\Gamma^2} \sum_{i=1}^{3} \{-T^i(\nabla\lambda^{i^T}\mathbf{V}) + [\nabla(T^i\lambda^i)^T\mathbf{n} + H(T^i\lambda^i)](\mathbf{V}^T\mathbf{n})\} d\Gamma. \qquad (49)
\end{aligned}$$

By Betti's reciprocal theorem [13],

$$\begin{aligned}
a(\mathbf{z}, \bar{\mathbf{z}}) &\equiv \iiint_\Omega \left[\sum_{i,j=1}^{3} \sigma^{ij}(\mathbf{z})\varepsilon^{ij}(\bar{\mathbf{z}})\right] d\Omega \\
&= \iiint_\Omega \left[\sum_{i,j=1}^{3} \sigma^{ij}(\bar{\mathbf{z}})\varepsilon^{ij}(\mathbf{z})\right] d\Omega \equiv a(\bar{\mathbf{z}}, \mathbf{z}), \quad \text{for all } \mathbf{z}, \bar{\mathbf{z}} \in [H^1(\Omega)]^3. \qquad (50)
\end{aligned}$$

Thus, $a(\dot{\mathbf{z}}, \boldsymbol{\lambda}) = a(\boldsymbol{\lambda}, \dot{\mathbf{z}})$ and equations (42), (48) and (50) yield

$$\begin{aligned}\psi' &= \iiint_\Omega \sum_{i,j=1}^{3} [\sigma^{ij}(\mathbf{z})\varepsilon^{ij}(\nabla\boldsymbol{\lambda}^T\mathbf{V}) + \sigma^{ij}(\boldsymbol{\lambda})\varepsilon^{ij}(\nabla\mathbf{z}^T\mathbf{V})]d\Omega \\ &- \iiint_\Omega \left[\sum_{i=1}^{3} f^i(\nabla\lambda^{i^T}\mathbf{V})\right] d\Omega - \iiint_\Omega \sum_{i,j=1}^{3} [g_{\sigma ij}(\mathbf{z})\sigma^{ij}(\nabla\mathbf{z}^T\mathbf{V})]m_p d\Omega \\ &- \iint_\Gamma \left[\sum_{i,j=1}^{3} \sigma^{ij}(\mathbf{z})\varepsilon^{ij}(\boldsymbol{\lambda})\right] (\mathbf{V}^T\mathbf{n})d\Gamma + \iint_{\Gamma^1\cup\Gamma^2} \left[\sum_{i=1}^{3} f^i\lambda^i\right] (\mathbf{V}^T\mathbf{n})d\Gamma \\ &+ \iint_{\Gamma^2} \sum_{i=1}^{3} \{-T^i(\nabla\lambda^{i^T}\mathbf{V}) + [\nabla(T^i\lambda^i)^T\mathbf{n} + H(T^i\lambda^i)](\mathbf{V}^T\mathbf{n})\}d\Gamma. \end{aligned} \quad (51)$$

As before, the variational identities of equations (8) and (47) may be used to transform the domain integrals of equation (51) to boundary integrals by identifying $\bar{\mathbf{z}}$ in equation (8) and $\bar{\boldsymbol{\lambda}}$ in equation (47) with $(\nabla\boldsymbol{\lambda}^T\mathbf{V})$ and $(\nabla\mathbf{z}^T\mathbf{V})$, respectively, in equation (51), obtaining

$$\begin{aligned}\psi' &= \iint_\Gamma \left[\sum_{i,j=1}^{3} \sigma^{ij}(\mathbf{z})n_j(\nabla\lambda^{i^T}\mathbf{V})\right] d\Gamma + \iint_\Gamma \{ \sum_{i,j=1}^{3} \sigma^{ij}(\boldsymbol{\lambda})n_j(\nabla z^{i^T}\mathbf{V}) \\ &- \sum_{i,j=1}^{3} \left[\sum_{k,\ell=1}^{3} g_{\sigma k\ell}(\mathbf{z})C^{k\ell ij}m_p\right] n_j(\nabla z^{i^T}\mathbf{V})\}d\Gamma \\ &- \iint_\Gamma \left[\sum_{i,j=1}^{3} \sigma^{ij}(\mathbf{z})\varepsilon^{ij}(\boldsymbol{\lambda})\right] (\mathbf{V}^T\mathbf{n})d\Gamma - \iint_{\Gamma^1\cup\Gamma^2} \left[\sum_{i=1}^{3} f^i\lambda^i\right] (\mathbf{V}^T\mathbf{n})d\Gamma \\ &+ \iint_{\Gamma^2} \sum_{i=1}^{3} \{-T^i(\nabla\lambda^{i^T}V) + [\nabla(T^i\lambda^i)^T\mathbf{n} + H(T^i\lambda^i)](\mathbf{V}^T\mathbf{n})\}d\Gamma. \end{aligned} \quad (52)$$

Since $\bar{\Omega}_p \subset \Omega, m_p = 0$ on Γ. Using boundary conditions of equations (7) and (46), equation (52) becomes

$$\begin{aligned}\psi' &= \iiint_{\Gamma^0} \sum_{i,j=1}^{3} [\sigma^{ij}(\mathbf{z})n_j(\nabla\lambda^{i^T}\mathbf{V}) + \sigma^{ij}(\boldsymbol{\lambda})n_j(\nabla z^{i^T}\mathbf{V})]d\Gamma \\ &- \iint_\Gamma \left[\sum_{i,j=1}^{3} \sigma^{ij}(\mathbf{z})\varepsilon^{ij}(\boldsymbol{\lambda})\right] (\mathbf{V}^T\mathbf{n})d\Gamma + \iint_{\Gamma^1\cup\Gamma^2} \left[\sum_{i=1}^{3} f^i\lambda^i\right] (\mathbf{V}^T\mathbf{n})d\Gamma \\ &+ \iint_{\Gamma^2} \sum_{i=1}^{3} [\nabla(T^i\lambda^i)^T\mathbf{n} + H(T^i\lambda^i)](\mathbf{V}^T\mathbf{n})d\Gamma. \end{aligned} \quad (53)$$

References pp. 54–55

On Γ^0, $\mathbf{z} = \boldsymbol{\lambda} = 0$ implies $\nabla z^i = (\nabla z^{i^T}\mathbf{n})\mathbf{n}$ and $\nabla \lambda^i = (\nabla \lambda^{i^T}\mathbf{n})\mathbf{n}$. Hence, equation (53) becomes

$$\psi' = \int\int_{\Gamma^0} \sum_{i,j=1}^{3} [\sigma^{ij}(\mathbf{z})n_j(\nabla \lambda^{i^T}\mathbf{n}) + \sigma^{ij}(\boldsymbol{\lambda})n_j(\nabla z^{i^T}\mathbf{n})](\mathbf{V}^T\mathbf{n})d\Gamma$$
$$- \int\int_{\Gamma} \left[\sum_{i,j=1}^{3} \sigma^{ij}(\mathbf{z})\varepsilon^{ij}(\boldsymbol{\lambda})\right](\mathbf{V}^T\mathbf{n})d\Gamma + \int\int_{\Gamma^1\cup\Gamma^2} \left[\sum_{i=1}^{3} f^i\lambda^i\right](\mathbf{V}^T\mathbf{n})d\Gamma$$
$$+ \int\int_{\Gamma^2} \sum_{i=1}^{3} [\nabla(T^i\lambda^i)]^T\mathbf{n} + H(T^i\lambda^i)](\mathbf{V}^T\mathbf{n})d\Gamma \tag{54}$$

which is the desired result.

Computational Considerations in Boundary Representation—To calculate design sensitivity information of equation (54) numerically, stresses, strain and/or normal derivatives of state and adjoint variables must be evaluated on the boundary. Thus, when a numerical method such as the finite element method is used for analysis, the accuracy of finite element results for state and adjoint variables and their derivatives on the boundary becomes critical. It is well known [17] that results of finite element analysis on the boundary may not be satisfactory for a system with nonsmooth load and for interface problems. This is of particular concern in stress design sensitivity analysis, since the adjoint load for an average stress constraint is a concentrated load on the subdomain Ω_p, over which stress is averaged.

Several methods might be considered to overcome this difficulty. The first choice is to use a finite element method that gives accurate results on the boundary. A second choice is to use a different numerical method, such as the boundary element method [18,19]. In the finite element method, the unknown function, e.g., displacement, is approximated by shape functions that do not satisfy the governing equations but usually satisfy kinematic boundary conditions. Nodal parameters z^i, e.g., nodal displacements, are then determined by approximate satisfaction of both differential equations and nonkinematic boundary conditions, in a domain integral mean sense. On the other hand, in the boundary element method, approximating functions satisfy the governing equations in the domain, but not the boundary conditions. Nodal parameters are determined by approximate satisfaction of boundary conditions in a weighted boundary integral sense. An important advantage of the boundary element method in the boundary representation of shape design sensitivity analysis is that it better represents boundary conditions and is usually more accurate in determining stress at the boundary.

Domain Representation—Another method to be investigated is the use of domain information to best utilize the basic character of finite element analysis. To develop a domain method [20], consider the material derivative formulas of Lemma 1. Instead of using equation (19), the result given in equation (20), which requires information on the domain rather than on the boundary, can be used. Taking the variation of equation (9), using equations (17), (20) and (22) plus the fact that $f^{i\prime} = T^{i\prime} = 0$,

$$\iiint_\Omega \sum_{i,j=1}^{3} [\sigma^{ij}(\mathbf{z}')\varepsilon^{ij}(\bar{\mathbf{z}}) + \sigma^{ij}(\mathbf{z})\varepsilon^{ij}(\bar{\mathbf{z}}')]d\Omega + \iiint_\Omega \nabla \left[\sum_{i,j=1}^{3} \sigma^{ij}(\mathbf{z})\varepsilon^{ij}(\bar{\mathbf{z}})\right]^T \mathbf{V} d\Omega$$
$$+ \iiint_\Omega \left[\sum_{i,j=1}^{3} \sigma^{ij}(\mathbf{z})\varepsilon^{ij}(\bar{\mathbf{z}})\right] \text{div } \mathbf{V} d\Omega$$
$$= \iiint_\Omega \left[\sum_{i=1}^{3} f^i \bar{z}^{i'}\right] d\Omega + \iiint_\Omega \nabla \left[\sum_{i=1}^{3} f^i \bar{z}^i\right]^T \mathbf{V} d\Omega$$
$$+ \iiint_\Omega \left[\sum_{i=1}^{3} f^i \bar{z}^i\right] \text{div } \mathbf{V} d\Omega + \iint_{\Gamma^2} \left[\sum_{i=1}^{3} T^i \bar{z}^{i'}\right] d\Gamma$$
$$+ \iint_{\Gamma^2} \left\{\nabla \left[\sum_{i=1}^{3} T^i \bar{z}^i\right]^T \mathbf{n} + H \left[\sum_{i=1}^{3} T^i \bar{z}^i\right]\right\} (\mathbf{V}^T \mathbf{n}) d\Gamma,$$
$$\text{for all } \bar{\mathbf{z}} \in Z. \tag{55}$$

Using equations (15) and (27), equation (55) can be rewritten as

$$\iiint_\Omega \sum_{i,j=1}^{3} [\sigma^{ij}(\dot{\mathbf{z}})\varepsilon^{ij}(\bar{\mathbf{z}}) - \sigma^{ij}(\mathbf{z})\varepsilon^{ij}(\nabla\bar{\mathbf{z}}^T \mathbf{V}) - \sigma^{ij}(\bar{\mathbf{z}})\varepsilon^{ij}(\nabla \mathbf{z}^T \mathbf{V})]d\Omega$$
$$+ \iiint_\Omega \nabla \left[\sum_{i,j=1}^{3} \sigma^{ij}(\mathbf{z})\varepsilon^{ij}(\bar{\mathbf{z}})\right]^T \mathbf{V} d\Omega$$
$$+ \iiint_\Omega \left[\sum_{i,j=1}^{3} \sigma^{ij}(\mathbf{z})\varepsilon^{ij}(\bar{\mathbf{z}})\right] \text{div } \mathbf{V} d\Omega$$
$$= \iiint_\Omega \sum_{i=1}^{3} \bar{z}^i (\nabla f^{i^T} \mathbf{V}) d\Omega + \iiint_\Omega \left[\sum_{i=1}^{3} f^i \bar{z}^i\right] \text{div } \mathbf{V} d\Omega$$
$$+ \iint_{\Gamma^2} \left\{ -\sum_{i=1}^{3} T^i (\nabla \bar{z}^{i^T} \mathbf{V}) + \left(\nabla \left[\sum_{i=1}^{3} T^i \bar{z}^i\right]^T \mathbf{n}\right.\right.$$
$$\left.\left. + H \left[\sum_{i=1}^{3} T^i \bar{z}^i\right]\right) (\mathbf{V}^T \mathbf{n}) \right\} d\Gamma, \quad \text{for all } \bar{\mathbf{z}} \in \mathrm{Z}. \tag{56}$$

It can be verified that

$$\sum_{i,j=1}^{3} \sigma^{ij}(\mathbf{z})\varepsilon^{ij}(\nabla\bar{\mathbf{z}}^T \mathbf{V}) = \sum_{i,j=1}^{3} \sigma^{ij}(\mathbf{z})(\nabla \bar{z}_j^{i^T} \mathbf{V} + \nabla \bar{z}^{i^T} \mathbf{V}_j) \tag{57}$$

and

$$\nabla\left[\sum_{i,j=1}^{3}\sigma^{ij}(z)\varepsilon^{ij}(\bar{z})\right]^{T}\mathbf{V}=\sum_{i,j=1}^{3}[\sigma^{ij}(z)(\nabla\bar{z}_{j}^{i^{T}}\mathbf{V})+\sigma^{ij}(\bar{z})(\nabla z_{j}^{i^{T}}\mathbf{V})] \tag{58}$$

where $\mathbf{V}_j = [V_j^1, V_j^2, V_j^3]^T$. Using the above results, equation (56) becomes

$$\begin{aligned}
a(\dot{z},\bar{z}) &\equiv \iiint_{\Omega}\left[\sum_{i,j=1}^{3}\sigma^{ij}(\dot{\mathbf{z}})\varepsilon^{ij}(\bar{\mathbf{z}})\right]d\Omega \\
&= \iiint_{\Omega}\sum_{i,j=1}^{3}[\sigma^{ij}(\mathbf{z})(\nabla\bar{z}^{i^{T}}\mathbf{V}_j)+\sigma^{ij}(\bar{\mathbf{z}})(\nabla z^{i^{T}}\mathbf{V}_j)]d\Omega \\
&\quad - \iiint_{\Omega}\left[\sum_{i,j=1}^{3}\sigma^{ij}(\mathbf{z})\varepsilon^{ij}(\bar{\mathbf{z}})\right]\text{div }\mathbf{V}d\Omega + \iiint_{\Omega}\sum_{i=1}^{3}\bar{z}^{i}(\nabla f^{i^{T}}\mathbf{V})d\Omega \\
&\quad + \iiint_{\Omega}\left[\sum_{i=1}^{3}f^{i}\bar{z}^{i}\right]\text{div }\mathbf{V}d\Omega \\
&\quad + \iint_{\Gamma^2}\left\{-\sum_{i=1}^{3}T^{i}(\nabla\bar{z}^{i^{T}}\mathbf{V}) + \left(\nabla\left[\sum_{i=1}^{3}T^{i}\bar{z}^{i}\right]^{T}\mathbf{n}\right.\right. \\
&\quad \left.\left. + H\left[\sum_{i=1}^{3}T^{i}\bar{z}^{i}\right]\right)(\mathbf{V}^{T}\mathbf{n})\right\}d\Gamma, \quad \text{for all } \bar{\mathbf{z}}\in Z.
\end{aligned} \tag{59}$$

As in equation (41), equation (59) is a variational equation for $\dot{\mathbf{z}} \in Z$.

Consider the mean stress functional of equation (39) in the form

$$\psi = \iiint_{\Omega} g(\boldsymbol{\sigma}(\mathbf{z}))m_p d\Omega = \frac{\iiint_{\Omega_p} g(\boldsymbol{\sigma}(\mathbf{z}))d\Omega}{\iiint_{\Omega_p} d\Omega}. \tag{60}$$

Taking the material derivative of equation (60) and using equation (20) [21],

$$\begin{aligned}
\psi' &= \left[\iiint_{\Omega_p}(g' + \nabla g^{T}\mathbf{V} + g\text{ div }\mathbf{V})d\Omega\iiint_{\Omega_p}d\Omega\right. \\
&\quad \left. - \iiint_{\Omega_p}g\,d\Omega\iiint_{\Omega_p}\text{div }\mathbf{V}d\Omega\right]\Big/\left(\iiint_{\Omega_p}d\Omega\right)^{2} \\
&= \iiint_{\Omega}\sum_{i,j=1}^{3}g_{\sigma ij}(\mathbf{z})[\sigma^{ij}(\dot{\mathbf{z}}) - \sigma^{ij}(\nabla\mathbf{z}^{T}\mathbf{V})]m_p d\Omega
\end{aligned}$$

$$+\iiint_\Omega \sum_{k=1}^{3}\left[\sum_{i,j=1}^{3} g_{\sigma ij}(\mathbf{z})\sigma_k^{ij}(\mathbf{z})V^k\right] m_p d\Omega + \iiint_\Omega g \text{ div } \mathbf{V} m_p d\Omega$$
$$-\iiint_\Omega g m_p d\Omega \iiint_\Omega m_p \text{ div } \mathbf{V} d\Omega. \tag{61}$$

It can be shown that

$$\sigma^{ij}(\nabla \mathbf{z}^T \mathbf{V}) = \sum_{k,\ell=1}^{3} C^{ijk\ell}(\nabla z_\ell^{k^T}\mathbf{V} + \nabla z^{k^T}\mathbf{V}_\ell) \tag{62}$$

and

$$\sum_{k=1}^{3} \sigma_k^{ij}(\mathbf{z})V^k = \sum_{k,\ell=1}^{3} c^{ijk\ell}(\nabla z_\ell^{k^T}\mathbf{V}). \tag{63}$$

Using these results, equation (61) becomes

$$\psi' = \iiint_\Omega \left[\sum_{i,j=1}^{3} g_{\sigma ij}(\mathbf{z})\sigma^{ij}(\dot{\mathbf{z}})\right] m_p d\Omega$$
$$-\iiint_\Omega \sum_{i,j=1}^{3}\left[\sum_{k,\ell=1}^{3} g_{\sigma ij}(\mathbf{z})C^{ijk\ell}(\nabla z^{k^T}\mathbf{V}_\ell)\right] m_p d\Omega$$
$$+\iiint_\Omega g \text{ div } \mathbf{V} m_p d\Omega - \iiint_\Omega g m_p d\Omega \iiint_\Omega m_p \text{ div } \mathbf{V} d\Omega. \tag{64}$$

As in the linear elasticity problem discussed in the section on adjoint variable formulations, the adjoint equation of equation (43) is defined. By the same method used earlier in this section, the sensitivity formula is obtained as

$$\psi' = \iiint_\Omega \sum_{i,j=1}^{3} [\sigma^{ij}(\mathbf{z})(\nabla \lambda^{i^T}\mathbf{V}_j) + \sigma^{ij}(\boldsymbol{\lambda})(\nabla z^{i^T}\mathbf{V}_j)]d\Omega$$
$$-\iiint_\Omega \left[\sum_{i,j=1}^{3} \sigma^{ij}(\mathbf{z})\varepsilon^{ij}(\boldsymbol{\lambda})\right] \text{ div } \mathbf{V} d\Omega + \iiint_\Omega \sum_{i=1}^{3} \lambda^i(\nabla f^{i^T}\mathbf{V})d\Omega$$
$$+\iiint_\Omega \left[\sum_{i=1}^{3} f^i\lambda^i\right] \text{ div } \mathbf{V} d\Omega$$
$$+\iint_{\Gamma^2} \left\{-\sum_{i=1}^{3} T^i(\nabla \lambda^{i^T}\mathbf{V}) + \left(\nabla\left[\sum_{i=1}^{3} T^i\lambda^i\right]^T \mathbf{n}\right.\right.$$
$$\left.\left. + H\left[\sum_{i+1}^{3} T^i\lambda^i\right]\right)(\mathbf{V}^T\mathbf{n})\right\} d\Gamma$$

References pp. 54–55

$$-\iiint_{\Omega} \sum_{i,j=1}^{3} \left[\sum_{k,\ell=1}^{3} g_{\sigma ij}(\mathbf{z}) C^{ijk\ell} (\nabla z^{k^T} \mathbf{V}_\ell) \right] m_p d\Omega$$
$$+ \iiint_{\Omega} g \operatorname{div} \mathbf{V} m_p d\Omega - \iiint_{\Omega} g m_p d\Omega \iiint_{\Omega} m_p \operatorname{div} \mathbf{V} d\Omega. \quad (65)$$

Computational Considerations in Domain Representation—Several comments may be made about the advantages and disadvantages of this domain method. One disadvantage is that a velocity field must be defined in the domain that satisfies regularity properties. There is no unique way of defining domain velocity fields for a given normal velocity field ($\mathbf{V}^T\mathbf{n}$) on the boundary. Also, numerical evaluation of the sensitivity result of equation (65) is more complicated than evaluation of equation (53), because equation (65) requires integration over the entire domain, whereas equation (53) requires integration only over the variable boundary. This problem can be alleviated by introducing a boundary layer [22] of finite elements that vary during perturbation of the shape of a structural component. This approach is illustrated schematically in Figure 4. The domain Ω is divided into subdomains Ω_1 and Ω_2, with Ω_1 held fixed and only the boundary layer Ω_2 modified. In this way, the velocity field need be defined only on Ω_2. The thickness of the boundary layer Ω_2 will depend on trade-offs between numerical accuracy and numerical efficiency.

There are several advantages associated with the domain method, in addition to its numerical accuracy. Variational identities are not required to transform domain integrals to boundary integrals. This is a significant advantage of the domain method in design sensitivity analysis of built-up structures, which are treated in [20]. Built-up structures are made up of combinations of structural components, with interface conditions. In the domain method, interface conditions are not required to obtain shape design sensitivity formulas. This greatly simplifies the derivation, since contributions from each component are simply added [20, 23]. As for numerical accuracy, results of finite element analysis on interface boundaries are often unsatisfactory for built-up structures, due to abrupt changes of boundary conditions. Using the domain method and careful finite element analysis, difficulties in stress evaluation at interfaces are avoided and accurate sensitivity results are obtained. Moreover, as seen in reference [23], interface boundaries for built-up structures are often straight lines and/or plane sections. Thus, a domain velocity field can be easily defined for a given normal velocity field ($\mathbf{V}^T\mathbf{n}$) on the boundary.

NUMERICAL EXPERIENCE

Substantial numerical experimentation has been carried out using the material derivative shape design sensitivity analysis formulation with boundary representation. Good results have been reported [2, 25] for a variety of single structural components. These studies have shown that great care must be taken in projecting stress information to the boundary to achieve acceptable design sensitivity accuracy. Higher-order elements and extrapolation from Gauss points have been shown

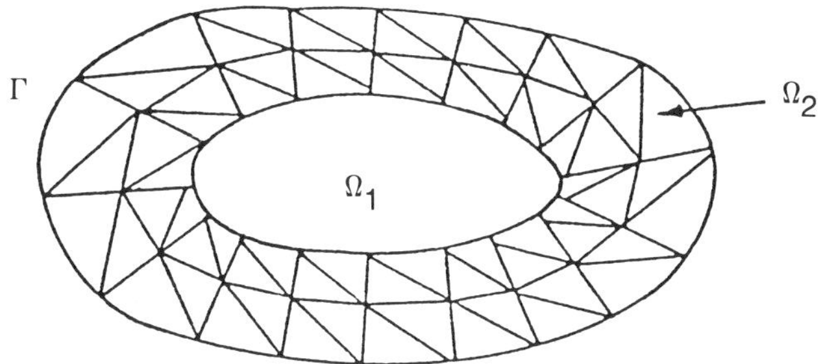

Figure 4. Boundary layer.

to be essential in achieving acceptable accuracy. Substantially inaccurate results have been observed when low-order elements are used and elementary boundary projection approaches are employed. A moderate amount of experience using the boundary representation and the boundary element method for analysis [26, 27] has shown consistently excellent results.

Numerical experimentation with the domain representation approach [20, 22-24] has indicated consistently good results for structural components, without the requirement for sophisticated elements, clever boundary projection methods or drastically refined grids. In order to be more quantitative, two examples are briefly discussed here to permit numerical comparison.

The classical fillet with optimized boundary profile Γ_1 in Figure 5 is used to study accuracy of stress design sensitivity. Sensitivity of von Mises stress averaged over individual finite elements is employed to test accuracy of the domain and boundary formulations. Numerical results obtained in [20] and [21] are quoted here to give an indication of differences in accuracy that may be expected. A 0.1% pertubation in boundary Γ_1 of Figure 5 is introduced. Averaged von Mises stresses for selected elements for the design of Figure 5 are denoted as OLD, and von Mises stresses for the same elements for the perturbed design are denoted as NEW in Tables 1 and 2 for the boundary and domain approaches, respectively. Actual changes in von Mises stress on the respective elements and the change predicted by the associated design sensitivity analysis method are also given. The ratio of predicted to actual change x 100 is tabulated as a percent agreement between the two. Note that results obtained with the domain approach are consistently superior to those obtained with the boundary approach, with the exception of element number 9 for the domain approach. Poor results with this element are associated with the very small change in von Mises stress, leading to inaccuracy in the difference between OLD and NEW values used in calculating the actual change.

As a more significant example, consider the plane stress problem of Figure 6 in which the interface boundary γ between materials of substantially different moduli of elasticity ($E_2/E_1 = 7.65$) is controlled by design variable b. The expression for design sensitivity associated with interface boundary movement with the domain approach is obtained by simply adding the results of equation (65) for both

References pp. 54–55

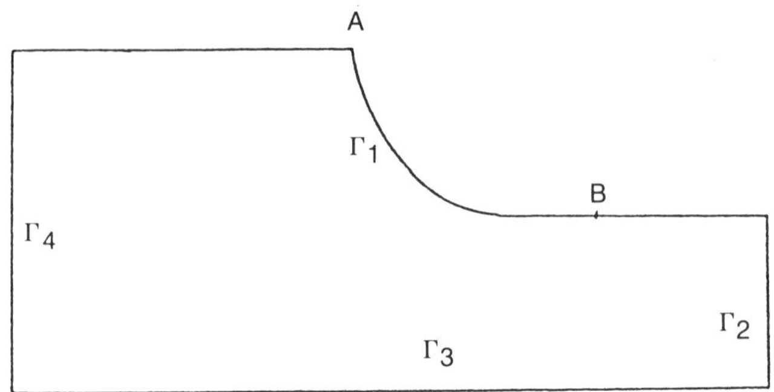

Figure 5. Fillet.

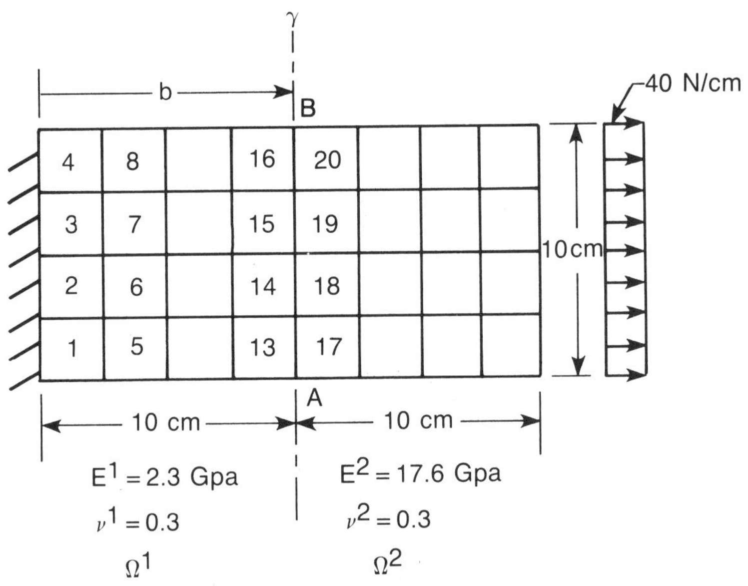

Figure 6. Interface problem.

segments of the structure. Additional computations are carried out in [21] to adapt the boundary representation of equation (54) to account for interface boundary conditions.

Holding the total length of the structural component as 20 centimeters in Figure 6 and perturbing the value of b by 3% leads to design sensitivity predictions using the boundary and domain methods presented in Tables 3 and 4, respectively. Apart from elements on which the actual change in von Mises stress is very small and hence inaccurate (e.g., element 22), design sensitivity predictions by the domain method in Table 4 are quite accurate. In stark contrast, results obtained using the boundary method in Table 3 are extremely poor. The difficulty encountered in this

Table 1
Shape sensitivity of fillet using boundary approach

Elt #	Von Mises stress OLD	Von Mises stress NEW	Actual change	Predicted change	Ratio x 100 %
1	7.1348E 02	7.1331E 02	−1.6878E-01	−1.4733E-01	87.3
2	6.5911E 02	6.5899E 02	−1.2209E-01	−1.0575E-01	86.6
3	5.5968E 02	5.5964E 02	−3.9794E-02	−3.3762E-02	84.8
4	4.4105E 02	4.4111E 02	5.5723E-02	4.8161E-02	86.4
5	3.4984E 02	3.4997E 02	1.3963E-01	1.1888E-01	85.1
6	7.2892E 02	7.2874E 02	−1.7547E-01	−1.5506E-01	88.4
7	6.7769E 02	6.7756E 02	−1.2937E-01	−1.1275E-01	87.2
8	5.7717E 02	5.7713E 02	−4.3532E-02	−3.6705E-02	84.3
9	4.4401E 02	4.4407E 02	5.6628E-02	4.9339E-02	87.1
10	3.1938E 02	3.1953E 02	1.5139E-01	1.2898E-01	85.2
11	7.5125E 02	7.5107E 02	−1.8421E-01	−1.6772E-01	91.0
12	7.0910E 02	7.0895E 02	−1.4457E-01	−1.2734E-01	88.1
13	6.0921E 02	6.0916E 02	−4.5148E-02	−3.6188E-02	80.2
14	4.4736E 02	4.4744E 02	7.7085E-02	6.7049E-02	87.0
15	2.6003E 02	2.6020E 02	1.6836E-01	1.4359E-01	85.3

Table 2
Shape sensitivity of fillet using domain approach

Elt #	Von Mises stress OLD	Von Mises stress NEW	Actual change	Predicted change	Ratio x 100 %
1	343.515150	343.759790	0.244640	0.249350	101.925
2	331.365630	331.620230	0.254610	0.259470	101.909
3	306.913410	307.186320	0.272910	0.278020	101.873
4	270.036600	270.332220	0.295620	0.300980	101.811
5	221.172710	221.487850	0.315140	0.320500	101.701
6	162.210080	162.529200	0.319120	0.323950	101.512
7	98.627940	98.915280	0.287340	0.290810	101.207
8	42.960540	43.112780	0.152240	0.155200	101.939
9	21.549280	21.546520	−0.002770	−0.006800	245.578
10	365.879780	366.077950	0.198170	0.201970	101.917
11	359.105100	359.306990	0.201890	0.205760	101.914
12	345.803600	346.012450	0.208850	0.212850	101.916
13	326.438890	326.657620	0.218730	0.222980	101.942
14	301.382400	301.615570	0.233160	0.237910	102.035
15	269.915720	270.174200	0.258480	0.264300	102.251
16	227.437390	227.742300	0.304910	0.312660	102.544

References pp. 54–55

example is associated with inaccuracy in finite element information projected to the interface boundary. While more refined finite element grids and higher order elements would improve results obtained using the boundary method, superiority of the domain method in this application is clear.

The drastically superior performance of the domain approach in the interface problem is repeated in a number of applications of design sensitivity theory to built-up structures that are composed of multiple components. Numerical experimentation with a beam-plate-truss built-up structure [23] and with a three-dimensional box structure [20] has illustrated the superiority of the domain method over the boundary approach.

As noted, even though the boundary and domain representations of shape design sensitivity are mathematically equivalent, their numerical performance is highly dependent upon the analysis method employed. Since the finite element analysis method generates information that tends to be more accurate in the interior of the domain, it appears to be best suited for implementation with the domain representation. In contrast, the boundary element analysis method provides greatest accuracy on the boundary and is most naturally suited to design sensitivity implementation using the boundary representation.

Table 3
Shape sensitivity of interface problem using boundary approach

Elt #	Von Mises stress OLD	NEW	Actual change	Predicted change	Ratio x 100 %
1	393.01304	393.17922	0.16618	0.20403	122.77621
2	364.37867	364.76664	0.38796	0.67218	173.25764
5	388.07514	388.36215	0.28701	0.56684	197.49952
6	402.26903	402.83406	0.56503	0.42080	74.47325
9	386.43461	386.84976	0.41515	−0.08520	−20.52248
10	407.14612	407.48249	0.33637	0.14159	42.09417
13	388.59634	388.95414	0.35780	−0.53089	−148.37419
14	379.04276	379.25247	0.20971	−1.90134	−906.64742
17	441.68524	442.25032	0.56507	−13.85905	−2452.60502
18	424.05820	425.22910	1.17089	−13.63066	−1164.12453
21	424.19015	424.70840	0.51825	−0.21408	−41.30779
22	378.85433	378.97497	0.12064	0.76770	636.37320
25	407.71528	408.23368	0.51840	0.49878	96.21538
26	387.87304	387.32342	−0.54962	−0.48837	88.85661

Table 4
Shape sensitivity of interface problem using domain approach

Elt #	Von Mises stress OLD	NEW	Actual change	Predicted change	Ratio x 100 %
1	393.013040	393.179220	0.166180	0.179540	108.036
2	364.378670	364.766640	0.387960	0.378400	97.535
5	388.075140	388.362150	0.287010	0.286710	99.898
6	402.269030	402.834060	0.565030	0.596340	105.540
9	386.434610	386.849760	0.415150	0.417480	100.562
10	407.146120	407.482490	0.336370	0.368570	109.573
13	388.596340	308.954140	0.357800	0.375490	104.940
14	379.042760	379.252470	0.209710	0.201590	96.125
17	441.685240	442.250320	0.565070	0.570690	100.994
18	424.058200	425.229100	1.170890	1.128710	96.397
21	424.190150	424.708400	0.518250	0.539190	104.042
22	378.854330	378.974970	0.120640	0.063960	53.017
25	407.715280	408.233680	0.518400	0.517100	99.749
26	387.873040	387.323420	−0.549620	−0.560830	102.040

ACKNOWLEDGEMENT

This research was supported by the National Science Foundation Project No. CEE 83-19871.

REFERENCES

1. R. T. Haftka and M. P. Kamat, *Elements of Structural Optimization.* Martinus Nijhoff Publishers, Boston (1985).
2. E. J. Haug, K. K. Choi and V. Komkov, *Design Sensitivity Analysis of Structural Systems.* Academic Press, New York, (1986).
3. M. E. Botkin, Shape optimization of plate and shell structures. *AIAA J.*, **20** (2), 268–273 (1982).
4. A. Francavilla, C. V. Ramakrishnan and O. C. Zienkiewicz, Optimization of shape to minimize stress concentration. *J. Strain Analysis* **10** (2), 63–70 (1975).
5. V. Braibant and C. Fleury, Shape optimal design using b-splines, *Computer Meth. Appl. Mech. Eng.* **44**, 247–267 (1984).
6. K. Dems and Z. Mroz, Variational approach by means of adjoint systems to structural optimization and sensitivity analysis – II Structural shape variation. *Int. J. Solids Struct.* **20** (6), 527–552 (1984).
7. J. Cea, Problems of shape optimal design, pp. 1005–1048 in *Optimization of Distributed Parameter Structures*, (Edited by E.J. Haug and J. Cea). Sijthoff and Noordhoff, The Netherlands (1981).
8. E.J. Haug, K.K. Choi, J.W. Hou and Y.M. Yoo, A variational method for shape optimal design of elastic structures, pp. 105–137 in *New Directions in Optimum Structural Design*, (Edited by E. Atrek, R. H. Gallagher, K. M. Ragsdell and O. C. Zienkiewicz). John Wiley and Sons (1984).
9. K. K. Choi and E. J. Haug, Shape design sensitivity analysis of elastic structures. *J. Struct. Mech.* **11** (2), 231–269 (1983).
10. N. V. Banichuk, Optimization of elastic bars in torsion. *Int. J. Solids Struct.* **12**, 275–286 (1976).
11. M. S. Na, N. Kikuchi and J. E. Taylor, Shape optimization for elastic torsion bars, pp. 216–223 in *Optimization Methods in Structural Design*, (Edited by H. Eschenauer and N. Olhoff). Bibliographisches Institut, Zurich, Switzerland (1983).
12. R. J. Yang and M. E. Botkin, The relationship between the variational approach and the implicit differentiation approach to shape design sensitivities. *The Optimum Shape: Automated Structural Design* (Edited by J. A. Bennett and M. E. Botkin). Plenum Press, New York (1986).
13. J. S. Sokolnikoff, *Mathematical Theory of Elasticity.* McGraw-Hill, New York (1956).
14. J-P. Zolesio, The material derivative (or speed) method for shape optimization, pp. 1089–1151 in *Optimization of Distributed Parameter Structures* (Edited by E. J. Haug and J. Cea). Sijthoff and Noordhoff, The Netherlands (1981).
15. J-P. Zolesio, Gradient des coûte governes par des problems de Neumann poses des wuverts anguleux en optimization de domain. CRMA-Report 116, University of Montreal, Canada (1982).
16. W. Kecs and P. P. Teodorescu, *Application of the Theory of Distributions in Mechanics.* Abacus Press, Tunbridge Wells, England, (1974).

17. I. Babuška and A. K. Aziz, Survey lectures on the mathematical foundations of the finite element method, pp. 1–359 in *The Mathematical Foundations of the Finite Element Method with Applications to Partial Differential Equations.* Academic Press (1972).

18. C. A. Brebbia and S. Walker, *Boundary Element Techniques in Engineering.* Newnes-Butterworths, Boston, MA (1980).

19. P. K. Banerjee and R. Butterfield, *Boundary Element Methods in Engineering Science.* McGraw-Hill, London (1981).

20. K. K. Choi and H. G. Seong, A domain method for shape design sensitivity analysis of built-up structures. *Computer Methods in Applied Mechanics and Engineering* (1986), to appear.

21. K. K. Choi, Shape design sensitivity analysis of displacement and stress constraints. *J. Struct. Mech.* **13** (1), 27–41 (1985).

22. H. G. Seong and K. K. Choi, Boundary layer approach to shape design sensitivity analysis. *J. Struct. Mech.* (1986), to appear.

23. K. K. Choi and H. G. Seong, Design component method for sensitivity analysis of built-up structures. *J. Struct. Mech.* (1986), to appear.

24. K. K. Choi and H. G. Seong, A numerical method for shape design sensitivity analysis and optimization of built-up structures. *The Optimum Shape: Automated Structural Design* (Edited by J. A. Bennett and M. E. Botkin). Plenum Press, New York (1986).

25. R. J. Yang and K. K. Choi, Accuracy of finite element based design sensitivity analysis. *J. Struct. Mech.*, **13** (2), 223–239 (1985).

26. C. A. Mota Soares, H. C. Rodrigues, L. M. Oliveira Faria and E. J. Haug, Optimization of the geometry of shafts using boundary elements. *J. Mech. Transm. Autom. Des.*, **106** (1), 199–202 (1984).

27. C. A. Mota Soares, H. C. Rodrigues and K. K. Choi, Shape optimal structural design using boundary element and minimum compliance techniques *J. Mech. Transm. Autom. Des.*, **106** (4), 516–521 (1984).

DISCUSSION

B. Prasad *(Electronic Data Systems)*

You commented that it only takes about one percent of the analysis cost to do the design sensitivity. If I recall, when you discretize the models and then you perform the design sensitivity, that cost was relatively much higher.

Haug

I'm sorry, I did not communicate that clearly. The cost of evaluating the integral was only about three percent of the analysis cost. We still have the cost of solving the basic finite element problem and the cost of solving the adjoint problem. Those are the dominant costs. In terms of just calculating the integral, the formulas look oppressive but they compute fast.

R. T. Haftka *(Virginia Polytechnic Institute and State University)*

Along the same line, doesn't this integral require the derivatives of the stress field? When you are using the divergence theorem, you have the divergence of the product of the lambda times phi. Wouldn't that produce the derivatives of the stresses associated with the adjoint?

Haug

No. Those formulas involve only stresses and strains, which are themselves derivatives of displacement fields. Anytime we have a divergence or gradient of something, the quantity being differentiated is only displacement related, so the additional gradients involve derivatives of displacement that are of the same order as stress.

E. J. Haug

D. Grierson *(University of Waterloo)*

Which is better, your estimated sensitivity or your actual sensitivity?

Haug

The actual sensitivity, by far.

Grierson

What method did you use to determine, for example, a perturbation of the actual solution?

Haug

We tend to make perturbations in design and then compare the differences in stress and the predicted sensitivity. Then we down the size of perturbations in design until we have no significant change in the sensitivities from the finite element result. In this comparison, we run into a problem because when we cut the perturbations way down, then the differences that are coming out—the difference between the original finite element solution and the new one—become very small and we lose significant figures. If we make the perturbations too large, then not only do the effects of nonlinearities creep in, but we don't get accurate derivatives. This is why we tend to make perturbations as small as we can.

Grierson

Then how dependent are those actual sensitivities on the method of finite element analysis that you use?

Haug

We've been pleasantly surprised in using two different finite element codes: one a displacement formulation and the other a hybrid method. We've gotten very fine accuracy from both of them. We are going to try another hybrid code shortly.

Grierson

So you are not using a boundary element technique when you are doing the boundary quantity evaluations?

Haug

No, we don't have to do boundary element analysis. You will have a chance to hear a very enthusiastic Carlos Mota Soares at this symposium, telling you that with boundary elements you get fine accuracy with the boundary integral scheme, and he's right.

J. Taylor *(The University of Michigan)*

When you state your criterion function, you have a measure m in there to weight the local stress. Would you elaborate on how one might represent a truly local criterion.

Haug

In displacement, there is no problem with using a Dirac delta measure. We've done that and, by the way, the adjoint load is a unit load.

Taylor

Computationally, how do we represent a local criterion?

Haug

Basically, by putting unit loads on nodes. But for stress, life is more complicated because in some sense we have no right to apply the Dirac delta measure to a stress-related quantity. I think it makes sense only to do some form of local averaging, but you can make that local averaging as tight as you want. However, the adjoint loads tend to have local couples and things like that, so now you get into some delicate technical questions.

Taylor

I think the delicacy becomes really serious when the order of magnitude of the local measure comes down to the size of an element.

Haug

We tend to work at the element level, averaging over elements, and we get very good results. If the area is smaller than the element, then you are somehow putting torques and loads on a small piece of an element. I think then you have very serious questions of procedure.

C. Fleury *(University of California - Los Angeles)*

I must strongly disagree with the last statement of your presentation. My feeling is that it's much better to discretize first and then to use some scheme to get the sensitivities. My main concern is that to get these results requires some degree of smoothness of the domain variation, but you do not mention this property. Also you should have some degree of smoothness for the displacement polynomials in the finite element mesh or you will not get the right answer.

Haug

Frankly, I've been wanting to talk about that and I appreciate the question. Doing the variational analysis, we know theoretically from the French mathematics school of Cea that the velocity field has to be as regular as the displacement field, in the Sobolev sense—it has to have a derivative in the Sobolev sense. These integrals make sense as long as the velocity field is in H^1 for the elasticity problem or H^2 for beams and plates. The fascinating thing is that the velocity field, or design movement field, must have precisely the same mathematical properties as the displacement field. We parameterize the boundary and make a small perturbation in each of the parameters that define the boundary displacement. We then calculate an internal velocity field, using the finite element coded displacement field, which is a very natural one. The crucial thing is to have the same regularity properties for the velocity field that you have for the displacement field. They have to be compatible.

Fleury

For stresses, for example, if you use displacement models for the finite elements, the stress will not be very good. When you transfer the domain integral to a boundary integral on interfaces between adjacent finite elements, you lose some accuracy.

Haug

That's what I pointed out. You lose accuracy when you project to the boundary. It is the traces that are being projected. In a sense, the projections to the boundaries are not accurate. It is well known that if you project from the interior, even using Gauss point projection and all that, it is difficult to get accurate values on the boundary.

Fleury

So you believe that for this type of approach it would be much better to use boundary element methods?

Haug

That's right. Say we have two formulations that are mathematically precise and equivalent. Numerically, it is more practical to use the boundary integral form of the result with boundary element methods, whereas the domain integral form of the result is more appropriate with the finite element method.

L. Schmit *(University of California - Los Angeles)*

In the domain integral method, the selection of the numerical integration scheme is at least open. In our previous discussion you suggested that a straightforward approach would be to use the same numerical integration scheme as in the finite element code that existed when the finite element stiffness matrix **K** was assembled.

Haug

One reason is that the stress information is readily available from the commercial code at those points.

Schmit

Could you comment on the possibility or the merits of other integration schemes. In other words, "How should I do this numerical integration?" not "What data do I have?"

Haug

I don't want to speak for Kyung Choi, who will be talking tomorrow, but I believe we have not investigated that fully. We have been trying to understand what this alternative approach involves. We want to compare it with a lot of work we have done in the past, so we just went to the Gauss points and did the numerical integration. A great number of decisions need to be made on how best to evaluate these quantities.

Kyung is going to talk about what he calls a domain element method in which you precompute some of these quantities, just as you form the stiffness matrices. I think there is a lot of potential for this approach, particularly with symbolic computation codes that are available. We have not yet thoroughly investigated the question of how best to evaluate these integrals. However, even with the crude computations we've done, we are very impressed with the kind of accuracy that can be obtained.

E. Atrek *(Engineering Mechanics Research Corp.)*

Since we're optimizing the shape of some body, aren't we also interested in whether the material can be removed from the inside as well? I cite as an extreme case the bone structure as an optimum. What applications does this method have in terms of change in the interior regions?

Haug

There is nothing to prevent your boundary from having a hole inside and moving the hole boundary around.

Atrek

Yes, but how do you first achieve that hole?

Haug

That's a good question, but I am incapable of coping with the topological problem involved in the creation of holes. I hope that some of our better mathematicians will wrestle with the question of connectivity of the optimum domain. We can select some subdomains inside where we think we might want holes, and optimize their shapes. But I am going to leave it to Bob Kohn and other good mathematicians to figure out how many holes we need in the optimum design.

THE RELATIONSHIP BETWEEN THE VARIATIONAL APPROACH AND THE IMPLICIT DIFFERENTIATION APPROACH TO SHAPE DESIGN SENSITIVITIES

R. J. YANG and M. E. BOTKIN

Engineering Mechanics Department
General Motors Research Laboratories
Warren, Michigan

Abstract

The most commonly used approach to the design sensitivity problem results from the implicit differentiation of the discretized equilibrium equations. The most general implementation of this technique requires that finite differences be used to differentiate the element stiffness matrices. Proper choice of the step size is necessary to obtain high levels of accuracy and to avoid round-off errors. Furthermore, since it is necessary to operate on the element matrices, this method is difficult to implement into a general purpose finite element program. A more recent shape design sensitivity formulation, based upon variational calculus, avoids having to differentiate the discretized equations and results in an analytical expression for the derivative. This approach is based upon the total derivative of the variational state equation and uses an adjoint variable technique for design sensitivity analysis. Only structural response data on the boundary of the structure are necessary, thereby making implementation into a general purpose program less difficult. This paper attempts to compare the two different techniques and point out the similarities. Two test problems demonstrate the accuracy of the variational approach.

INTRODUCTION

Shape optimal design is an important class of structural design problems in which the shape of a two- or three-dimensional structural component is to be determined, subject to constraints involving natural frequencies, displacements and stresses of the structure. In general, the performance derivative, which is essential in the use of any direct optimization technique, cannot be obtained analytically. A finite difference scheme, using the implicit differentiation method, is widely used in the structural optimization world and has been applied to shape optimization problems [1–4]. In this scheme, the discretized equilibrium equation of the structure is first differentiated. The performance change is then obtained by varying each design variable by a specific amount. The advantages of this approach are its generality and simplicity. A disadvantage is that one has to choose a proper step size for the change in design parameters. The proper choice of step size is a process of trial and error; a poor choice may cause numerical difficulty due to round-off error and/or nonlinearities of the structural response. An additional drawback of using the implicit differentiation approach for shape variables is the difficulty of implementing the technique into a general purpose finite element program.

Haug et al. [5–8] developed a unified theory of structural design sensitivity analysis for linear elastic structures, using a variational formulation of the structural equations. This theory allows one to take the total derivative, or material derivative, of the variational state equation and to use an adjoint variable technique for design sensitivity analysis. The main attraction of this approach is that one can compute the derivatives of structural performances analytically. No discretization approximations are involved during the derivation, and step size need not be specified in the calculation. However, the formulation requires evaluating accurate stress quantities on the boundaries, and these are often difficult to obtain.

In this chapter the variational design sensitivity formulation is interpreted, and the formulation is related to the implicit differentiation approach. Two examples are given to illustrate the use of the variational approach and to compare the results of the two different approaches.

IMPLICIT DIFFERENTIATION APPROACH (IDA)

A structural system can be discretized to obtain the equilibrium equation as

$$\mathbf{K}\mathbf{z} = \mathbf{F} \tag{1}$$

where $\mathbf{K}$ is the reduced global stiffness matrix, and $\mathbf{z}$ and $\mathbf{F}$ are the displacement and force vectors, respectively. If one implicitly differentiates the equilibrium equation with respect to a shape design variable vector $\mathbf{b}$, and assumes that force vector $\mathbf{F}$ is independent of $\mathbf{b}$, the following result is obtained:

$$\frac{\partial \mathbf{z}}{\partial b_i} = -\mathbf{K}^{-1}\frac{\partial \mathbf{K}}{\partial b_i}\mathbf{z}$$

or

$$\Delta \mathbf{z} = -\mathbf{K}^{-1}\Delta\mathbf{K}\mathbf{z}. \tag{2}$$

Traditionally, the calculation of $\partial(\mathbf{K})/\partial b_i$ is done either by analytically carrying out the differentiation [9] or by performing a numerical finite difference calculation.

When the finite difference calculation is performed for shape design sensitivity, one is actually comparing two separate finite element meshes and effectively evaluating changes in the finite element mesh discretization as well as the boundary change. In general, the stiffness matrix $\mathbf{K}$ or strain recovery matrix $\mathbf{B}$ is sufficiently complex to preclude analytical evaluation of the derivative for shape optimization. Also, the analytical derivative tends to be element dependent, whereas the finite difference approach is not. For conventional or nonshape structural optimization; i.e., where area, thickness or material properties are chosen as design variables, analytical derivatives may be obtained in some cases.

VARIATIONAL DESIGN SENSITIVITY ANALYSIS (VDSA)

Material Derivative—Since the shape of domain Ω of a structural component is treated as the design variable, it is convenient to think of Ω as a continuous medium and utilize the material derivative idea from continuum mechanics. The pointwise material derivative (if it exists) at $\mathbf{x} \in \Omega$ is defined as [7]

$$\begin{aligned}\dot{\mathbf{z}}(\mathbf{x}) &\equiv \frac{d}{d\tau}\mathbf{z}_\tau(\mathbf{x}, \tau\mathbf{V}(\mathbf{x}))\,|_{\tau=0} \\ &= \lim_{\tau\to 0}\frac{\mathbf{z}_\tau(\mathbf{x}+\tau\mathbf{V}(\mathbf{x})) - \mathbf{z}(\mathbf{x})}{\tau}\end{aligned} \tag{3}$$

where τ is the parameter defining the transformation between initial and current shapes and can be thought of as time, and $\mathbf{V}(\mathbf{x})$ is the design perturbation and may be thought of as a design deformation velocity. If $\mathbf{z}_\tau$ has a regular extension in a neighborhood of $\bar{\Omega}$, then one has [7]

$$\dot{\mathbf{z}}(\mathbf{x}) = \mathbf{z}'(\mathbf{x}) + \nabla\mathbf{z}^T\mathbf{V}(\mathbf{x}) \tag{4}$$

where

$$\mathbf{z}'(\mathbf{x}) \equiv \lim_{\tau\to 0}\frac{\mathbf{z}_\tau(\mathbf{x}) - \mathbf{z}(\mathbf{x})}{\tau} \tag{5}$$

is the partial derivative of $\mathbf{z}$ and $\boldsymbol{\nabla}$ is a gradient symbol.

The material derivative of a functional ψ, which is usually taken as a structural performance measure, also can be defined as

$$\delta\psi \equiv \frac{d}{d\tau}\psi(\tau)|_{\tau=0} = \lim_{\tau\to 0}\frac{\psi(\tau) - \psi(0)}{\tau}. \tag{6}$$

Specifically, one defines ψ as

$$\psi = \int_{\Omega_\tau} f_\tau(\mathbf{x}_\tau)\, d\Omega_\tau \tag{7}$$

where f_τ is a regular function defined in Ω_τ, and subscript τ denotes the deformed configuration. Using equation (6), the material derivative of ψ is [6, 7]

$$\delta\psi = \int_\Omega f'(\mathbf{x})\, d\Omega + \int_\Gamma f(\mathbf{x})\mathbf{V}^T\mathbf{n}\, d\Gamma \tag{8}$$

where $\mathbf{n}$ is a outward unit normal vector of boundary Γ.

In equation (6) one sees that the term material derivative is a misnomer and is actually a differential. This is not obvious from the definition of the material derivative. To better understand this, one may first consider the derivative of a real valued function of a real variable, and then extend it to the derivative of a functional of a function.

Let u map a real set $\mathbf{R}$ to another set $\mathbf{R}$, i.e., $\mathbf{R} \rightarrow \mathbf{R}$, and the definition of the derivative is as follows:

$$u'(x_0) = \lim_{a \to 0} \frac{u(x_0 + a) - u(x_0)}{a}$$

or

$$u'(x_0)(x - x_0) = \lim_{a \to 0} \left\{ \frac{u(x_0 + a) - u(x_0)}{a} (x - x_0) \right\}. \tag{9}$$

Writing $x - x_0 = \eta$, and replacing a by $a\eta$, one then has

$$u'(x_0)\eta = \lim_{a \to 0} \frac{u(x_0 + a\eta) - u(x_0)}{a}. \tag{10}$$

Then it is clear that the right side of equation (10) is a differential, not a derivative, since $u'(x_0)$ is the derivative at point x_0, and η is a difference of $x - x_0$. In equation (10), if a is replaced by τ, the right side of the equation is the same as the definition of material derivative of equation (3). This interpretation can be expanded to function space and to the so-called Gateaux derivative [10–11]. The material derivative is in fact a kind of Gateaux derivative; thus it is a differential quantity and not an ordinary derivative.

Another way to see this fact is to simply define the volume of a solid as a functional:

$$\psi = \int_\Omega d\Omega. \tag{11}$$

When taking the material derivative of equation (11) by the form of equation (8), one has

$$\delta\psi = \int_\Gamma \mathbf{V}^T \mathbf{n} \, d\Gamma. \tag{12}$$

Then, one clearly observes that the right side of equation (12) represents the change of volume instead of the rate of volume change.

Variational Formulation—Consider a displacement functional at an isolated fixed point $\bar{\mathbf{x}}$; i.e.,

$$\psi = \int_\Omega \delta(\mathbf{x} - \bar{\mathbf{x}}) \mathbf{z} \, d\Omega \tag{13}$$

where δ is the Dirac measure at zero, and the state variable or displacement $\mathbf{z}$ is governed by the following variational equation,

$$a(\mathbf{z}, \bar{\mathbf{z}}) \equiv \int_\Omega \sigma^{ij}(\mathbf{z}) \varepsilon^{ij}(\bar{\mathbf{z}}) \, d\Omega = \int_{\Gamma^2} T_i \bar{z}_i \, d\Gamma \tag{14}$$

where $a(\mathbf{z},\bar{\mathbf{z}})$ is the energy bilinear form, σ^{ij} and ε^{ij} are stress and strain tensors respectively, Γ^2 is loaded boundary, T_i is traction vector, and $\bar{\mathbf{z}}$ is a kinematically admissible virtual displacement. Note that summation convention for a repeated index is used throughout this paper.

The differentiability results for the energy bilinear forms, static response, and constraint functional were used to develop the shape design sensitivity formulas. Since the proofs found in reference [7] are tedious and do not contribute insight into the adjoint variable technique, the derivations are not repeated here.

One can take material derivatives of both equations (13) and (14) by assuming that the actual load and the corresponding boundary are unchanged during deformation, and applying the adjoint variable method [5–8] to obtain

$$\delta\psi = \int_{\Gamma^0} \{\sigma^{ij}(\boldsymbol{\lambda})\nabla z_i^T \mathbf{V} + \sigma^{ij}(\mathbf{z})\nabla\lambda_i^T \mathbf{V}\}\, n_j\, d\Gamma - \int_{\Gamma} \sigma^{ij}(\mathbf{z})\varepsilon^{ij}(\boldsymbol{\lambda})\mathbf{V}^T\mathbf{n}\, d\Gamma \tag{15}$$

where Γ^0 and Γ denote kinematically constrained and moving boundaries, respectively, and the adjoint variable $\boldsymbol{\lambda}$ is governed by

$$a(\boldsymbol{\lambda},\bar{\boldsymbol{\lambda}}) = \int_{\Omega} \delta(\mathbf{x}-\bar{\mathbf{x}})\bar{\boldsymbol{\lambda}}\, d\Omega \tag{16}$$

where $\bar{\boldsymbol{\lambda}}$ is a kinematically admissible virtual displacement. Note that the discretized stiffness matrices for the equilibrium equation of (14) and the adjoint equation of (16) are identical since the energy bilinear forms are the same, and both $\bar{\mathbf{z}}$ and $\bar{\boldsymbol{\lambda}}$ are in the same space of kinematically admissible displacements. Physically, the adjoint equation of (16) is interpreted by simply applying a unit load at the point where the displacement is of interest. The detailed derivation can be found in references [5], [6] and [7].

Numerical Intepretation—For simplicity, one can assume that the kinematically constrained boundary is fixed; then equation (15) becomes

$$\delta\psi = -\int_{\Gamma} \sigma^{ij}(\mathbf{z})\varepsilon^{ij}(\boldsymbol{\lambda})\mathbf{V}^T\mathbf{n}\, d\Gamma. \tag{17}$$

Note that in equation (17) only the boundary integral appears, and it can be computed once boundary stresses, strains and velocity are available.

When the shape design parameter vector $\mathbf{b}$ is defined, it can be linearized in tems of $\delta\mathbf{b}$ by

$$\mathbf{b} = \mathbf{b}^0 + \tau\delta\mathbf{b} \tag{18}$$

where $\mathbf{b}^0$ is the nominal design at the given iteration. Presume that points on the boundary Γ are specified by a position vector $\mathbf{r}(\mathbf{b})$; then the velocity field at the boundary is defined by using equation (18) as [5]

$$\mathbf{V} \equiv \frac{d}{d\tau}(\mathbf{r}(\mathbf{b})) = \frac{\partial\mathbf{r}}{\partial\mathbf{b}}\delta\mathbf{b}. \tag{19}$$

Substituting **V** into equation (17) and rewriting the left-hand side, one obtains

$$\begin{aligned}\frac{\partial\psi}{\partial\mathbf{b}}\delta\mathbf{b} &= -\int_\Gamma \sigma^{ij}(\mathbf{z})\varepsilon^{ij}(\boldsymbol{\lambda})\mathbf{n}^T\frac{\partial\mathbf{r}}{\partial\mathbf{b}}\delta\mathbf{b}\,d\Gamma \\ &= -\left(\int_\Gamma \sigma^{ij}(\mathbf{z})\varepsilon^{ij}(\boldsymbol{\lambda})\mathbf{n}^T\frac{\partial\mathbf{r}}{\partial\mathbf{b}}d\Gamma\right)\delta\mathbf{b}.\end{aligned} \tag{20}$$

Finally, the gradient of a displacement can be rewritten as

$$\frac{\partial\psi}{\partial\mathbf{b}} = -\int_\Gamma \sigma^{ij}(\mathbf{z})\varepsilon^{ij}(\boldsymbol{\lambda})\mathbf{n}^T\frac{\partial\mathbf{r}}{\partial\mathbf{b}}d\Gamma. \tag{21}$$

Note that in equation (21), once the physical boundary is parameterized, i.e., $\partial\mathbf{r}/\partial\mathbf{b}$ is known, the derivative of a displacement can be computed using boundary information only. The design variable step size δb disappears in equation (21), while it is needed and crucial in the finite difference approach for the implicit differentiation method.

RELATIONSHIP BETWEEN IDA AND VDSA

To obtain the design sensitivity of equation (17), one has to solve equation (14) for the state variable $\mathbf{z}$ and equation (16) for the adjoint variable $\boldsymbol{\lambda}$. As mentioned before, $\boldsymbol{\lambda}$ is physically obtained by applying a unit load at the nodal point where the displacement is of interest. If all the displacement derivatives are desired, then $\boldsymbol{\lambda}$ is an $N \times N$ matrix, instead of an $N \times 1$ vector, where N is the total number of degrees of freedom of the system. The discretized adjoint equation for variable $\boldsymbol{\lambda}$ is then written as

$$\mathbf{K}\boldsymbol{\lambda} = \mathbf{I} \tag{22}$$

where **I** is the identity matrix. One may notice that the $\boldsymbol{\lambda}$ is simply the flexibility matrix or the inverse of the stiffness matrix $\mathbf{K}^{-1}$ from finite element analysis. A similar interpretation also can be found in [12]. However, the definition for the adjoint variable is different by a minus sign. Applying this observation to equation (17), a formula similar to equation (2) can be obtained.

First, the actual stresses and adjoint strains in equation (17) can be expressed as

$$\begin{aligned}\boldsymbol{\sigma}(\mathbf{z}) &= \mathbf{DBz}_e \\ \boldsymbol{\varepsilon}(\boldsymbol{\lambda}) &= \mathbf{B}\boldsymbol{\lambda}_e\end{aligned} \tag{23}$$

where **D** is the elasticity matrix, **B** is the strain recovery matrix and the subscript e refers to elemental displacement, which is related to the global displacement vector by a transformation matrix $\boldsymbol{\beta}$, as

$$\mathbf{z}_e = \boldsymbol{\beta}\mathbf{z}. \tag{24}$$

Equation (17) then is discretized to a computable finite element form as

$$\delta\psi = -\sum_{N_b} \int_{\Gamma} \mathbf{z}^T \mathbf{C} \boldsymbol{\lambda} \mathbf{V}^T \mathbf{n} \, d\Gamma \tag{25}$$

where $\mathbf{C} = \boldsymbol{\beta}^T \mathbf{B}^T \mathbf{D} \mathbf{B} \boldsymbol{\beta}$ and N_b is the boundary element number. In equation (25) $\mathbf{C}$ is a $N \times N$ symmetric matrix; thus $\boldsymbol{\lambda}$ and $\mathbf{z}$ are interchangable. Replacing $\boldsymbol{\lambda}$ by $\mathbf{K}^{-1}$, one obtains

$$\delta\psi = -\mathbf{K}^{-1} \left(\sum_{N_b} \int_{\Gamma} \mathbf{C} \mathbf{V}^T \mathbf{n} \, d\Gamma \right) \mathbf{z}. \tag{26}$$

Comparing equations (26) and (2), one obtains

$$\left(\Delta\mathbf{K} - \sum_{N_b} \int_{\Gamma} \mathbf{C} \mathbf{V}^T \mathbf{n} \, d\Gamma \right) \mathbf{z} = 0. \tag{27}$$

Note that from equation (1)

$$\mathbf{z} = \mathbf{K}^{-1} \mathbf{F}. \tag{28}$$

Since $\mathbf{K}$ is a nonsingular matrix, one can make $\mathbf{z}$ arbitrary in the space R^n by arbitrarily varying $\mathbf{F}$. This result implies that

$$\Delta\mathbf{K} = \sum_{N_b} \int_{\Gamma} \mathbf{C} \mathbf{V}^T \mathbf{n} \, d\Gamma. \tag{29}$$

One should notice that equation (29) is valid only when the finite element mesh remains unperturbed. In other words, only boundary nodes are allowed to move, since in equation (27) all information is transformed to the boundary, and as a result the internal node effects were excluded automatically. Note that in equation (29) $\mathbf{V}^T \mathbf{n}$ is the normal movement of the solid boundary, and thus it can be interpreted as a domain change $d\Omega_+$, in an infinitesimal sense. Equation (29) then can be rewritten as

$$\Delta\mathbf{K} \simeq \sum_{N_b} \int_{\Omega} \mathbf{C} \, d\Omega_+. \tag{30}$$

This approximation to $\Delta\mathbf{K}$ has been proposed by Zienkiewicz and Campbell [13], who pointed out that the major effect on stiffness change is due to the change of the area in integration rather than the change of $\boldsymbol{\beta}^T \mathbf{B}^T \mathbf{D} \mathbf{B} \boldsymbol{\beta}$ or the $\mathbf{C}$ matrix, in equation (25). Thus, the variational formulation can be related to the implicit differentiation approach for change of traction-free boundaries. As for kinematically constrained boundaries, equation (17) is no longer valid. Instead, one should employ equation (15) which includes an additional term on Γ^0. However, it has not been proven here that both approaches are equivalent in this case.

References pp. 72–73

It is worthwhile noting that the derivatives computed by the variational approach, namely equations (15) and (17), are based on boundary information only. This implies that the accuracy of design sensitivity depends on accurate boundary information. A better estimation of boundary stresses leads to a more accurate design sensitivity estimate. Conversely, if the stiffness matrix is to be differentiated as in the IDA method, accurate calculation of the stiffness matrix is required. Numerical results are shown in the next section to demonstrate this fact.

Another way to relate the two approaches is to directly take the material derivative of the discretized equilibrium equation of (1) to obtain

$$\frac{D\mathbf{z}}{D\tau} = -\mathbf{K}^{-1}\frac{D\mathbf{K}}{D\tau}\mathbf{z} \tag{31}$$

and, evaluating $D\mathbf{K}/D\tau$ using equation (8),

$$\frac{D\mathbf{K}}{D\tau} = \sum_{N_e}\left\{\int_\Omega \frac{\partial \mathbf{C}}{\partial \tau}\,d\Omega + \int_\Gamma \mathbf{C}\mathbf{V}^T\mathbf{n}\,d\Gamma\right\} \tag{32}$$

where N_e is the number of finite elements. Using the fact that $\partial\mathbf{C}/\partial\tau = 0$ and assuming the interelemental boundary is fixed, equation (29) becomes

$$\Delta\mathbf{z} = -\mathbf{K}^{-1}\left(\sum_{N_b}\int_\Gamma \mathbf{C}\mathbf{V}^T\mathbf{n}\,d\Gamma\right)\mathbf{z}. \tag{33}$$

Note that equation (33) is identical to equation (26), which is obtained from variational design sensitivity theory. Recall that equation (26) is valid only for traction-free boundaries, while equation (32) does not imply this condition. However, equation (33) is derived from the discretized equilibrium equation, in which the kinematical boundary conditions are already imposed to obtain the reduced stiffness matrix **K**. This may automatically exclude the possibility of moving a kinematically constrained boundary.

The relationship between IDA and VDSA shown in this section is valid only for problems with homogeneous kinematical boundary conditions.

NUMERICAL EXAMPLES

Two examples are discussed in this section. The results are obtained based on the assumption that the interelement boundary is fixed.

A Cantilever Beam—A simple cantilever beam is considered as an example to demonstrate the use of the variational design sensitivity theory and the boundary parameterization, and to compare the two approaches discussed in the previous sections. The finite element configuration, dimensions, material properties, loading condition and design variable are shown in Figure 1. Design variable **b** is chosen to

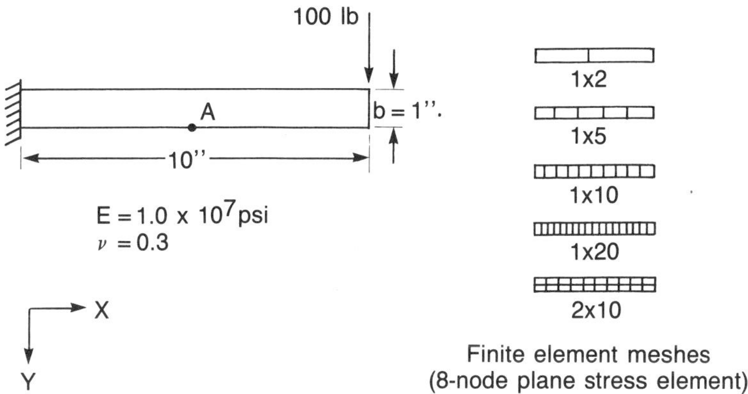

Figure 1. Cantilever beam.

move the upper traction-free boundary. Thus, $\mathbf{n}^T \partial \mathbf{r}/\partial \mathbf{b}$ of equation (21) is equal to 1, and equation (21) is reduced to

$$\frac{\partial \psi}{\partial \mathbf{b}} = -\int_\Gamma \sigma^{ij}(\mathbf{z})\varepsilon^{ij}(\boldsymbol{\lambda})\, d\Gamma. \tag{34}$$

An eight-node isoparametric plane stress element is employed for analysis. The boundary stresses and strains that appear in equation (34) are computed by linear extrapolation from the stresses at Gauss points, where the optimal or the best approximate stresses are located [15]. The external load of 100 lbs. is parabolically applied at the right end of the solid.

Numerical results for design sensitivity of point A in the Y direction for 1×2, 1×5, 1×10, 1×20 and 2×10 meshes are shown in Table 1. The exact solution is obtained from classical beam theory. In Table 1, column 2 represents the displacement in the Y direction for the initial design, b. The percentage of error shown in column 3 is compared to the solution from classical beam theory. Column 4 shows the value of the displacement sensitivity at point A for the implicit approach (IDA) which evaluates the stiffness derivative analytically, and column 5 is the error estimate with respect to beam theory. Columns 6 and 7 are the results for the variational approach (VDSA).

From columns 5 and 7 of Table 1, one observes that the displacement sensitivities for the implicit and variational approaches are very close. This numerical implementation implies that these two approaches are equivalent. However, both approaches depend on the accuracy of the finite element solution.

References pp. 72–73

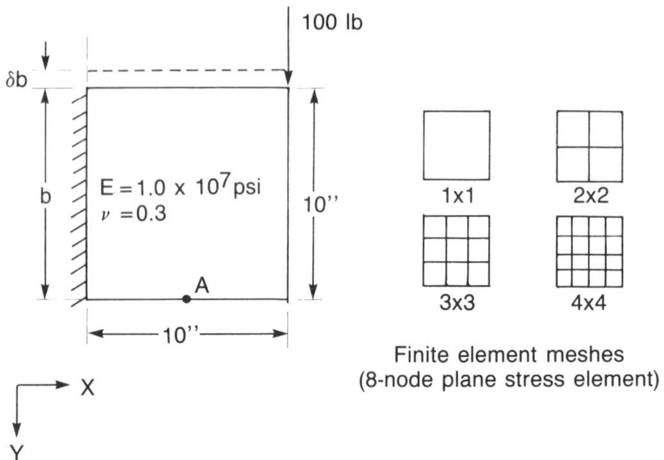

Figure 2. Cantilever thin plate.

Table 1
Accuracy of design sensitivity for cantilever beam

Mesh	v	Error (%)	IDA (analytical) dv/db	IDA (analytical) Error (%)	VDSA dv/db	VDSA Error (%)
1x2	1.188E-2	5.0	-3.533E-2	5.4	-3.532E-2	5.4
1x5	1.233E-2	1.4	-3.671E-2	1.7	-3.670E-2	1.7
1x10	1.249E-2	0.1	-3.718E-2	0.5	-3.716E-2	0.5
1x20	1.257E-2	-0.6	-3.741E-2	-0.2	-3.740E-2	-0.1
2x10	1.259E-2	-0.7	-3.738E-2	-0.0	-3.747E-2	-0.3
exact	1.250E-2		-3.735E-2			

A Two-Dimensional Thin Plate—The second example is a simple two-dimensional thin plate [14]. The geometric configuration is the same as the first example except that the height is larger. The finite element configuration, dimensions, material properties, loading condition and design variable are shown in Figure 2.

Numerical results for design sensitivity of point A in the Y direction for 1×1, 2×2, 3×3, 4×4, 5×5 and 6×6 meshes are shown in Table 2. Column 2 represents the displacement in the Y direction for the initial design, b. Column 3 shows the value of the displacement derivative at point A for the implicit approach (IDA) which evaluates the stiffness derivative analytically. Column 4 is the IDA result of the finite difference method which uses 0.001 b as the step size, and column 5 is the result for the variational approach (VDSA).

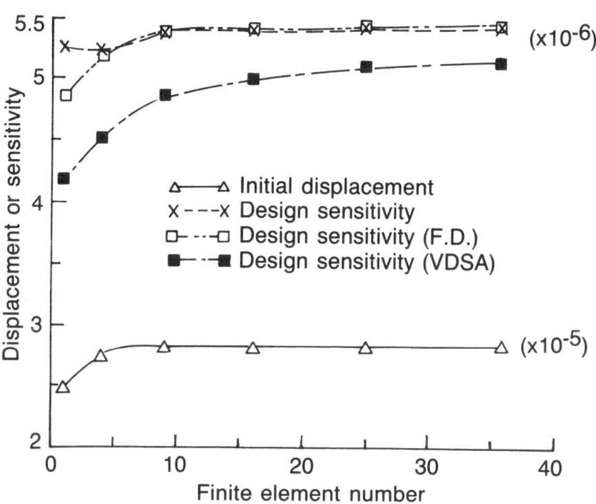

Figure 3. Accuracy of design sensitivity.

Figure 3 shows the same results as Table 2. From Figure 3 and Table 2, one observes that the displacements and the sensitivities for the implicit approach, either from the analytical or finite difference methods, do not change much after the 3 × 3 finite element mesh. However, the design sensitivity for the variational approach (VDSA) is still increasing at the refinement limit of the finite element mesh. This implies that the variational approach is more sensitive to the finite element results, although it provides the analytical formulation for sensitivities.

From this simple example, one concludes that the variational approach tends to yield better gradient estimates once a more accurate analysis is used, and better boundary stresses are obtained. The same conclusion is also found in reference [16].

Table 2
Accuracy of design sensitivity for thin plate

		IDA	IDA(F.D.)	VDSA
Element	v	dv/db	dv/db	dv/db
1x1	2.495E-5	-5.248E-6	-4.845E-6	-4.196E-6
2x2	2.760E-5	-5.235E-6	-5.173E-6	-4.518E-6
3x3	2.841E-5	-5.375E-6	-5.381E-6	-4.856E-6
4x4	2.845E-5	-5.394E-6	-5.409E-6	-4.995E-6
5x5	2.854E-5	-5.413E-6	-5.436E-6	-5.093E-6
6x6	2.856E-5	-5.426E-6	-5.456E-6	-5.158E-6

References pp. 72–73

SUMMARY

Two approaches for obtaining sensitivities of structural response with respect to shape variation have been reviewed and compared. One well-known method is obtained from implicit differentiation of the finite element equations of equilibrium. This formulation requires either the analytical derivative of an element stiffness matrix with respect to shape variation or a finite difference approximation of the derivative. Either of these quantities is difficult to obtain from a general purpose finite element program for which the source is not available. More recently, a new theory has been developed in which one takes the material derivative of the variational state equation and uses an adjoint variable technique for design sensitivity analysis. The advantage of this approach is that an analytical expression for the sensitivities is obtained which is dependent only upon boundary solution quantities that are generally available from general purpose programs. The primary disadvantage is that these quantities are difficult to obtain accurately.

It was shown in this paper that by introducing the discrete finite element expressions for stresses and strains into the variational formulation for sensitivities, the two approaches result in the same matrix equation. Certain assumptions were made that limit this relationship to unconstrained and unloaded boundaries. The term in the matrix equation for the variational approach, which represents the stiffness change in an element along a moving boundary, comes only from the change in the area of integration; it does not include the change in the terms of the strain recovery matrix.

Two demonstration problems were presented in which comparisons were made between the accuracy of the two approaches. In general it was observed that although the variational formulation does not directly depend upon the finite element method and is considered to be the more accurate of the two approaches, it does depend on the finite element results for the needed boundary solution information. As a result the accuracy tends to be indirectly affected by the finite element solution. The primary deficiency in the implicit differentiation approach comes when finite differences are employed in the computation of the stiffness matrix derivatives, in that it is difficult to select the correct step size to avoid inaccuracies.

REFERENCES

1. J. A. Bennett and M. E. Botkin, Structural shape optimization with geometric description and adaptive mesh refinement. *AIAA, J.*, **23** (3), 458–464 (1985).

2. M. E. Botkin, Adaptive finite element technique for plate structures. *AIAA, J.*, **23** (5), 812–814 (1985).

3. M. E. Botkin and J. A. Bennett, Shape optimization of three-dimensional folded plate structures. *AIAA J.*, **23** (11), 1804–1810 (1985).

4. M. E. Botkin, Shape optimization of plate and shell structures. *AIAA, J.*, **20** (2), 268–273 (1982).

5. E. J. Haug, K. K. Choi, J. W. Hou and Y. M. Yoo, A variational method for shape optimal design of elastic structures, *New Directions in Optimum Structural Design* (Edited by E. Atrek, et al.). John Wiley & Sons, New York (1984).

6. K. K. Choi and E. J. Haug, Shape design sensitivity analysis of elastic structures. *J. Struct. Mech.* **11**(2), 231–269 (1983).

7. E. J. Haug, K. K. Choi and V. Komkov, *Design Sensitivity Analysis of Structural Systems*. Academic Press (1985).

8. K. K. Choi, Shape design sensitivity analysis of displacement and stress constraints. *J. Struct. Mech.* **13**(1), 27–41 (1985).

9. C. V. Ramakrishnan and A. Francavilla, Structural shape optimization using penalty functions. *J. Struct. Mech.* **3**(4), 403–422 (1974–1975).

10. K. K. Choi, Personal communication.

11. R. F. Curtain and A. J. Pritchard, *Functional Analysis in Modern Applied Mathematics*. Academic Press (1977).

12. A. D. Belegundu and J. S. Arora, A sensitivity interpretation of adjoint variables in optimal design. *Comput. Meth. Appl. Mech. Eng.* **48**, 81–89 (1985).

13. O. C. Zienkiewicz and J. S. Campbell, Shape optimization and sequential linear programming, in *Optimum Structural Design* (Edited by R. H. Gallagher and O. C. Zienkiewicz). John Wiley & Sons (1973).

14. L. A. Schmit, Personal communication.

15. J. Barlow, Optimal stress locations in finite element models. *Int. J. Num. Meth. Eng.* **10**, 243–251 (1976).

16. R. J. Yang and K. K. Choi, Accuracy of finite element based shape design sensitivity analysis. *J. Struct. Mech.* **13**(2) (1985).

DISCUSSION

E. Atrek *(Engineering Mechanics Research Corporation)*

What if the response of the interior region is critical? Let's not talk about holes (as we did in the discussion of Dr. Haug's paper), but I noticed in your summary that only structural response on the boundary is necessary. Suppose that I have body forces or some loading (or constraint) that makes the interior region more critical. This information is not reflected in your boundary response, so it would seem that the method is not well equipped to deal with this problem. Am I correct in assuming this?

Yang

I didn't quite understand your point.

Atrek

Let's suppose I have a problem where the shear stress near the neutral axis of some beam (modeled, say, by plane stress elements) is an important consideration in design. Since you are only interested in boundary response data, that would not be reflected in the solution to the problem, would it?

E. J. Haug *(University of Iowa)*

You *can* get sensitivity of internal stresses with no problem by evaluating boundary integral expressions.

Atrek

Maybe I am misunderstanding the wording in the summary: "Only structural response data on the boundary of the structure is necessary."

Haug

You can evaluate your sensitivity using only boundary information, but that doesn't mean that you have calculated sensitivity of stresses at the boundary—you can get sensitivity of stresses on the interior. The key is that adjoint loads are applied at that interior high-stress zone, and then you pick up information at the boundary before evaluating sensitivity.

Moderator—R. T. Haftka

What were the design variables in the problems you have shown? Did you have boundary variation or size variation?

Yang

In the beam problem, it is the height of the beam. We only consider boundary variation for this problem.

J. Taylor *(The University of Michigan)*

I think we share the feeling that the adjoint load automatically will emphasize the part of the current version of the structural domain which needs to have the sensitivity evaluated.

I have another question that may be related to refining the issue slightly. You showed a factor, M_p, defined to equal the inverse of the area of the elements. This is clearly a case where there is a distinctly local criterion, which means the measure grows large. I wonder if there is a way to be sure that this remains bounded. In other words, is there assurance that the computational procedure will function properly?

Yang

Actually we did not do too much research on that because we are using finite elements as our measure for M_p. So if the area approaches a very small number, we are not going to get a very good finite element solution. I think we can use finite element results to find out which element is small enough compared to the original one and to ensure M_p is bounded.

R. J. Yang

B. Prasad *(Electronic Data Systems)*

Perhaps I do not understand the conclusions. You pointed out that in the variational approach the idea is independent of the finite element types. But is it not true that VDSA in fact requires the shape function which is element dependent?

Yang

It is true that VDSA requires the shape function, but I do not think it is critical if we use different shape functions.

Prasad

No, I am talking about the shape function, the **C** matrix which you use in the VDSA.

Yang

I don't use the **C** matrix in calculating sensitivities for the variational approach. The only information I need is stress—real stress, adjoint strain, velocity and normal movement on the boundary.

Prasad

Don't you use the elements' domain information, too?

Yang

I don't need domain information.

K. W. Brown *(Pratt & Whitney Aircraft)*

How do the two methods compare in terms of computer execution time— for the adjoint method versus the implicit method—for an equivalent level of accuracy?

Yang

I did not take the computation time aspect into account, but it may depend here on how many active constraints there are, and how many design variables you have. For the adjoint variable method, if you have an overdesign, you do not have any adjoint loads to calculate in the first several iterations. That probably saves some time. On the other hand, if you use the implicit approach, you have to evaluate a number of artificial loads in each iteration no matter how many active constraints you have.

Z. Mróz *(Polish Academy of Science)*

I think that you have not considered loaded boundaries. Then you may also have loading variation due to boundary variation (e.g., pressure loading). The sensitivity expressions should account for more complex cases.

Yang

You're right, but our automotive components usually have free boundaries which may change. If we want to change the other boundaries, we are also able to do that.

Mróz

With free boundaries, there is another problem with the class of variations. If you introduce a small notch or crack, how could you evaluate your sensitivity to handle stress concentration? And your computation effort would certainly increase. If you have smooth boundaries, you could consider a small perturbation in the form of a v-notch. Could you use your approach in this case?

Yang

I think that is a different problem. We are trying to compare these two approaches in a general sense. The boundaries are assumed to be regular and do not have any crack.

Mróz

What is the precise definition of the class of boundaries that are used in your sensitivity analysis?

Moderator

Ed [Haug], would you like to take that question?

Haug

I disagree that we can't handle loaded boundaries. I showed the formulas that account for them. I think Ren-Jye [Yang] simply did not encounter such terms in his problems. Even if you have corners on the boundary, there are jump terms or discrete terms that come into the sensitivity. Correct me if I am wrong, but roughly, with C^1 boundaries you don't have corners and no such jumps arise. If you have even a corner with a finite slope jump, you can get discrete contributions to sensitivities. As you get cusps, then you really have a mess on your hands. We are basically presuming a locally C^1 boundary in most of our work.

C. Fleury *(University of California – Los Angeles)*

With respect to the beam example, you mention that you get the derivative analytically. Do you do this by differentiating the element stiffness matrix?

Yang

Yes.

VARIATIONAL APPROACH TO SHAPE SENSITIVITY ANALYSIS AND OPTIMAL DESIGN

ZENON MRÓZ
Institute of Fundamental Technological Research
Warsaw, Poland

Abstract

A uniform variational approach to sensitivity analysis of linear and nonlinear structures with both material and geometrical nonlinearities is presented. Arbitrary stress, strain or displacement functionals are considered, and they are augmented by the associated bilinear functionals expressed in terms of primary and adjoint fields. Their variation with respect to stress or displacement fields provides compatability or equilibrium conditions for primary and adjoint structures, whereas variation with respect to material variables or shape of the body provides sensitivity of the functional.

General results can easily be particularized to the case of beams, disks, plates or shells with external boundary or interface variation, and can be used to solve optimal design problems by iterative procedures based on optimality conditions.

INTRODUCTION

The present paper discusses uniform variational approaches to problems of sensitivity analysis and optimal design of structures, including shape as one of the design options. By the term *shape* we understand not only the shape of external structure surfaces but also the shape of interfaces in composite materials or the shape of reinforcing layers introduced in three-dimensional or surface structures (e.g., shell or plate stiffeners). We could also imagine, by analogy to hinges in beam structures, the displacement discontinuity surfaces introduced into a structure in order to reduce stress level for imposed displacements or initial distortions of a structure. Therefore the present formulation may provide a broader perspective on shape design problems and constitute a more uniform treatment.

First, the concept of sensitivity with respect to design functions or parameters is introduced. Next the shape variation is considered. Both direct and adjoint structure approaches are discussed. Finally some particular examples and applications are presented.

SENSITIVITY ANALYSIS OF STRUCTURES WITH FIXED SHAPE

When analytical or numerical solution is determined for a specified mathematical model, there is an interest in assessing variations or derivatives of a physical field with respect to parameters of the problem. Such information is important in evaluating the accuracy of the mathematical model, in identification procedures or in optimization of system response with respect to some parameters. The methods of sensitivity analysis have therefore been explored in various fields of science and engineering, for instance, in [1].

In structural mechanics, sensitivity analysis is even more important since any redesign process requires assessment of the variation of local or global structural response characteristics due to structure modification. This modification may involve structure stiffness parameters, shape, support and loading action or structure topology. In this section we shall discuss only variation of stiffness parameters or functions within specified structure configuration, loading and support conditions. The analysis will next be extended by considering varying the shape of the structure.

To formulate the problem, consider a functional G depending on a set of stiffness parameters b_k. The sensitivity analysis is aimed at expressing the variation of G explicitly in terms of variations of b_k, for instance, in a form of polynomial expansion

$$\Delta G = S_i \delta b_i + \frac{1}{2} R_{ij} \delta b_i \delta b_j + \ldots = \delta G + \frac{1}{2} \delta^2 G + \ldots \tag{1}$$

where S_i and R_{ij} are respectively the first order sensitivity vector and the second order sensitivity matrix. This class of problems was studied in a number of papers [3–13] and recently presented in a book [2]. In this section, we briefly outline the problem in a way discussed in detail in papers [3–6].

Sensitivity Analysis for Linear Elastic Structures — Assuming small strain theory, consider a linear elastic structure of specified shape and boundary conditions with tractions and displacements prescribed on boundary portions S_T and S_u. Assume the stiffness or compliance matrices $\mathbf{D}$ and $\mathbf{E}$ to depend on a set of design functions $\phi_k = \phi_k(\mathbf{x})$ so that the relations between stress $\boldsymbol{\sigma}$ and strain $\boldsymbol{\varepsilon}$ are

$$\boldsymbol{\sigma} = \mathbf{D}(\phi_k)\boldsymbol{\varepsilon} \qquad \boldsymbol{\varepsilon} = \mathbf{E}(\phi_k)\boldsymbol{\sigma} \tag{2}$$

where $\mathbf{D}\boldsymbol{\varepsilon} = D_{ijkl}\varepsilon_{kl}$ denotes the matrix product of tensors of different order, and the scalar product of tensors or vectors of the same order is obtained by placing two symbols in juxtaposition, thus: $\boldsymbol{\sigma}\boldsymbol{\varepsilon} = \sigma_{ij}\varepsilon_{ij}$. When small variations $\delta\phi_k$ of design functions are imposed, we obtain from (2)

$$\begin{aligned}\delta\boldsymbol{\sigma} &= \mathbf{D}\delta\boldsymbol{\varepsilon} + \delta\mathbf{D}\boldsymbol{\varepsilon} = \mathbf{D}\delta\boldsymbol{\varepsilon} + \frac{\partial\mathbf{D}}{\partial\phi_k}\delta\phi_k\boldsymbol{\varepsilon} = \delta\boldsymbol{\sigma}' + \delta\boldsymbol{\sigma}'' \\ \delta\boldsymbol{\varepsilon} &= E\delta\boldsymbol{\sigma} + \delta\mathbf{E}\boldsymbol{\sigma} = \mathbf{E}\delta\boldsymbol{\sigma} + \frac{\partial\mathbf{E}}{\partial\phi_k}\delta\phi_k\boldsymbol{\sigma} = \delta\boldsymbol{\varepsilon}' + \delta\boldsymbol{\varepsilon}''.\end{aligned} \tag{3}$$

(Also see Figure 1.) The equations of equilibrium and compatibility are expressed through the virtual work equations

$$\int \boldsymbol{\sigma}\delta\boldsymbol{\varepsilon}\,dV = \int \mathbf{T}^0\delta\mathbf{u}\,dS_T + \int \mathbf{f}\delta\mathbf{u}\,dV \tag{4}$$

$$\int \delta\boldsymbol{\sigma}\boldsymbol{\varepsilon}\,dV = \int \delta\mathbf{T}\mathbf{u}^0\,dS_u + \int \delta\mathbf{f}\mathbf{u}\,dV \tag{5}$$

where $\mathbf{T}^0$ and $\mathbf{u}^0$ are the specified tractions and displacements on S_T and S_u whereas $\mathbf{f}$ denotes the body force vector, depending in general on the design functions, $\mathbf{f} = \mathbf{f}(\phi_k)$, for instance, in the case of disks or plates of varying thickness with account for self-weight.

Consider a general functional form

$$G = \int \psi(\boldsymbol{\sigma}, \mathbf{u}, \boldsymbol{\phi})\,dV + \int h(\mathbf{u}, \mathbf{T})\,dS \tag{6}$$

or its more particular version discussed in [3, 4]

$$G = \int \psi(\boldsymbol{\sigma}, \boldsymbol{\phi})\,dV + \int \Phi(\mathbf{u}, \boldsymbol{\phi})\,dV + \int f(\mathbf{T})\,dS_u + \int g(\mathbf{u})\,dS_T \tag{7}$$

where the integrands are continuous and differentiable functions of their arguments. Since for any statically admissible stress field $\boldsymbol{\sigma}^s$ that is in equilibrium with body forces $\mathbf{f}^s$ and surface tractions $\mathbf{T}^s$, and for kinematically admissible fields $\mathbf{u}^k$ and $\boldsymbol{\varepsilon}^k$, the following virtual work equation holds:

$$\int \boldsymbol{\sigma}^s\boldsymbol{\varepsilon}^k\,dV - \int \mathbf{f}^s\mathbf{u}^k\,dV - \int \mathbf{T}^s\mathbf{u}^k\,dS = 0. \tag{8}$$

This equation can be added to (6) as a constraint condition. Let the stress field $\boldsymbol{\sigma}^s$ be identified with the actual field $\boldsymbol{\sigma}$. Let the displacement and strain fields $\mathbf{u}^k$ and $\boldsymbol{\varepsilon}^k$ be identified as the fields associated with the adjoint structure of the same shape and elastic stiffness moduli but with different boundary conditions and initial strain or stress fields within the structure. Denote the state fields within the adjoint structure by $\boldsymbol{\sigma}^a$, $\boldsymbol{\varepsilon}^a$ and $\mathbf{u}^a$, and combine equation (6) with (8) which provides the augmented functional

$$\bar{G} = \int \psi\,dV + \int h\,dS - \int \boldsymbol{\sigma}\boldsymbol{\varepsilon}^a\,dV + \int \mathbf{f}\mathbf{u}^a\,dV + \int \mathbf{T}\mathbf{u}^a dS. \tag{9}$$

For simplicity the Lagrange multiplier was assumed to be unity, as it only scales boundary conditions for the adjoint structure. In what follows, we shall show that considering the first variation of equation (9) and requiring stationarity with respect to the fields $\mathbf{u}$ and $\mathbf{u}^a$, we obtain the equilibrium conditions for both primary and adjoint structures together with respective boundary conditions. The variation of equation (9) with respect to ϕ_k then provides the first order sensitivity of the structure. The first variation of (9) is expressed as follows:

$$\begin{aligned}\delta\bar{G} = &\int \left(\frac{\partial\psi}{\partial\boldsymbol{\sigma}}\delta\boldsymbol{\sigma} + \frac{\partial\psi}{\partial\mathbf{u}}\delta\mathbf{u} + \frac{\partial\psi}{\partial\phi_k}\delta\phi_k\right) dV \\ &+ \int \left(\frac{\partial h}{\partial\mathbf{u}}\delta\mathbf{u} + \frac{\partial h}{\partial\mathbf{T}}\delta\mathbf{T}\right) dS - \int (\boldsymbol{\sigma}\delta\boldsymbol{\varepsilon}^a + \delta\boldsymbol{\sigma}\boldsymbol{\varepsilon}^a)\, dV \\ &+ \int (\delta\mathbf{f}\mathbf{u}^a + \mathbf{f}\delta\mathbf{u}^a)\, dV + \int (\delta\mathbf{T}\mathbf{u}^a + \mathbf{T}\delta\mathbf{u}^a)\, dS.\end{aligned} \tag{10}$$

Assume that the constitutive relations for the adjoint structure take the following form (also see Figure 1):

$$\boldsymbol{\sigma}^a = \mathbf{D}(\boldsymbol{\varepsilon}^a - \boldsymbol{\varepsilon}^{ai}) \tag{11}$$

where $\boldsymbol{\varepsilon}^{ai}$ is the initial strain field to be specified later. Assuming that $\delta\mathbf{u}^0 = 0$ on S_u and $\delta\mathbf{T}^0 = 0$ on S_T, expression (10) can be transformed to the following form:

$$\begin{aligned}\delta\bar{G} = &\int \left(\frac{\partial\psi}{\partial\phi_k}\delta\phi_k + \delta\mathbf{f}\mathbf{u}^a\right) dV \\ &- \left\{\int (\boldsymbol{\sigma}\delta\boldsymbol{\varepsilon}^a - \mathbf{f}\delta\mathbf{u}^a)\, dV - \int \mathbf{T}\delta\mathbf{u}^a dS\right\} \\ &- \left\{\int [(\boldsymbol{\varepsilon}^a - \boldsymbol{\varepsilon}^{ai})\delta\boldsymbol{\sigma} - \mathbf{f}^a\delta\mathbf{u}]\, dV - \int \mathbf{T}^{a0}\delta\mathbf{u}\, dS_T\right\} \\ &- \left\{\int \left[\left(\boldsymbol{\varepsilon}^{ai} - \frac{\partial\psi}{\partial\boldsymbol{\sigma}}\right)\delta\boldsymbol{\sigma} + \left(\mathbf{f}^a - \frac{\partial\psi}{\partial\mathbf{u}}\right)\delta\mathbf{u}\right] dV\right. \\ &+ \int \left(\mathbf{T}^{a0} - \frac{\partial h}{\partial\mathbf{u}}\right)\delta\mathbf{u}\, dS_T \\ &\left.- \int \left(\mathbf{u}^{a0} + \frac{\partial h}{\partial\mathbf{T}}\right)\delta\mathbf{T}\, dSu.\right\}\end{aligned} \tag{12}$$

Let us note that the expression within the first pair of braces can be regarded as the equilibrium equation of the primary structure expressed through the virtual work equation

$$\int \boldsymbol{\sigma}\delta\boldsymbol{\varepsilon}^a\, dV = \int \mathbf{f}\delta\mathbf{u}^a\, dV + \int \mathbf{T}\delta\mathbf{u}^a\, dS. \tag{13}$$

By virtue of equations (3) and (11) the expression in the second pair of braces can be transformed as follows:

$$\int [(\boldsymbol{\varepsilon}^a - \boldsymbol{\varepsilon}^{ai})\delta\boldsymbol{\sigma} - \mathbf{f}^a\delta\mathbf{u}]\, dV - \int \mathbf{T}^{a0}\delta\mathbf{u}\, dS_T$$

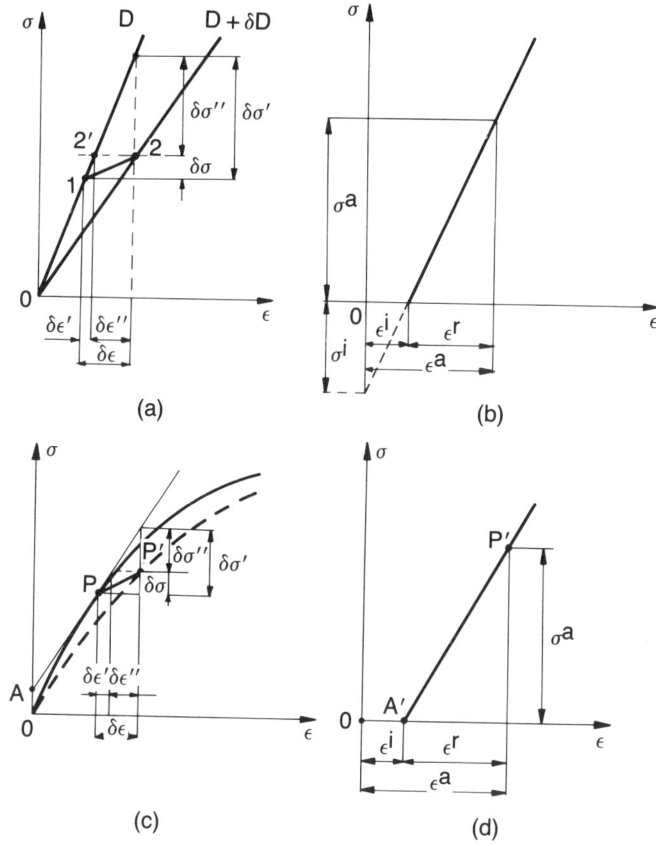

Figure 1. (a) Variation of stress and strain due to stiffness modulus variation, (b) stress and strain in the adjoint structure, (c) nonlinear stress-strain relation, and (d) stress-strain relation in the adjoint structure.

$$= \left\{ \int (\boldsymbol{\sigma}^a \delta \boldsymbol{\varepsilon} - \mathbf{f}^a \delta \mathbf{u})\, dV - \int \mathbf{T}^{a0} \delta \mathbf{u}\, dS_T \right\}$$
$$+ \int (\boldsymbol{\varepsilon}^a - \boldsymbol{\varepsilon}^{ai}) \frac{\partial \mathbf{D}}{\partial \phi_k} \boldsymbol{\varepsilon} \delta \phi_k \, dV. \tag{14}$$

Since the equilibrium conditions for the adjoint structure are expressed by the virtual work equation

$$\int \boldsymbol{\sigma}^a d\boldsymbol{\varepsilon}\, dV = \int \mathbf{f}^a \delta \mathbf{u}\, dV + \int \mathbf{T}^{a0} \delta \mathbf{u}\, dS_T, \tag{15}$$

only the last term of equation (14) does not vanish. Let us note that the expression in the third pair of braces of (12) vanishes when the following loading and boundary conditions are satisfied for the adjoint structure:

$$\begin{aligned} \boldsymbol{\varepsilon}^{ai} &= \frac{\partial \psi}{\partial \boldsymbol{\sigma}}, \qquad \mathbf{f}^a = \frac{\partial \psi}{\partial \mathbf{u}} \qquad \text{within V}, \\ \mathbf{T}^{a0} &= \frac{\partial h}{\partial \mathbf{u}} \qquad \text{on } S_T, \qquad u^{a0} = -\frac{\partial h}{\partial \mathbf{T}} \qquad \text{on } S_u. \end{aligned} \tag{16}$$

In view of equations (13) through (16), the variation δG is then expressed as follows:

$$\delta G = \delta \bar{G} = \int \left[\frac{\partial \psi}{\partial \phi_k} + \frac{\partial \mathbf{f}}{\partial \phi_k} \mathbf{u}^a - (\boldsymbol{\varepsilon}^a - \boldsymbol{\varepsilon}^{ai}) \frac{\partial \mathbf{D}}{\partial \phi_k} \boldsymbol{\varepsilon} \right] \delta \phi_k \, dV, \tag{17}$$

that is, expressed explicitly in terms of the variations $\delta \phi_k$ of the design functions. Noting that

$$\mathbf{DE} = \mathbf{I}, \quad \frac{\partial D}{\partial \phi_k} \mathbf{E} = -\mathbf{D} \frac{\partial \mathbf{E}}{\partial \phi_k}, \tag{18}$$

expression (18) can be presented in the equivalent form,

$$\delta G = \int \left[\frac{\partial \psi}{\partial \phi_k} + \frac{\partial \mathbf{f}}{\partial \phi_k} \mathbf{u}^a + \boldsymbol{\sigma}^a \frac{\partial \mathbf{E}}{\partial \psi_k} \boldsymbol{\sigma} \right] \delta \psi_k \, dV. \tag{19}$$

It is seen that augmenting the functional G by the bilinar form (8), we generate equilibrium and boundary conditions for both structures and also generate the sensitivity expression with respect to the design functions. Considering the complementary bilinear form, that is,

$$\tilde{G} = \int \psi \, dV + \int h \, dS + \int \boldsymbol{\sigma}^a \boldsymbol{\varepsilon} \, dV - \int \mathbf{f}^a \mathbf{u} \, dV - \int \mathbf{T}^a \mathbf{u} \, dS \tag{20}$$

the identical sensitivity expression, i.e., equation (19), is obtained.

The case when the functional (6) depends also on strain $\boldsymbol{\varepsilon}$ can be treated in a similar way. Assume that $\psi = \psi(\boldsymbol{\sigma}, \boldsymbol{\varepsilon}, \mathbf{u}, \boldsymbol{\phi})$ and the stress-strain relation for the adjoint structure takes the form

$$\boldsymbol{\sigma}^a = \mathbf{D}(\boldsymbol{\varepsilon}^a - \boldsymbol{\varepsilon}^{ai}) - \boldsymbol{\sigma}^{ai} \tag{21}$$

where $\boldsymbol{\sigma}^{ai}$ is the initial stress introduced independently of the initial strain. Following the previous derivation, we obtain

$$\boldsymbol{\sigma}^{ai} = \frac{\partial \psi}{\partial \boldsymbol{\varepsilon}} \tag{22}$$

and thc variation of G from equation (17) or (19) is still valid.

When the design functions ψ_k depend on a set of design parameters $b_m (m = 1, 2, \ldots p)$ so that

$$\psi_k = \psi_k(x, b_m), \quad \delta \psi_k = \frac{\partial \psi_k}{\partial b_m} \delta b_m, \tag{23}$$

then the expressions (17) or (19) take the form

$$\delta G = \left\{ \int \left[\frac{\partial \psi}{\partial b_m} + \frac{\partial f}{\partial b_m} \mathbf{u}^a - (\boldsymbol{\varepsilon}^a - \boldsymbol{\varepsilon}^{ai}) \frac{\partial \mathbf{D}}{\partial b_m} \boldsymbol{\varepsilon} \right] dV \right\} \delta b_m \tag{24}$$

and

$$\delta G = \left\{ \int \left[\frac{\partial \psi}{\partial b_m} + \frac{\partial f}{\partial b_m} \mathbf{u}^a + \boldsymbol{\sigma}^a \frac{\partial \mathbf{E}}{\partial b_m} \boldsymbol{\sigma} \right] dV \right\} \delta b_m. \tag{25}$$

The presented approach, which will be called the *adjoint structure approach*, is based on the concept of an adjoint structure and its stress, strain and displacement fields. Having determined solutions for both primary and adjoint structures, the first variations or derivatives with respect to design functions or parameters are explicitly expressed. The other alternative approach, which will be called the *direct method*, is possible by direct calculation of variations (or derivatives) of stress, strain and displacement fields due to variation of design parameters. In fact, starting from decomposition in equation (3), the stress or strain variations $\delta\boldsymbol{\sigma}''$ and $\delta\boldsymbol{\varepsilon}''$ can be regarded as the *initial* variations in the elastic body of stiffness or compliance matrices $\mathbf{D}$ and $\mathbf{E}$. The virtual work equation

$$\int \delta\boldsymbol{\sigma}' \delta\boldsymbol{\varepsilon} \, dV = - \int \delta\boldsymbol{\sigma}'' \delta\boldsymbol{\varepsilon} \, dV \tag{26}$$

can be regarded as the equilibrium condition for the stress field $\delta\boldsymbol{\sigma}'(\mathbf{x})$ with the specified initial stress field $\delta\boldsymbol{\sigma}''(\mathbf{x})$. The boundary conditions for $\delta\boldsymbol{\sigma}'$ and $\delta\mathbf{u}$ are

$$\delta\boldsymbol{\sigma}'\mathbf{n} = -\delta\boldsymbol{\sigma}''\mathbf{n} \quad \text{on } S_T, \qquad \delta\mathbf{u} = 0 \quad \text{on } S_u. \tag{27}$$

When there are n stiffness parameters, the direct approach requires $n+1$ solutions, whereas the adjoint structure approach requires only two solutions and is more economical. However, when there are m functionals and n stiffness parameters, then the direct approach would require $n+1$ solutions whereas the adjoint structure approach would need $m+1$ solutions in order to generate all variations. Thus, the choice between the two approaches depends on the ratio of m to n and on the convenience of generating solutions associated with one or the other approach.

Sensitivity Analysis for Nonlinear Structures — The analysis for a linear elastic structure can easily be generalized for a physically nonlinear material within small strain theory or for geometrically nonlinear theory. Consider first a general nonlinear stress-strain relation for the small strain theory following from the elastic potential:

$$\boldsymbol{\sigma} = \mathbf{S}(\boldsymbol{\varepsilon}, \mathbf{b}) = \frac{\partial U(\boldsymbol{\varepsilon}, \mathbf{b})}{\partial \boldsymbol{\varepsilon}} \tag{28}$$

where $\mathbf{b}$ denotes the vector of stiffness parameters and U is the specific strain energy per unit volume. The variation of stress is now specified as follows:

$$\delta\boldsymbol{\sigma} = \frac{\partial S}{\partial \boldsymbol{\varepsilon}} \delta\boldsymbol{\varepsilon} + \frac{\partial S}{\partial \mathbf{b}} \delta\mathbf{b} = \frac{\partial^2 U}{\partial \boldsymbol{\varepsilon} \partial \boldsymbol{\varepsilon}} \delta\boldsymbol{\varepsilon} + \frac{\partial S}{\partial \mathbf{b}} \delta\mathbf{b} = \delta\boldsymbol{\sigma}' + \delta\boldsymbol{\sigma}''. \tag{29}$$

It is seen that the decomposition in equation (29) is identical to equation (3), provided the tangent stiffness matrix $\mathbf{D}_t$ is specified by the relation

$$\mathbf{D}_t(\boldsymbol{\varepsilon}, \mathbf{b}) = \frac{\partial \mathbf{S}}{\partial \boldsymbol{\varepsilon}} = \frac{\partial^2 U}{\partial \boldsymbol{\varepsilon} \partial \boldsymbol{\varepsilon}}. \tag{30}$$

The adjoint structure is now described by the linear relationship in equation (11) between stress and strain with the tangent stiffness matrix $\mathbf{D}_t$:

$$\boldsymbol{\sigma}^a = \mathbf{D}_t(\boldsymbol{\varepsilon}^a - \boldsymbol{\varepsilon}^{ai}). \tag{31}$$

The sensitivity expressions (17) and (19) remain valid in this case, that is,

$$\delta G = \left\{ \int \left[\frac{\partial \psi}{\partial \mathbf{b}} + \frac{\partial \mathbf{f}}{\partial \mathbf{b}} \mathbf{u}^a - (\boldsymbol{\varepsilon}^a - \boldsymbol{\varepsilon}^{ai}) \frac{\partial \mathbf{S}}{\partial \mathbf{b}} \right] dV \right\} \delta \mathbf{b}. \tag{32}$$

Figures 1(c) and 1(d) illustrate relations (29) and (31).

The case of nonlinear geometric theory can be treated along similar lines. Assume that the Cartesian coordinates x_i are interpreted as material coordinates and introduce the displacement field $\mathbf{u}(\mathbf{x})$ whose deformation gradient is $F_{ij} = \delta_{ij} + u_{i,j}$, where δ_{ij} is the Kronecker symbol. The material is assumed to be hyperelastic, obeying the constitutive law of the form

$$t_{ij} = \frac{\partial U}{\partial F_{ij}} = \mathbf{S}(\mathbf{F}, \mathbf{b}) \tag{33}$$

where $\mathbf{t}$ is the first (nonsymmetric) Piola-Kirchhoff stress tensor, related to the symmetric Cauchy stress $\boldsymbol{\sigma}$ by the formula $t_{ij} F_{kj} = \det(\mathbf{F}) \sigma_{ik}$ and $U(\mathbf{F}, \mathbf{b})$ is the specific strain energy per unit volume in the undeformed state. Instead of (29), we now have

$$\delta t_{ij} = \frac{\partial^2 U}{\partial F_{ij} \partial F_{kL}} \delta F_{kL} + \frac{\partial^2 U}{\partial F_{ij} \partial b_k} \delta b_k = \delta t'_{ij} + \delta t''_{ij}. \tag{34}$$

The functional (6) is now expressed in the undeformed configuration in terms of $\mathbf{t}, \mathbf{u}, \boldsymbol{\phi}$, and boundary tractions or displacements. The adjoint structure is linear with the constitutive relation and the initial displacement gradient specified as follows:

$$\begin{aligned} t^a_{ij} &= \frac{\partial^2 U}{\partial F_{ij} \partial F_{k\ell}} (u^a_{k,\ell} - u^i_{k,\ell}) = D^t_{ijk\ell}(u^a_{k,\ell} - u^i_{k,\ell}) \\ u^i_{k,\ell} &= \frac{\partial \psi}{\partial t_{k\ell}} \end{aligned} \tag{35}$$

whereas the body forces $\mathbf{f}^a$, surface tractions and displacements $\mathbf{T}^{a0}$, $\mathbf{u}^{a0}$, are specified by (16). The formula (32) remains valid provided $\boldsymbol{\varepsilon}^a$ and $\boldsymbol{\varepsilon}^{ai}$ are replaced by $u^a_{i,j}$ and $u^{ai}_{i,j}$. These relations can also be expressed in terms of the symmetric Piola-Kirchhoff stress tensor s_{ij} and the conjugate Green strain tensor e_{ij}. A detailed discussion of sensitivity analysis and optimal design of nonlinear beams and plates was presented in reference [6]. The second-order sensitivity analysis was considered in [5, 6, 13, 14] and resulted in expressions for second-order variations and derivatives of functionals.

SENSITIVITY ANALYSIS OF STRUCTURES WITH SHAPE TRANSFORMATION

General Relations — In the previous section, we considered the case of variation of material stiffness or cross-sectional functions or parameters of a primary structure of specified shape. In this section, a more complex case will be considered when the shape of the external boundary of a body is not specified in advance but can vary in order to attain the desired properties of a structure. Besides the external boundary, we can also investigate the variation of shape of interfaces between different materials, of the shape of reinforcing layers or of displacement discontinuity surfaces within the structure. The concept of shape variation can therefore be considered in a broader context, not necessarily referring to the shape of external surfaces.

Consider an elastic body occupying a domain Υ of volume V and bounded by the surface S. On boundary portions S_u and S_T the displacements $\mathbf{u}^0(x)$ and tractions $\mathbf{T}^0(x)$ are specified (see Figure 2). The body undergoes a deformation process specified by the displacement field $\mathbf{u}(\mathbf{x}, t)$ where t is a time-like parameter. In a finite deformation process the initial configuration C_0 is deformed into C_d, $C_0 \to C_d$, $\mathbf{x}^d = \mathbf{x} + \mathbf{u}$. Consider now the transformation process of the body resulting in variation of boundary shape of the initial configuration. To describe boundary variation, the transformation vector function $\boldsymbol{\phi}(\mathbf{x}, t)$, $\mathbf{x} \in S_0$ should be specified on the boundary surface S. The transformation of each boundary point will thus be specified, $\mathbf{x}^t = \mathbf{x} + \boldsymbol{\phi}, C_0 \to C_t$. However, it is more convenient to specify the transformation field not only on S but also in the interior and exterior of the body as the space field. In this way, for any instantaneous shape of the structure, its further modification is specified. As we are interested in a continuing process of modification (for instance, in the incremental redesign process) we will consider the transformation rate field $\dot{\boldsymbol{\phi}}(\mathbf{x}, t)$ or the infinitesimal variation field $\delta\boldsymbol{\phi}(\mathbf{x}, t) = \dot{\boldsymbol{\phi}} dt$ specified within a domain enclosing the instantaneous structure configuration. Similarly to spatial description of fluid motion, now $\mathbf{x}$ denotes the instantaneous position of a structure point. Obviously, the continuation of the transformation field and its rate from the surface into the enclosed domain is quite arbitrary, provided the proper continuity conditions are satisfied. In fact, it will be shown that the first variations of functionals can be expressed as surface integrals in terms of surface data only.

Consider an infinitesimal transformation field $\delta\boldsymbol{\phi}(x)$ from the assumed configuration, so that a typical point P, initially placed at $\mathbf{x}$ passes to the position x^*, so that

$$P \to P^* : \mathbf{x}^* = \mathbf{x} + \delta\boldsymbol{\phi}(\mathbf{x}) \tag{36}$$

and $\delta\boldsymbol{\phi}$ is assumed as the differentiable field. The volume and surface elements of a structure and the unit normal vector to the boundary surface are then transformed

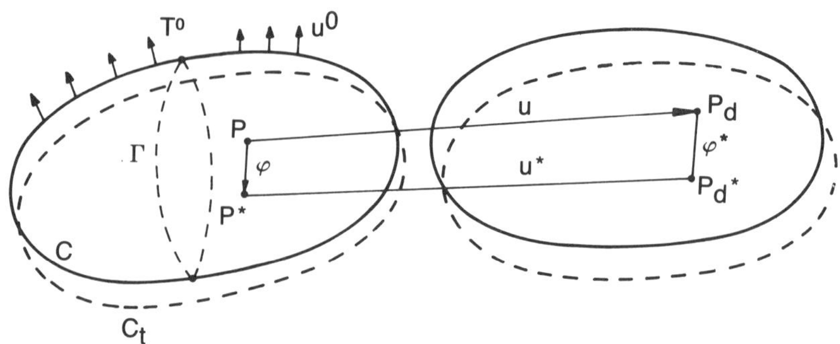

Figure 2. **Transformation and deformation processes in a structure.**

as follows (see [4] for detailed derivation):

$$\begin{aligned} \delta(dV) &= \delta\phi_{k,k}\, dV, \\ \delta(dS) &= (\delta_{k\ell} - n_k n_i)\delta\phi_{k,\ell}\, dS, \\ \delta n_j &= (n_j n_\ell - \delta_{j\ell}) n_k \delta\phi_{k,\ell}, \end{aligned} \tag{37}$$

where the comma preceding an index denotes partial differentiation. The variations of state fields within a structure are now expressed in a fixed reference system as follows:

$$\delta\mathbf{u} = \delta\bar{\mathbf{u}} + \mathbf{u}_{,k}\,\delta\psi_k, \qquad \delta\boldsymbol{\varepsilon} = \delta\bar{\boldsymbol{\varepsilon}} + \boldsymbol{\varepsilon}_{,k}\,\delta\phi_k, \qquad \delta\boldsymbol{\sigma} = \delta\bar{\boldsymbol{\sigma}} + \boldsymbol{\sigma}_{,k}\,\delta\phi_k \tag{38}$$

whereas the variations of body forces and surface tractions are

$$\begin{aligned} \delta T_i = \delta(\sigma_{ij} n_j) &= \delta\sigma_{ij} n_j + \sigma_{ij}\delta n_j \\ &= \delta\bar{\sigma}_{ij} n_j + \sigma_{ij,k}\delta\phi_k n_j \\ &\quad + \sigma_{ij}(n_j n_\ell - \delta_{j\ell}) n_k \delta\phi_{k,\ell} \\ \delta\mathbf{f} = \delta\bar{\mathbf{f}} + \mathbf{f}_{,k}\,\delta\phi_k. \end{aligned} \tag{39}$$

The dashed state variations in equation (38) and (39) denote the local variation at the initial configuration of the body (local variations), and the remaining terms represent the variations due to transformation on the material point. Similarly to variations, the rates are expressed as a sum of local time derivatives and convective terms:

$$\dot{\mathbf{u}} = \dot{\bar{\mathbf{u}}} + \mathbf{u}_{,k}\,\dot{\boldsymbol{\phi}}_k, \quad \dot{\boldsymbol{\varepsilon}} = \dot{\bar{\boldsymbol{\varepsilon}}} + \boldsymbol{\varepsilon}_{,k}\,\dot{\boldsymbol{\phi}}_k, \quad \dot{\boldsymbol{\sigma}} = \dot{\bar{\boldsymbol{\sigma}}} + \boldsymbol{\sigma}_{,k}\,\dot{\boldsymbol{\phi}}_k. \tag{40}$$

Consider now the first variations or derivatives of volume and surface integrals. Consider the volume integral

$$I_v = \int f\, dV \tag{41}$$

and its variation

$$\delta I_v = \int \delta f \, dV + \int f\delta(dV) = \int (\delta f + f\delta\phi_{k,k})\, dV = \int \delta \bar{f} dV + \int f\delta\phi_n \, dS \quad (42)$$

where $\delta\phi_n = \delta\phi_k n_k$ is the normal component of $\delta\boldsymbol{\phi}$. The time derivative of I_v is expressed similarly as

$$\begin{aligned} \frac{d}{dt}(I_v) &= \int \left(\frac{Df}{Dt} + f\dot{\phi}_{k,k}\right) dV = \int \left[\frac{\partial f}{\partial t} + (f\dot{\phi}_k)_{,k}\right] dV \\ &= \int \frac{\partial f}{\partial t} dV + \int f\dot{\phi}_n \, dS \end{aligned} \quad (43)$$

and is familiar from continuum mechanics. Here Df/Dt denotes the material time derivative and $\partial f/\partial t$ is the local derivative with fixed position of the material point.

Consider now the surface integral

$$I_s = \int f \, dS \quad (44)$$

over the regular surface S. In view of equation (37), its variation can be expressed as follows:

$$\begin{aligned} \delta I_s &= \int \delta f \, dS + \int f\delta(dS) = \int [\delta f + f(\delta_{k\ell} - n_k n_\ell)\delta\phi_{k,\ell}]\, dS \\ &= \int [\delta f + f(\delta\phi_{\alpha,\alpha} - 2K_m \delta\phi_n]\, dS \end{aligned} \quad (45)$$

where $\delta\phi_\alpha(\alpha = 1, 2)$ are the tangential components of $\delta\boldsymbol{\phi}$ referred to a curvilinear reference system within the surface, and K_m is the mean surface curvature (now the comma denotes the covariant derivative). The last expression can next be transformed as follows [7]:

$$\begin{aligned} \delta I_s &= \int (\delta f - f_{,\alpha}\delta\phi_\alpha - 2fK_m\delta\phi_n)\, dS + \oint_\Gamma f d\phi_\mu \, d\ell \\ &= \int (\delta_n f - 2fK_m\delta\phi_n)\, dS + \oint_\Gamma f\delta\phi_\mu \, d\ell \end{aligned} \quad (46)$$

where we applied the Green theorem for a continuous and differentiable vector field $\mathbf{v}$ specified on a regular surface S bounded by a piecewise smooth closed curve:

$$\int v_{\alpha,\alpha} \, dS = \int_\Gamma v_\alpha \mu_\alpha \, d\ell = \int_\Gamma v_\mu \, d\ell \quad (47)$$

where ℓ is the arc length and $\boldsymbol{\mu}$ denotes the unit vector normal to Γ and tangent to S, pointing towards the outside of S. The symbol $\delta_n f$ denotes the variation along the direction normal to S; thus

$$\delta_n f = \delta \bar{f} + f_{,n}\delta\phi_n \quad (48)$$

References pp. 104–105

and $\delta\phi_\mu = \delta\phi_\alpha \mu_\alpha = \delta\phi_i \mu_i$ is the component of the transformation vector $\delta\boldsymbol{\phi}$ in the direction of $\boldsymbol{\mu}$. To calculate the variation of a surface integral over a piecewise regular surface, formula (46) can be directly applied over each regular surface with subsequent addition of consecutive variations. In particular, the variation of the surface integral over a *closed* piecewise regular surface S takes the form

$$\delta I_s = \int (\delta_n f - 2fK_m \delta\phi_n)\, dS + \sum \int_\Gamma (f^+ \delta\phi_\mu^+ + f^- \delta\phi_\mu^-)\, d\ell \tag{49}$$

where the sum of the line integrals is taken over all edges of the surface S, and + and − signs refer to quantities evaluated on the two regular surface sections intersecting along the edge Γ. Note that the pair $(\delta\phi_\mu^+, \delta\phi_\mu^-)$ of tangential transformation variation at the edge is uniquely defined by the pair ($\delta\phi_n^+$ and $\delta\phi_n^-$) of the normal surface variations on the edge. The time derivative of I_s is expressed similarly:

$$\frac{dI_s}{dt} = \int \left(\frac{\delta f}{\delta t} - 2fK_m \dot{\phi}_n \right) dS + \oint_\Gamma f \dot{\phi}_\mu \, d\ell \tag{50}$$

where $\delta f/\delta t$ is the *transformation derivative* of a spatial field $f(x,t)$ specified by the relation

$$\frac{\delta f}{\delta t} = \frac{\partial f}{\partial t} + f_{,n} \dot{\phi}_n. \tag{51}$$

Note that the transformation derivative describes the variation along the normal direction to the boundary.

Variation of Regular and Piecewise Regular Shapes — Consider first the case of a regular boundary, (Figure 3a), and derive the variation of the functional G specified by (6), following the adjoint structure approach. For the augmented functional (9) we have

$$\begin{aligned} \delta\bar{G} = &\int \delta\psi\, dV + \int \psi \delta(dV) + \int \delta h\, dS + \int h \delta(dS) \\ &- \int \delta(\boldsymbol{\sigma}\boldsymbol{\varepsilon}^a)\, dV - \int \boldsymbol{\sigma}\boldsymbol{\varepsilon}^a \delta(dV) + \int \delta(\mathbf{f}\mathbf{u}^a)\, dV \\ &+ \int (\mathbf{f}\mathbf{u}^a) \delta(dV) \\ &+ \int \delta(\mathbf{T}\mathbf{u}^a)\, dS + \int (\mathbf{T}\mathbf{u}^a) \delta(dS). \end{aligned} \tag{52}$$

In view of equations (37) through (39), we obtain

$$
\begin{aligned}
\delta \bar{G} = & \int \left(\frac{\partial \psi}{\partial \boldsymbol{\sigma}} \delta \bar{\boldsymbol{\sigma}} + \frac{\partial \psi}{\partial \mathbf{u}} \delta \bar{\mathbf{u}} \right) dV + \int \psi n_k \delta \phi_k \, dS \\
& + \int \left[\frac{\partial h}{\partial \mathbf{u}} \delta \mathbf{u} + \frac{\partial h}{\partial \mathbf{T}} \delta \mathbf{T} + h(\delta_{k\ell} - n_k n_\ell) \delta \phi_{k,\ell} \right] dS \\
& - \int \left[\delta \bar{\boldsymbol{\sigma}} \varepsilon^a + \boldsymbol{\sigma} \delta \bar{\boldsymbol{\varepsilon}}^a - \delta \bar{\mathbf{f}} \mathbf{u}^a - \mathbf{f} \delta \bar{\mathbf{u}}^a \right] dV \\
& - \int [\boldsymbol{\sigma} \boldsymbol{\varepsilon}^a - \mathbf{f} \mathbf{u}^a] n_k \delta \phi_k \, dS \\
& + \int [\delta \mathbf{T} \mathbf{u}^a + \mathbf{T} \delta \mathbf{u}^a + \mathbf{T} \mathbf{u}^a (\delta_{k\ell} - n_k n_\ell) \delta \phi_{k,\ell}] \, dS.
\end{aligned}
\tag{53}
$$

Note that the local variations $\delta \bar{\boldsymbol{\sigma}}$ and $\delta \bar{\mathbf{u}}$ occur in the volume integral whereas the total variations $\delta \boldsymbol{\sigma}$ and $\delta \mathbf{u}$ occur in the surface integrals. Formula (53) can be rearranged as follows:

$$
\begin{aligned}
\delta \bar{G} = & \int [(\psi - \boldsymbol{\sigma} \boldsymbol{\varepsilon}^a + \mathbf{f} \mathbf{u}^a) n_k \delta \phi_k + (\mathbf{T}^a \mathbf{u}_{,k} + \mathbf{T} \mathbf{u}^a_{,k}) \delta \phi_k \\
& + (h + \mathbf{T} \mathbf{u}^a)(\delta_{k\ell} - n_k n_\ell) \delta \phi_{k,\ell}] \, dS + \int \delta \bar{\mathbf{f}} \mathbf{u}^a \, dV \\
& + \int \left(\frac{\partial h}{\partial \mathbf{T}} + \mathbf{u}^a \right) \delta \mathbf{T}^0 \, dS_T + \int \left(\frac{\partial h}{\partial \mathbf{u}} - \mathbf{T}^a \right) \delta \mathbf{u}^0 \, dS_u \\
& - \left\{ \int (\boldsymbol{\sigma} \delta \bar{\boldsymbol{\varepsilon}}^a - \mathbf{f} \delta \bar{\mathbf{u}}^a) \, dV - \int \mathbf{T} \delta \bar{\mathbf{u}}^a \, dS \right\} \\
& - \left\{ \int [(\boldsymbol{\varepsilon}^a - \boldsymbol{\varepsilon}^{ai}) \delta \bar{\boldsymbol{\sigma}} - \mathbf{f}^a \delta \bar{\mathbf{u}}] \, dV \right. \\
& \left. - \int \mathbf{T}^a \delta \bar{\mathbf{u}} \, dS \right\} - \left\{ \int \left[\left(\boldsymbol{\varepsilon}^{ai} - \frac{\partial \psi}{\partial \boldsymbol{\sigma}} \right) \delta \bar{\boldsymbol{\sigma}} + \left(\mathbf{f}^a - \frac{\partial \psi}{\partial \mathbf{u}} \right) \delta \bar{\mathbf{u}} \right] dV \right. \\
& \left. + \int \left(\mathbf{T}^{a0} - \frac{\partial h}{\partial \mathbf{u}} \right) \delta \mathbf{u} \, dS_T - \int \left(\mathbf{u}^{a0} + \frac{\partial h}{\partial \mathbf{T}} \right) \delta \mathbf{T} \, dS_u \right\}.
\end{aligned}
\tag{54}
$$

Consider a linear elastic structure for which the stress-strain relations (2) occur, whereas for the adjoint structure the relations in equation (11) are valid. Then for a homogeneous material, we have

$$
\delta \bar{\boldsymbol{\sigma}} = \delta \boldsymbol{\sigma} - \boldsymbol{\sigma}_{,k} \delta \phi_k = \mathbf{D} \delta \boldsymbol{\varepsilon} - \boldsymbol{\sigma}_{,k} \delta \phi_k = \mathbf{D}(\delta \boldsymbol{\varepsilon} - \boldsymbol{\varepsilon}_{,k} \delta \phi_k) = \mathbf{D} \delta \bar{\boldsymbol{\varepsilon}} \tag{55}
$$

and

$$
\int (\boldsymbol{\varepsilon}^a - \boldsymbol{\varepsilon}^{ai}) \delta \bar{\boldsymbol{\sigma}} \, dV = \int (\boldsymbol{\varepsilon}^a - \boldsymbol{\varepsilon}^{ai}) \mathbf{D} \delta \bar{\boldsymbol{\varepsilon}} \, dV = \int \boldsymbol{\sigma}^a \delta \bar{\boldsymbol{\varepsilon}} \, dV. \tag{56}
$$

Substituting (56) into (54), one obtains

$$
\begin{aligned}
\delta\bar{G} = &\int [(\psi - \boldsymbol{\sigma}\boldsymbol{\varepsilon}^a + \mathbf{f}\mathbf{u}^a)n_k\delta\phi_k + (\mathbf{T}\mathbf{u}^a_{,k} + \mathbf{T}^a\mathbf{u}_{,k})\delta\phi_k \\
&+ (h + \mathbf{T}\mathbf{u}^a)(\delta_{k\ell} - n_k n_\ell \delta\phi_{k,\ell}]\, dS + \int \left(\frac{\partial h}{\partial \mathbf{T}} + \mathbf{u}^a\right)\delta\mathbf{T}^0\, dS_T \\
&+ \int \left(\frac{\partial h}{\partial \mathbf{u}} - \mathbf{T}^a\right)\delta\mathbf{u}^0\, dS_u + \int \delta\bar{\mathbf{f}}\mathbf{u}^a\, dV \\
&- \left\{\int (\boldsymbol{\sigma}\delta\bar{\boldsymbol{\varepsilon}}^a - \mathbf{f}\delta\bar{\mathbf{u}}^a)\, dV - \int \mathbf{T}\delta\bar{\mathbf{u}}^a\, dS\right\} \\
&- \left\{\int (\boldsymbol{\sigma}^a\delta\bar{\boldsymbol{\varepsilon}} - \mathbf{f}^a\delta\bar{\mathbf{u}})\, dV - \int \mathbf{T}^a\delta\bar{\mathbf{u}}\, dS\right\} \\
&- \left\{\int \left[\left(\boldsymbol{\varepsilon}^{ai} - \frac{\partial\psi}{\partial\boldsymbol{\sigma}}\right)\delta\bar{\boldsymbol{\sigma}} + \left(\mathbf{f}^a - \frac{\partial\psi}{\partial\mathbf{u}}\right)\delta\bar{\mathbf{u}}\right] dV \right. \\
&\left. + \int \left(\mathbf{T}^{a0} - \frac{\partial h}{\partial\mathbf{u}}\right)\delta\mathbf{u} dS_T - \int \left(\mathbf{u}^{a0} + \frac{\partial h}{\partial\mathbf{T}}\right)\delta\mathbf{T}\, dS_u\right\}
\end{aligned}
\tag{57}
$$

The expression is similar to (12) where the terms in the last pair of braces specifies the data for boundary conditions, initial strain and body force fields within the adjoint structure identical to (16). On the other hand, the terms in the first and the second pair of braces in equation (57) constitute the equilibrium conditions of both primary and adjoint structures expressed through the virtual work equations at their initial configurations. The remaining terms of (57) represent the first variation of G due to infinitesimal transformation $\delta\boldsymbol{\phi}$; thus,

$$
\begin{aligned}
\delta\bar{G} = &\int \{[(\psi - \boldsymbol{\sigma}\boldsymbol{\varepsilon}^a + \mathbf{f}\mathbf{u}^a)n_k + \mathbf{T}\mathbf{u}^a_{,k} + \mathbf{T}^a\mathbf{u}_{,k}]\delta\phi_k \\
&+ (h + \mathbf{T}\mathbf{u}^a)(\delta_{k\ell} - n_k n_\ell)\,\delta\phi_{k,\ell}\}\, dS \\
&+ \int \left(\frac{\partial h}{\partial\mathbf{T}} + \mathbf{u}^a\right)\delta\mathbf{T}^0\, dS_T \\
&+ \int \left(\frac{\partial h}{\partial\mathbf{u}} - \mathbf{T}^a\right)\delta\mathbf{u}^0\, dS_u + \int \delta\bar{f}\mathbf{u}^a\, dV.
\end{aligned}
\tag{58}
$$

It should be stressed that for specified loading, support conditions, and body force field, their variations $\delta\mathbf{T}^0$ on S_T, $\delta\mathbf{u}^0$ on S_u, and $\delta\bar{\mathbf{f}}$ within V are known and can be expressed in terms of the transformation field variation $\delta\boldsymbol{\phi}$ and its gradients.

Expression (58) contains gradients of the transformation vector on S. In view of equation (46), it can further be transformed to the following form containing only components of $\delta\boldsymbol{\phi}$, namely,

$$\begin{aligned}\delta\bar{G} = &\int [\psi - \boldsymbol{\sigma}\boldsymbol{\varepsilon}^a + \mathbf{f}\mathbf{u}^a + (h + \mathbf{T}\mathbf{u}^a)_{,n} - (h + \mathbf{T}\mathbf{u}^a)2K_m]\,\delta\phi_n\,dS \\ &+ \int \left(\frac{\partial h}{\partial \mathbf{T}} + \mathbf{u}^a\right)(\delta\mathbf{T}^0 - \mathbf{T}^0_{,k}\delta\phi_k)\,dS_T \\ &+ \int \left(\frac{\partial h}{\partial \mathbf{u}} - \mathbf{T}^a\right)(\delta\mathbf{u}^0 - \mathbf{u}^0_{,k}\delta\phi_k)\,dS_u \\ &+ \int \delta\bar{\mathbf{f}}\mathbf{u}^a\,dV + \oint_\Gamma [\![h + \mathbf{T}\mathbf{u}^a]\!]\,\delta\phi_\mu\,d\ell\end{aligned} \tag{59}$$

where the line integral can be selected along the curve Γ separating the boundary portions S_T and S_u. Since $\delta\phi_\mu^+ = -\delta\phi_\mu^- = \delta\phi_\mu$, the integrand $[\![h + \mathbf{T}\mathbf{u}^a]\!] = (h+\mathbf{T}\mathbf{u}^a)^+ - (h+\mathbf{T}\mathbf{u}^a)^-$ represents the discontinuity of $h+\mathbf{T}\mathbf{u}^a$ along Γ. Therefore, this line integral describes the interaction between boundaries with discontinuous boundary conditions.

Consider now a case when the boundary is formed by a set of regular surfaces intersecting at edges L_k, (Figure 3). Following formula (49), the line integrals along L_k in (59) will appear in the form

$$\sum_k \int_{\Gamma_k} [(h + \mathbf{T}\mathbf{u}^a)^+\delta\phi_\mu^+ + (h + \mathbf{T}\mathbf{u}^a)^-\delta\phi_\mu^-]d\ell_k \tag{60}$$

and (59) should be completed by the contribution of equation (60) along the edges of intersection of regular surface sections.

Variation of Interface Shapes — Consider now the case when the stress, strain or displacement fields may suffer discontinuities on the interface S^c (Figure 4). The interface divides the body into two parts, $V^{(1)}$ and $V^{(2)}$, within which stress and strain in fields are continuous and have continuous derivatives. Assume the transformation field to modify only S^c whereas the external boundary remains unaltered; thus $\delta\phi = 0$ on S. Three types of discontinuity surfaces have mechanical significance and they will now be briefly discussed.

For a composite structure with two different materials occupying subdomains $V^{(1)}$ and $V^{(2)}$ the stiffness moduli vary discontinuously on the *composite interface* S^c (see Figure 4(a)). However, the displacements $\mathbf{u} = \mathbf{u}^c$ and the surface tractions $\mathbf{T}^c = \boldsymbol{\sigma}^c\mathbf{n}$ are continuous, but their gradients and both stress and strain components exhibit discontinuities on S^c. A jump across S^c will be denoted by $[\![\ ]\!]$, so that

$$[\![f]\!] = f^{(2)} - f^{(1)} \tag{61}$$

where the superscripts (1) and (2) refer to the quantities defined on S^c regarded as the boundary of $V^{(1)}$ and $V^{(2)}$ respectively. Moreover, $\mathbf{n}^{(1)} = \mathbf{n}^{(2)} = \mathbf{n}$, so we can write

$$\begin{aligned}&[\![\mathbf{u}]\!] = 0, \qquad [\![\mathbf{T}]\!] = [\![\boldsymbol{\sigma}]\!]\mathbf{n} = 0 \\ &[\![\mathbf{u}_{,k}]\!] = [\![\mathbf{u}_{,n}]\!]\mathbf{n}_k, \qquad [\![\mathbf{T}_{,k}]\!] = [\![\mathbf{T}_{,n}]\!]\mathbf{n}_k\end{aligned} \tag{62}$$

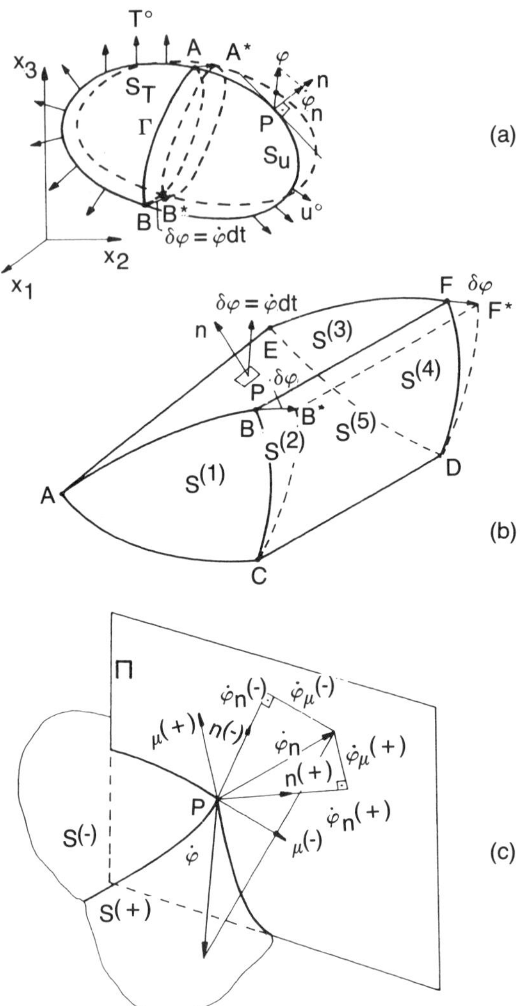

Figure 3. **(a) Regular external shape and its transformation, (b) piecewise regular external shape, and (c) decomposition of the transformation rate vector at the edge of intersection of regular surfaces S^+ and S^-.**

and

$$
\begin{aligned}
[\![\delta \mathbf{u}]\!] &= [\![\delta \bar{\mathbf{u}}]\!] + [\![\mathbf{u}_{,n}]\!]\delta\phi_n = \mathbf{0} \\
[\![\delta \mathbf{T}]\!] &= [\![\delta \bar{\mathbf{T}}]\!] + [\![\mathbf{T}_{,n}]\!]\delta\phi_n = [\![\boldsymbol{\sigma}]\!]\delta\mathbf{n} + [\![\delta\bar{\boldsymbol{\sigma}}]\!]\mathbf{n} + [\![\boldsymbol{\sigma}_{,k}]\!]\delta\phi_k\mathbf{n} = \mathbf{0}
\end{aligned}
\tag{63}
$$

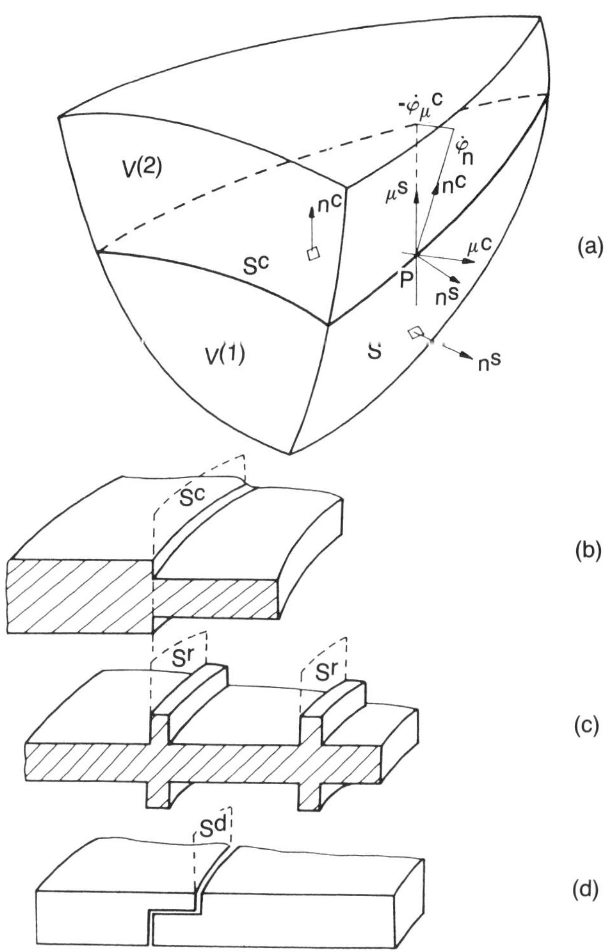

Figure 4. (a) Interface S^c separating two domains $V^{(1)}$ and $V^{(2)}$ in a structure, (b) interface S^c in a plate with stepwise varying thickness, (c) interface with reinforcing ring in a plate, and (d) discontinuity interface S^d in a plate.

From equations (63) and (37) it follows that

$$\begin{aligned} [\![\delta u_i]\!] &= -[\![u_{i,n}]\!]\delta\phi_n \\ [\![\delta \bar{T}_i]\!] &= [\![\delta\bar{\sigma}_{ij}]\!]n_j \\ &= [\![\sigma_{ij}]\!]n_k\delta\phi_{k,j} - [\![\sigma_{ij,k}]\!]n_j\delta\phi_k. \end{aligned} \tag{64}$$

Applying expression (59) to $V^{(1)}$ and $V^{(2)}$ and then adding, the variation G can be found in the case of moving interface S^c and fixed external boundaries. Setting

References pp. 104–105

$\delta\bar{\mathbf{f}} = 0$ within V, $h = 0$, $\delta\phi_k = 0$ on S, and $\delta\mathbf{T} = \delta\mathbf{u} = 0$ on S^c, we obtain

$$\begin{aligned}\delta\bar{G} = \int \{ [\![\psi]\!] - [\![\boldsymbol{\sigma}\boldsymbol{\varepsilon}^a]\!] + [\![\mathbf{f}]\!]\mathbf{u}^a + \mathbf{T}[\![\mathbf{u}^a_{,n}]\!] \\ -\mathbf{T}^a[\![\mathbf{u}_{,n}]\!]\}\, \delta\phi_n \, dS^c \\ + \int_\Gamma [\![\mathbf{T}\mathbf{u}^a]\!] \delta\phi_\mu \, d\ell\end{aligned} \tag{65}$$

where the line term refers to the curve of intersection of S^c with the boundary surface S.

The interface S^c may also refer to cases where a stepwise continuous thickness variation occurs within a plate, disk or beam (Figure 4(b)). Using generalized stress and strain components with a stiffness matrix depending on thickness, the varation δG associated with the variation of S^c can be expressed by equation (65). Examples of optimal design for unspecified interface are presented in reference [8].

The interface may also be a surface of discontinuity of tractions and displacements. The first case occurs when a reinforcing membrane is introduced on the interface S^r within the three-dimensional body, or a stiffener is placed onto a plate or shell, as in Figure 4(c). The traction discontinuity is then related to tangential forces through proper equilibrium equations.

The other case occurs when some displacement components suffer discontinuities on the interface S^d as in Figure 4(d). Such discontinuities are introduced in order to reduce stresses induced by distributed or localized distortion fields, for instance, those due to a thermal field or an imposed support displacement. A simple model of the displacement discontinuity is obtained by assuming that

$$[\![\mathbf{u}]\!] = \mathbf{f}(\mathbf{T}), \tag{66}$$

that is, by relating the discontinuity to tractions on S^d. The variation δG will also contain the expressions

$$-\int 2[\![\mathbf{T}]\!]\mathbf{u}^a K_m \delta\phi_n \, dS^r, \qquad -\int 2\mathbf{T}[\![\mathbf{u}^a]\!] K_m \delta\phi_n \, dS^d \tag{67}$$

as well as the terms presented in (65).

Direct Method of Sensitivity Analysis — In some cases, it may be practical to generate variations or derivatives of specified functionals by calculating variations of stress, strain and displacement fields from properly posed boundary-value problems. Starting from the expressions

$$\delta\boldsymbol{\sigma} = \delta\bar{\boldsymbol{\sigma}} + \boldsymbol{\sigma}_{,k}\,\delta\phi_k, \qquad \delta\boldsymbol{\varepsilon} = \delta\bar{\boldsymbol{\varepsilon}} + \boldsymbol{\varepsilon}_{,k}\,\delta\phi_k, \qquad \delta\mathbf{u} = \delta\bar{\mathbf{u}} + \mathbf{u}_{,k}\,\delta\phi_k, \tag{68}$$

it is seen that the total variations can easily be calculated from (68), providing the local variations $\delta\bar{\boldsymbol{\sigma}}$, $\delta\bar{\boldsymbol{\varepsilon}}$ and $\delta\bar{\mathbf{u}}$ are specified. Let us note that these variations satisfy the usual linear elasticity relation

$$\delta\bar{\boldsymbol{\sigma}} = \mathbf{D}\delta\bar{\boldsymbol{\varepsilon}}, \qquad \delta\bar{\sigma}_{ij,j} + \delta\bar{f}_i = 0, \qquad \delta\bar{\varepsilon}_{ij} = \frac{1}{2}(\delta\bar{u}_{ij,j} + \delta\bar{u}_{j,i}). \tag{69}$$

To specify boundary conditions for local variations, assume that surface tractions $\mathbf{T}^0(x)$ and displacements $\mathbf{u}^0(x)$ are specified as fields in the neighborhood of the actual body configuration. Then we have

$$\begin{aligned} \delta\mathbf{T}^0 &= \delta\bar{\boldsymbol{\sigma}}\mathbf{n} + \boldsymbol{\sigma}_{,k}\,\delta\phi_k\mathbf{n} + \boldsymbol{\sigma}\delta\mathbf{n} \qquad \text{on } S_T \\ \delta\mathbf{u}^0 &= \delta\bar{\mathbf{u}} + \mathbf{u}_{,k}\,\delta\phi_k \qquad \text{on } S_u \end{aligned} \tag{70}$$

from which the boundary conditions are specified in the form

$$\begin{aligned} \delta\bar{\boldsymbol{\sigma}}n &= \delta\mathbf{T}^0 - \boldsymbol{\sigma}_{,k}\,\delta\phi_k\mathbf{n} - \boldsymbol{\sigma}\delta\mathbf{n} \qquad \text{on } S_T \\ \delta\bar{\mathbf{u}} &= \delta\mathbf{u}^0 - \mathbf{u}_{,k}\,\delta\phi_k \qquad \text{on } S_u. \end{aligned} \tag{71}$$

Solving the boundary value problem in equations (69) and (71), the variation δG can be determined by direct calculus. This method may have some advantage over the adjoint structure approach when boundary variation is dependent on a set of parameters $b_k (k = 1, 2, 3, \dots m), \boldsymbol{\phi} = \boldsymbol{\phi}(\mathbf{x}, b_k)$ and when there are p functionals whose variations are to be determined.

Variation of Local Stress, Strain or Displacement Components — In the previous analysis, we considered the problem of variation of an arbitrary functional G defined over the whole domain Υ and/or the boundary surface of the structure. Thus the functional G has been regarded as a global structural response of a body. Consider now a closely related class of problems associated with variation of local stress, strain or displacement components within a body. It will be shown that they can be expressed as bilinear functionals with proper qualification of the adjoint body.

Assume first that the stress component σ_{qs} at a typical point x_0 within a structure is to be determined. Using the well-known property of the Dirac function $\delta(\mathbf{x} - \mathbf{x}_0), x \in \Upsilon$, the stress component σ_{qs} is expressed as follows:

$$\sigma_{qs}(\mathbf{x}_0) = G = \int \delta(\mathbf{x} - \mathbf{x}_0)\delta_{qk}\delta_{s\ell}\sigma_{k\ell}\,dV = \int \psi(\boldsymbol{\sigma})\,dV. \tag{72}$$

Comparing (72) with (6), we have

$$h(\mathbf{u}, \mathbf{T}) = 0 \qquad \text{on } S, \qquad \psi(\boldsymbol{\sigma}) = \delta(\mathbf{x} - \mathbf{x}_0)\delta_{qk}\delta_{s\ell}\sigma_{k\ell}. \tag{73}$$

Hence, the stress component $\sigma_{qs}(\mathbf{x}_0)$ can be expressed in the form of the bilinear functional (9) as

$$\sigma_{qs} = \bar{G} = \int \psi\,dV - \int \boldsymbol{\sigma\varepsilon}^a\,dV + \int \mathbf{f}\mathbf{u}^a\,dV + \int \mathbf{T}\mathbf{u}^a\,dS. \tag{74}$$

In view of (16), the adjoint structure is now specified as follows:

$$\begin{aligned} \varepsilon^{ai}_{k\ell} &= \frac{\partial\psi}{\partial\sigma_{k\ell}} = \delta(\mathbf{x} - \mathbf{x}_0)\delta_{qk}\delta_{s\ell}, \qquad \mathbf{f}^a = 0 \qquad \text{within } V \\ \mathbf{T}^{a0} &= 0 \qquad \text{on } S_T, \qquad \mathbf{u}^{a0} = 0 \qquad \text{on } S_u. \end{aligned} \tag{75}$$

Thus, for specified q and s, the first expression in (75) defines a unit *edge* or *screw dislocation* at $\mathbf{x}_0$.

The variation of any displacement component at the typical point $\mathbf{x}_0$ can be derived in a similar way. Introducing the integrand ψ in the form

$$\psi(\mathbf{u}) = \delta(\mathbf{x} - \mathbf{x}_0)\delta_{qk}u_k, \tag{76}$$

the displacement component u_q is expressed in the form

$$u_q(x_0) = \int \psi(\mathbf{u})\, dV = \int \delta(\dot{\mathbf{x}} - \mathbf{x}_0)\delta_{qk}u_k\, dV. \tag{77}$$

The adjoint structure with vanishing boundary tractions and displacement on S_t and S_u, respectively, and vanishing field of initial strain within V is now loaded by the body force f^a at $\mathbf{x}_0$, which is given as

$$f_k^a = \frac{\partial \psi}{\partial u_k} = \delta(\mathbf{x} - \mathbf{x}_0)\delta_{qk} \qquad \text{within } V, \tag{78}$$

and the associated bilinear functional is

$$\bar{G} = u_q(\mathbf{x}_0) - \int \boldsymbol{\sigma}\boldsymbol{\varepsilon}^a\, dV + \int \mathbf{T}\mathbf{u}^a\, dS = \int \boldsymbol{\sigma}^a\boldsymbol{\varepsilon}\, dV. \tag{79}$$

The expressions (17) and (59) for the first variation of G remain valid in this case.

APPLICATIONS TO OPTIMAL SHAPE DESIGN

The general analysis of previous sections can easily be particularized to any specific example. In this paper, we shall briefly discuss the application to optimal shape problems and derivation of proper optimality conditions.

Stationarity Conditions in Optimal Shape Design — One immediate application of sensitivity analysis is to derive the stationarity conditions in optimal shape design including both loaded, supported or free boundaries and piecewise regular shapes with interaction terms along edges of intersection.

Consider first the case of regular shape and assume the cost of the structure to be proportional to the material volume; thus

$$C = c\int dV, \qquad \delta C = c\int n_k\, \delta\phi_k\, dS \tag{80}$$

where c is a constant parameter. The problem is then reduced to minimizing or maximizing the objective functional G with a specified upper bound on the structure cost:

$$\text{min or max of } G, \qquad \text{subject to } C \leq C_0. \tag{81}$$

Introducing the functional

$$G' = G + \lambda(C - C_0) \tag{82}$$

where λ is the Lagrange multiplier, the condition of stationarity of G is expressed in the form

$$\delta G' = \delta G + \lambda \delta C + \delta\lambda(C - C_0) = 0 \tag{83}$$

and

$$\delta G = \lambda \delta C, \qquad \delta\lambda(C - C_0) = 0. \tag{84}$$

The second equality requires either $C = C_0$ or $\delta\lambda = 0$. The first stationarity condition can be expressed explicitly in terms of states of primary and adjoint structures on the boundary surface. Starting from the formulas (59) and (80), the general stationarity condition takes the form

$$\begin{aligned}
&\int [\psi - \boldsymbol{\sigma\varepsilon}^a + \mathbf{f}\mathbf{u}^a + (h + \mathbf{T}\mathbf{u}^a)_{,n} - (h + \mathbf{T}\mathbf{u}^a) 2K_m] \delta\phi_n \, dS \\
&\quad + \int \left(\frac{\partial h}{\partial \mathbf{T}} + \mathbf{u}^a \right) (\delta \mathbf{T}^0 - \mathbf{T}^0_{,k} \delta\phi_k) \, dS_T \\
&\quad + \int \left(\frac{\partial h}{\partial \mathbf{u}} - \mathbf{T}^a \right) (\delta \mathbf{u}^0 - \mathbf{u}^0_{,k} \delta\phi_k) \, dS_u \\
&\quad + \int \delta \bar{\mathbf{f}} \mathbf{u}^a dV + \sum_k \int_\Gamma [(h + \mathbf{T}\mathbf{u}^a)^+ \delta\phi_\mu^+ + (h + \mathbf{T}\mathbf{u}^a)^- \delta\phi_\mu^-] \, d\ell_\mu \\
&= -\lambda \int C \delta\phi_n \, dS.
\end{aligned} \tag{85}$$

This general condition can be applied to various particular cases. When only the loaded boundary is varied and when $\delta \bar{f} = 0$, $h = h(\mathbf{u})$ on S_T, and $h = 0$ on S_u, then for a regular boundary we obtain

$$\begin{aligned}
&\int [\psi - \boldsymbol{\sigma\varepsilon}^a + \mathbf{f}\mathbf{u}^a + (h + \mathbf{T}\mathbf{u}^a)_{,n} - (h + \mathbf{T}\mathbf{u}^a) 2K_m] \, \delta\phi_n \, dS_T \\
&\quad + \int \mathbf{u}^a (\delta \mathbf{T}^0 - \mathbf{T}^0_{,k} \delta\phi_k) \, dS_T = -\lambda C \int \delta\phi_n \, dS_T.
\end{aligned} \tag{86}$$

Let us note that when the specified traction field $\mathbf{T}^0$ does not depend on boundary configuration $\mathbf{T}^0 = \mathbf{T}^0(\mathbf{x})$, then the last term of (86) vanishes and the optimality condition takes the form

$$\psi - \boldsymbol{\sigma\varepsilon}^a + \mathbf{f}\mathbf{u}^a + (h + \mathbf{T}\mathbf{u}^a)_{,n} - (h + \mathbf{T}\mathbf{u}^a) 2K_m = -\lambda C \qquad \text{on } S_T. \tag{87}$$

In the case of free boundary variation with $h = 0$ within V, condition (8) provides

$$\boldsymbol{\sigma\varepsilon}^a - \psi = \alpha C = \text{ constant} \qquad \text{on } S^0. \tag{88}$$

For the variation of the interface S^c, in view of (65), the optimality condition is

$$[\![\psi]\!] - [\![\boldsymbol{\sigma\varepsilon}^a]\!] + [\![\mathbf{f}]\!]\mathbf{u}^a - [\![\sigma_{k\ell}]\!]\varepsilon^a_{k\ell} = -\lambda(C_1 - C_2) \qquad \text{on } S^c \tag{89}$$

References pp. 104–105

where C_1 and C_2 are the specific costs of materials occupying domains $V^{(1)}$ and $V^{(2)}$. Here $\sigma_{k\ell}$ and $\varepsilon_{k\ell}(k, \ell = 1, 2)$ are the internal stress and strain components within the plane tangential to S^c. In fact, on S^c, we may write

$$[\![\sigma_{ij}\varepsilon_{ij}]\!] = [\![\sigma_{k\ell}]\!]\varepsilon_{k\ell} + \sigma_{in}[\![\varepsilon_{in}]\!] \tag{90}$$

where σ_{in} are traction components and the conjugate strains.

The applicability of such optimality conditions was illustrated in the references [4, 8]. These conditions can also be used in generating evolution rules for the boundary surface during the iterative redesign procedure. Note that the local optimality conditions can generally be expressed in the form

$$\Phi(\mathbf{u}, \mathbf{u}^a, \boldsymbol{\sigma}, \boldsymbol{\sigma}^a, \boldsymbol{\varepsilon}, \boldsymbol{\varepsilon}^a) = \text{ constant} \qquad \text{on } S. \tag{91}$$

Assume n nodal points on the boundary that specify its evolution. Let the mean value of Φ on the instantaneous boundary be $\bar{\Phi}$. A subsequent boundary evolution can be specified by a rule providing normal transformation of each nodal point, for instance.

$$\delta\phi_n^{(p)} = k\left[\left(\frac{\Phi^{(p)}}{\bar{\Phi}}\right)^n - 1\right] \tag{92}$$

where k and n are positive parameters. Note that when $\Phi^{(p)} < \bar{\Phi}$, the boundary point p moves inside the body; and for $\Phi^{(p)} > \bar{\Phi}$, it moves to the exterior. The evolution process terminates when $\Phi^{(p)} = \bar{\Phi}$. However, such boundary evolution may not provide a solution corresponding to the global minimum of the functional because the stationarity point may not correspond to the optimal solution.

Mean Compliance Design —The case of mean compliance design was investigated in numerous papers. The mean structure compliance is measured by the work of surface tractions on S_T for a rigidly supported boundary portion $S_u(\mathbf{u}^0 = 0$ on $S_u)$. In the linear case the mean compliance C can therefore be identified with the complementary energy of the structure Π_σ and the mean stiffness S with the potential energy Π_u, so that

$$\begin{aligned} \Pi_\sigma &= \int W(\boldsymbol{\sigma})\, dV - \int \mathbf{T}\mathbf{u}^0\, dS_u = \int W(\boldsymbol{\sigma})\, dV = \frac{1}{2}\int \mathbf{T}^0\mathbf{u}\, dS_T \\ \Pi_u &= \int U(\boldsymbol{\varepsilon})\, dV - \int \mathbf{T}^0\mathbf{u}\, dS_T = -\int U(\boldsymbol{\varepsilon})\, dV = -\frac{1}{2}\int \mathbf{T}^0\mathbf{u}\, dS_T \end{aligned} \tag{93}$$

and

$$C = 2\Pi_\sigma, \qquad S = -C = 2\Pi_u. \tag{94}$$

As the states of primary and adjoint structures coincide, the optimality conditions are considerably simplified. For instance, for the free boundary S_0, the condition (88) provides

$$U = \frac{1}{2}\boldsymbol{\sigma\varepsilon} = U^0 = \text{ constant} \qquad \text{on } S_0. \tag{95}$$

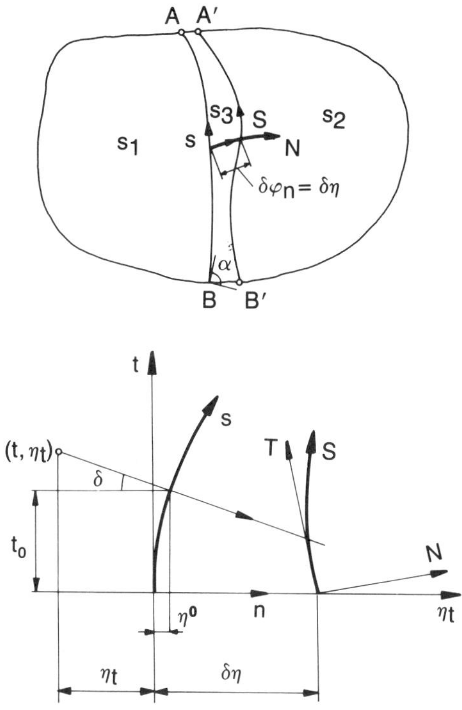

Figure 5. Stiffener AB in a plate and its two positions.

and for the interface S^c we obtain from (89)

$$\tilde{U} = [\![U]\!] - [\![\mathbf{f}]\!]\mathbf{u} - \mathbf{T}[\![\mathbf{u}_{,n}]\!] = \lambda(C_1 - C_2) = \tilde{U}^0. \tag{96}$$

The strong minimum theorem requires the specific strain energy to be a decreasing function when moving from the interior to the exterior of body domain in the neighborhood of S_0. As this condition is not satisfied in most cases, only stationarity is assured by the optimality conditions (95) and/or (96).

Consider now a problem of optimal shape of the interface S^r. This problem is related to stiffener design within the in-plane disc or plate subjected to tension and flexure. Considering the mean compliance design, the potential energy

$$\Pi_u = \int U^- \, dS_1 + \int U^+ \, dS_2 + \int U^s \, d\ell - \int \mathbf{T}^0 \mathbf{u} \, dS_T \tag{97}$$

should be maximized subject to the constraint set on volume or cost of the structure. Here $U^-(\boldsymbol{\varepsilon}, h), U^+(\boldsymbol{\varepsilon}, h)$ and $U^s(\boldsymbol{\varepsilon}_t, A_s)$ are the specific elastic strain energies of the plate in domains S_1 and S_2, and of the stiffener per unit length (Figure 5). The variation of Π_u for a disc is expressed as follows:

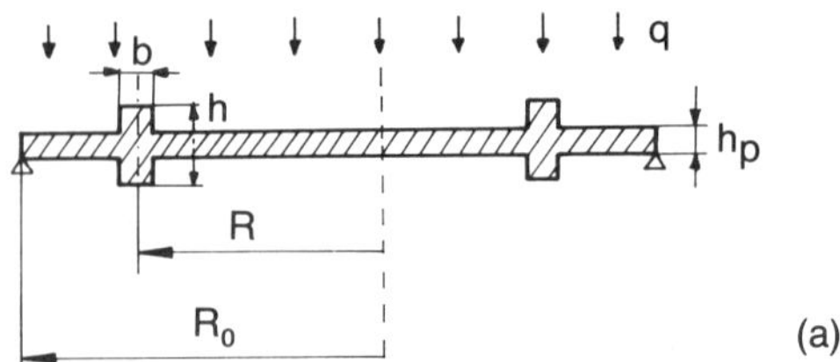

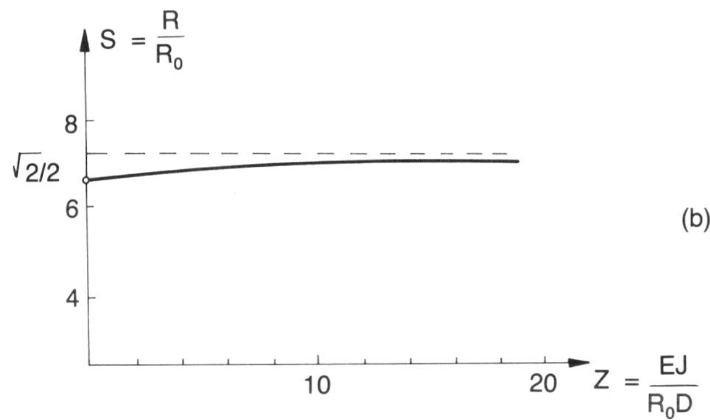

Figure 6. (a) Reinforcing ring within a circular plate, and (b) its optimal position.

$$\begin{aligned}\delta\Pi_u = \int \delta U^- \, dS_1 + \int \delta U^+ \, dS_2 + \int [\![U]\!]\delta\phi_n \, d\ell + \int \delta U^s \, d\ell \\ + \int U^s k_s \delta\phi_n \, d\ell = \int \frac{\partial U^s}{\partial A_s}\delta A_s \, d\ell + \int U^s k_s \delta\phi_n \, d\ell \\ - \frac{1}{2}\int [\![\mathbf{T}]\!](\mathbf{u}_{,n}^- + \mathbf{u}_{,n}^+)\delta\phi_n \, d\ell + \int N u_{n,t}\delta\phi_{n,t} \, d\ell\end{aligned} \tag{98}$$

where k_s is the stiffener curvature, N denotes the axial stiffener force (bending stiffness being neglected), and A_s is the cross-sectional area. A more general model accounting for axial, flexural and torsional stiffness of the stiffener within a disc or plate is considered in a separate paper [16].

We conclude this section by considering two simple examples. Consider first a circular, simply supported plate of radius R_0 with a circular stiffener of rectangular cross-section of fixed area $A_s = bh$ shown in Figure 6(a). Let us determine an optimal location of the stiffener within the plate. The optimality condition now takes the form

$$K_r^+ + K_r^- = -\frac{1}{r} w_{1r} \tag{99}$$

where w is the lateral deflection of the plate and K_r^+, K_r^- is the radial curvature on both sides of the stiffener. Figure 6(b) presents the optimal solution for varying ratio of the stiffener flexural stiffness EJ to the plate stiffness D. It is seen that the stiffener position tends asymptotically to $R_s = \frac{1}{2}\sqrt{2}R_0$ for large values of EJ.

Figure 7 presents the optimal solution for a simply supported and uniformly loaded rectangular plate with two stiffeners symmetrically situated with respect to the central axis. The optimality condition now takes the form

$$\int [T] w_{,n} \, d\ell = 0 \tag{100}$$

where [T] denotes the discontinuity of the transverse force and $w_{,n}$ is the deflection gradient in normal direction to the stiffener. It is seen from Figure 7 that for small rib stiffness EJ the two stiffeners coincide with the symmetry axis, $\bar{t} = t/a = 0.5$, and for large EJ the stiffeners tend to their limiting position $\bar{t} = 0.313$.

CONCLUDING REMARKS

In this chapter, we discussed uniform approaches to sensitivity analysis for both linear and nonlinear structures, accounting for material parameter or shape variation. Potential applications of the general results are numerous, first in optimal shape design problems but also in identification and shape recognition problems. Other applications relate to physical metallurgy or fracture mechanics where phase transitions or crack propagation can be treated as a shape transformation process. Variational formulation and the use of bilinear functionals result in the generation of field equations for both primary and adjoint structures and also in sensitivity expressions from the same functional. Such formulation may therefore be used in developing numerical procedures for analysis and in iterative redesign procedures including optimal design as one of the objectives.

References pp. 104–105

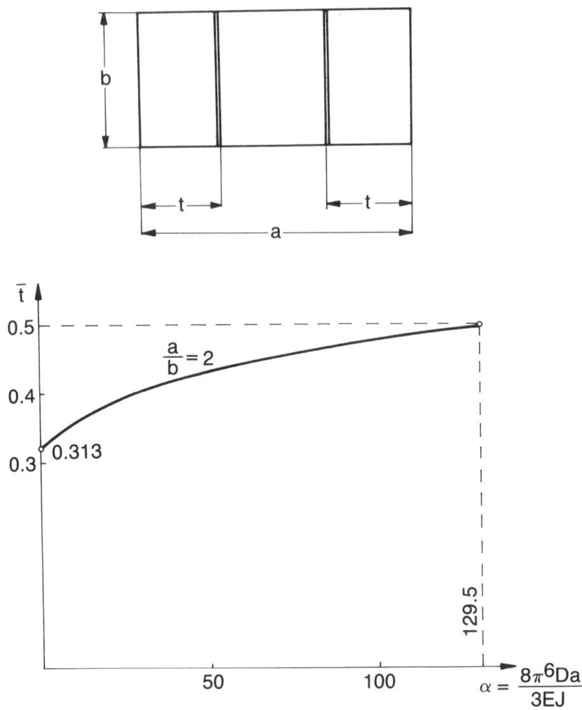

Figure 7. Rectangular plate with two stiffeners and their optimal position $\bar{t} = t/a$.

REFERENCES

1. R. Tomovic and M. Vukabratovic, *General Sensitivity Theory.* American Elsevier (1972).
2. E. J. Haug, K. K. Choi and V. Komkov, *Design Sensitivity Analysis of Structural Systems.* Academic Press (1985).
3. K. Dems and Z. Mróz, Variational approach by means of adjoint systems to structural optimization and sensitivity analysis—Part I: Variation of material parameters within fixed domain. *Int. J. Solids Struct.* **19**, 677–692 (1983).
4. K. Dems and Z. Mróz, Variational approach by means of adjoint systems to structural optimization and sensitivity analysis—Part II: Structure shape variation. *Int. J. Solids Struct.* **20**, 527–552 (1984).
5. K. Dems and Z. Mróz, Variational approach to first and second order sensitivity analysis of elastic structures. *Int. J. Numer. Meth. Eng.* **21**, 637–661 (1984).
6. R. T. Haftka and Z. Mróz, First and second order sensitivity analysis of linear and non-linear structures. Submitted for publication (1985).
7. H. Petryk and Z. Mróz, Time derivatives of integrals and functionals defined on varying volume and surface domains. *Arch. Mech.* (1986), submitted for publication.
8. K. Dems and Z. Mróz, Optimal shape design of multicomposite structures. *J. Struct. Mech.* **8**, 309–329 (1980).

9. Z. Mróz and A. Mironov, Optimal design of structures for global mechanical constraints *Arch. Mech.* **32**, 505–516 (1980).

10. Z. Mróz, M. P. Kamat and R. H. Plaut, Sensitivity analysis and optimal design of non-linear beams and plates. *J. Struct. Mech.* **13**, 245–266 (1985).

11. E. J. Haug and B. Rousselet, Design sensitivity analysis in structural mechanics–I: Static response. *J. Struct. Mech.* **8**, 17–41 (1980).

12. E. J. Haug, Second-order design sensitivity analysis of structural systems. *AIAA J.* **19**, 1087–1088 (1981).

13. R. T. Haftka, Second-order sensitivity derivatives in structural analysis. *AIAA J.* **20**, 1765–1766 (1982).

14. K. K. Choi and E. J. Haug, Shape design sensitivity analysis of elastic structures. *J. Struct. Mech.* **11**, 231–269 (1983).

15. K. Dems and Z. Mróz, Application of bilinear functionals in sensitivity analysis. *Arch. Mech.* (1985), submitted for publication.

16. Z. Mróz, K. Dems and D. Szeląg, Optimal stiffener design in discs and plates. *Arch. Mech.* (1986), submitted for publication.

DISCUSSION

B. Prasad *(Electronic Data Systems)*

When you are using a gradient-based optimizing technique and you have a discontinuity of the gradients, then the gradient-based optimization techniques wouldn't be applicable. Do you get gradient discontinuity when fields are discontinuous? If not, could you elaborate on this?

Mróz

You also have discontinuity in field quantities like tractions. We handle this problem by adding discontinuity relations along singular surfaces to the field equations. This surface may have three different interpretations. If we have the field equations plus discontinuity equations along the singular surface, then we can study the sensitivity of the whole performance of the structure due to the variation of our singular surface. So we get analytical expressions for sensitivities, and we have an optimality condition that relates the discontinuities on both sides of the cost function. Now we can solve directly by the Newton-Raphson technique to obtain the optimal interface shape, or by an incremental process of modification of interface (this is what we used).

Moderator—R. Haftka

I'm not sure that was the question. There is no discontinuity in the sensitivities. It is the original problem – the stress analysis problem – which has the discontinuities, but the sensitivities themselves can be continuous so that there won't be any problem in using gradient-based techniques.

Mróz

There is a continuous expression of functional variation or gradient associated with interface variation, though we have strong discontinuities in our field. We may have displacement and traction discontinuity and structural gradient continuity.

Z. Mróz

L. Schmit *(University of California - Los Angeles)*

Then the theory you used addresses the possibility of a smooth boundary in which there would be a notch?

Mróz

Yes.

Schmit

In other words, you could in fact incorporate the presence of a crack, if the original analysis could be carried out satisfactorily?

Mróz

Yes. We can admit edges on the free boundary and there is no singularity in the field, provided there is a convex domain. But if you have a concave notch or crack, then you have some singular fields at the notch root. Then, at the crack tip you have a square root singularity. With this, one has to be extremely careful in calculating surface integrals because they include a singularity, which is a finite contribution of zero boundary area. You have to know more about singular fields.

R. Kohn *(Courant Institute of Mathematics)*

Concerning the introduction of a crack, it seems that sensitivity analysis near the point where the crack is introduced may be difficult. But sensitivity analysis away from the introduced crack seems related to Griffith's crack growth criterion, which is analyzed using Rice's J integral. What relation does that topic bear to your theory?

Mróz

We elaborated some path-independent integrals for sensitivity which are based on both adjoint and primary fields. We generalized the J integral, so one can calculate sensitivity of any global functional by contour integral and there is no singularity problem. The problem is more delicate if you want to calculate local stress or local strain sensitivity.

Kohn

Does your theory allow computation of the variation on the introduction of a stiffener?

Mróz

Yes, I can examine the limit of a zero stiffener. This is an odd situation, however, because if we tried to find the optimal placement of six stiffeners in the plate for some stiffness ratios, they coalesce and we have only two. For some, they coalesce into one. So you may institute an infinite number of stiffeners but they have a tendency to polarize in some discrete number. I don't understand this mathematically, but you can start with many and end up with one or two by this process.

E. Haug *(University of Iowa)*

I suspect that when it comes to adding a rib, which is really a finite material addition, that may in fact be a relatively smooth change on the structure of the problem. However, if you go to hinges, or from a continuum to a hinge, or if you introduce new holes that weren't there before, then I suspect that there is a fundamental change in the structure of the problem. Your kinematically admissible set of displacement changes, and my guess is that this is not differentiable.

Mróz

It is differentiable because you have analytical expressions for both hinge and reinforcing interfaces. In the case of beams, we showed by a set of examples how to generate sensitivities and design hinges in the best way for structures which are subjected both to loading and specified displacements.

Haug

If you already have the hinge in your model and you move the hinge around, you are OK. But if, for example, you go from a beam model with no hinge and try to converge to a hinge and keep the continuity of slope in your formulation, then I think you have a difficulty.

Mróz

Perhaps I did not make it clear. We produced an elastic hinge with some local rules between displacement or rotation discontinuity and local tractions or bending moments. If this elasticity of the hinge tends to elasticity of the beam, we have a smooth transition. But if it is only a hinge with zero traction, of course there will be a discontinuous transition, from continuum to discontinuous, so this is an elastic hinge between tractions and displacement discontinuities.

J. Taylor *(The University of Michigan)*

Without detracting from a masterful exposure on this generalization in variational modeling and the many useful results brought to our attention in the portions presented, I believe the idea could be described in an alternate way. In particular, in what you called the formalism at the beginning of your paper, the definition of that functional can be given an interpretation where the governing equations are adjoined with multipliers. In that case, the multipliers come out to have the same physical meaning as your adjoint variables. The advantage to this is that it becomes possible, simply by following a formal procedure, to generate the system of governing equations.

Mróz

I agree that formally you can use variational functionals with constraints and you could introduce multipliers, integrate by parts, and get to an adjoint system. However, I think the approach I used is much easier because this so-called generating functional integrates the field equations and the sensitivities into one functional.

A. D. Belegundu *(GMI Engineering and Management Institute)*

This is interesting because I recently completed a paper which derives all sensitivity equations using Lagrange multipliers. In fact, Washizu has also derived equations in mechanics using multipliers. The variational principles and the Lagrange multiplier approach can be shown to be equivalent to each other depending on the interpretation of the Lagrange multiplier. Still, the approach you have used is good because sometimes in the Lagrange multiplier approach some physical significance is hard to obtain—which it will not be if the work is based on a variational principle. I have shown that Lagrange multipliers and adjoint variables are the same and represent sensitivity with respect to forcing functions.

D. Vasilopoulos *(General Motors Research Laboratories)*

Professor Mróz, you mentioned that this approach can also be used in nonlinear problems. In previous presentations in this symposium, certain problems were mentioned that arise from our inability to accurately compute the adjoint problems because of the singular loads that we have to use. Do you expect these difficulties to become more dramatic in nonlinear problems?

Mróz

I do not totally agree with the statement that you face great difficulty with a singularity if you have to calculate sensitivity of local stress or strain, or local deflection. Professor Haug said that the local deflection is determined by introducing a concentrated force in the adjoint system, generating the solution and calculating deflection by an integral expression. The local stress calculation was explained on my slides, but I didn't discuss this because time was short. If you introduce the unit dislocation into the adjoint system at the selected location, you have to find a solution for this dislocation problem. The respective stress component is expressed by an integral over the body domain of the adjoint strain and primary stress fields. The same for the strain component. As the adjoint field is self-equilibrated, it is sufficient to calculate a local integral around the respective point. As the local dislocation solutions are available in the analytical form, they can be used in representing a singularity. A similar procedure can be applied for local deflection problems in both linear and nonlinear structures.

Vasilopoulos

Do you mean that if this structure is available at the beginning, we do not solve the adjoint problem by doing another finite element solution with concentrated loads? Do you mean that we know this function before we even start? Is this function also available for the displacements?

Mróz

You have to solve the unit force problem in the same finite body, take the stress field and multiply it by the primal strain field to get the formal calculation. Next you take dislocations to solve the problem in the finite body, with strain multiplied by the primary stress expression.

Vasilopoulis

To use this dislocation, do we do a numerical analysis?

Mróz

Yes, because the expression for the dislocation solution in an infinite body is a closed form solution in dislocation theory. However, the dislocation solution in the finite body is not a closed form solution, so you have to use a singular field and solve the boundary value problem. But you can superpose both singular and regular parts of the solution so that only the regular part is to be determined numerically.

SESSION II
ANALYSIS AND MODELING FOR SHAPE OPTIMIZATION

Session Chairman
B. A. SZABO
Washington University
St. Louis, Missouri

AUTOMATIC FINITE ELEMENT MODELING FOR USE WITH THREE-DIMENSIONAL SHAPE OPTIMIZATION

M. S. SHEPHARD
Center for Interactive Computer Graphics
Rensselaer Polytechnic Institute
Troy, New York

M. A. YERRY
ARIES Technology, Inc.
Lowell, Massachusetts

Abstract

A key aspect of the shape optimization process is the generation and control of the numerical analysis models used to evaluate the constraint and gradient information required during the optimization. In general, the numerical analysis discretization, assumed here to be a finite element mesh, must be changed as the object evolves from its original to optimal shape. This paper discusses the capabilities needed to automatically generate and control these finite element models.

The paper considers the general questions of shape control as well as numerical model generation and control for three-dimensional objects. A specific approach to automatic mesh generation, the modified octree technique, is presented which is well suited for both three-dimensional shape optimization and adaptive finite element analysis. Finally, the integration of the component modeling capabilities needed for automatic three-dimensional shape optimization is discussed.

INTRODUCTION

Approaches to structural shape optimization employing finite element analysis results were established over a decade ago [1]. However, it is only recently that serious consideration has been given to the practical use of shape optimization [2–5]. Although there continue to be improvements in design sensitivity calculation and numerical optimization procedures, it is primarily because of other advances that shape optimization is becoming a practical design tool. The most obvious is the advancement in computer hardware that has yielded a manifold increase in the number of computations per computing dollar. Important advancements have also been made in procedures to generate and control finite element models. The purpose of this paper is to discuss the finite element and geometric modeling tools needed for the automatic optimization of the three-dimensional shapes.

Before addressing specific modeling capabilities, it is important to state the premise underlying the material present in this paper. Production shape optimization software will be used primarily by designers rather than by finite element or shape optimization software developers. Therefore these users will expect to specify the shape optimization problem in terms of a geometric description of the object. It should also be assumed that these users will not question the accuracy of the results produced by the numerical analysis models employed during the optimization.

Based on the above premise, the modeling concerns that must be addressed for structural shape optimization are: the description and control of geometric shape; the generation and control of numerical models; the determination of the information needed for shape optimization; and then the integration of all these capabilities into an operational procedure. The next section presents a discussion of the possibility of defining and controlling three-dimensional shapes using recently developed geometric modeling procedures. The subsequent section establishes the need for automatic mesh generation and adaptive analysis, while the fourth section discusses a particular mesh generator that is well suited for use for general three-dimensional shape optimization. The final section considers the possible integration of all these capabilities into a system for automatic shape optimization.

GEOMETRIC REPRESENTATIONS FOR THREE-DIMENSIONAL SHAPE OPTIMIZATION

The definition and control of three-dimensional objects for the purpose of optimizing their shape is much more complex than it is in two dimensions. Therefore it is impractical to assume that the "finite element" based geometric descriptions commonly used in two dimensions will be satisfactory in three dimensions. Typically, the amount of shape freedom desired in three dimensions is limited to a specific number of parameters controlling the size and position of geometric features that make up the object. This, combined with the desire to maintain control of the number of variables in the optimization process, indicates that it is not desirable to use individual nodal positions as unknowns in the optimization, but rather to use a limited number of geometrically based parameters.

A number of the recent contributions to the field of shape optimization have recognized this and have employed only specific geometric parameters in the optimization [2–5]. However, with the exception of the work of Botkin and Bennett on plate structures [2], the geometric shape parameters used have had a direct correspondence to the mesh patches used to generate the finite element model. The complexities of generating meshes for general three-dimensional objects and the desire to allow large shape changes will not allow such a direct correspondence to always be maintained.

Geometric Modeling of Three-Dimensional Objects—The three classes of geometric modeling systems available for the computerized description of three-dimensional objects are wireframe, surface and solid modelers. The early CAD systems that supported primarily drafting operations employed wireframe definitions in which the object is defined in terms of points and edges with additional annotation information appended. These definitions are not geometrically complete and can only be properly interpreted by a human operator using both the geometric and annotation information available.

Surface models contain the additional geometric information required to specify the surfaces bounded by the edges of the object. Although a surface representation contains all the geometric information required to define an object, these representations are not complete. Specifically, representation of only the surface does not allow the unique classification of any given point in space to be either inside, outside or on the surface of the object.

Solid modeling procedures are able to uniquely classify every point in space and thus yield a complete and unique representation of an object [6, 7]. This functionality is attained by adding the appropriate associativity information to the basic geometric information present in a surface representation. A number of systematic procedures have been developed for providing this information [6]. The most intuitive of these is the boundary representation, which adds associativities that tie vertices to edges, edges to loops of edges bounding surface patches, surface patches bounding shells and shells defining an object.

The most popular form of solid model representation used today is constructive solid geometry (CSG) in which an object is defined in terms of a number of simple subobjects, called primitives, which are scaled, positioned and combined using the Boolean operations of union, intersection and difference to produce the desired solid object. The process is demonstrated in Figure 1 where Figure 1(a) shows six superquadric primitives [8] which are combined by means of the Boolean operators to produce the object shown in Figure 1(b).

Shape Optimization Using Solid Models—There are a number of advantages to be gained if a solid model representation is used to define the geometry of a three-dimensional object in a shape optimization process. The most important is that a complete and unique representation, as employed by solid models, must be used if the shape optimization process is to be carried out in an automated manner for general shape changes. As will be indicated later, complete representations are also required for use with automatic mesh generation.

There are other advantages related to the geometric functionality a solid representation can provide. The first is that a solid model representation gives, at least to some extent, a parameterized representation of the object which can be

References pp. 132–134

used as the shape control parameters in the optimization. Consider, for example, an object defined in terms of a set of cylinders and blocks in a constructive solid model. The dimensions of the blocks, the length and radius of the cylinders and the relative positions of each of the primitives are natural parameters that could be used for shape control. In some problems they may represent a satisfactory set of parameters, while in others there would be a desire to allow more general shape modification.

Whatever level of generality is desired in the control of an object's shape, it is desirable to tie all the shape control parameters to the solid model representation so that the functionality of the solid modeler could be used to carry out the geometric calculations required to account for shape changes. If this is not done, the shape optimization procedures would be responsible for carrying out all shape control, which would mean the creation of the required geometric functionality within the shape optimization procedure. Considering the amount of effort required to develop a solid modeling capability, this approach is not justified.

To demonstrate the level of geometric shape flexibility that could be obtained by employing a solid modeling capability in shape optimization, consider the object shown in Figure 2(a) which represents the union of a block and two cylinders. Assuming that the shape design parameters for this example include only the diameter of the cylinders, Figure 2(b) shows a possible optimum shape for this object. In going from Figure 2(a) to Figure 2(b) the topology of the object (in this case number of curve segments and their associativities) has changed because the two cylinders no longer intersect each other. The solid model representation automatically accounts for this change and can communicate that information to the other aspects of the shape optimization process such as the automatic mesh generation.

There are a number of questions that must be addressed and capabilities that must be developed before three-dimensional shape optimization based on solid representations can be realized. However, the complexities involved in dealing with three-dimensional models indicate that it is necessary to use the functionality of solid modelers if general three-dimensional procedures are to be developed.

AUTOMATIC GENERATION AND CONTROL OF ANALYSIS MODELS FOR SHAPE OPTIMIZATION

Current procedures for finite element model generation typically rely on the use of mapped mesh generators that are capable of producing the finite elements in a region with specific fixed topologies [9–11]. Therefore, the analyst is required to partition the object of interest into a set of disjoint regions of the required topologies. The main drawback to basing finite element mesh generation for shape optimization on this approach is that it is a time-consuming process that requires manual intervention. Thus the shape optimization process must be carried out using the same mesh patches throughout the process, or the process must be interrupted for the manual intervention required to generate a new mesh. A second drawback of this approach is that the user is discouraged from creating mesh gradations

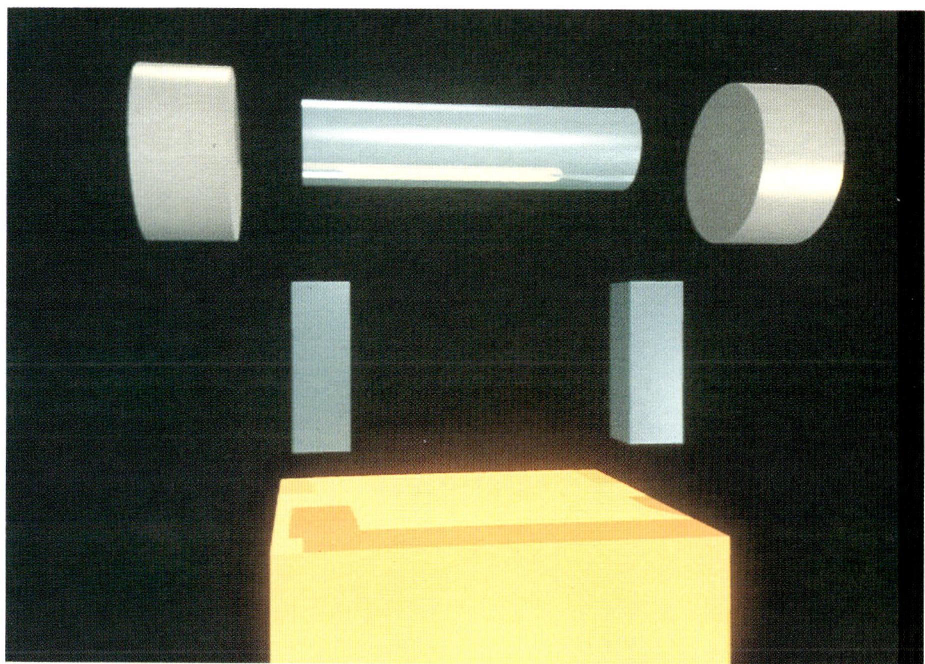

(a) Six superquadric primitives

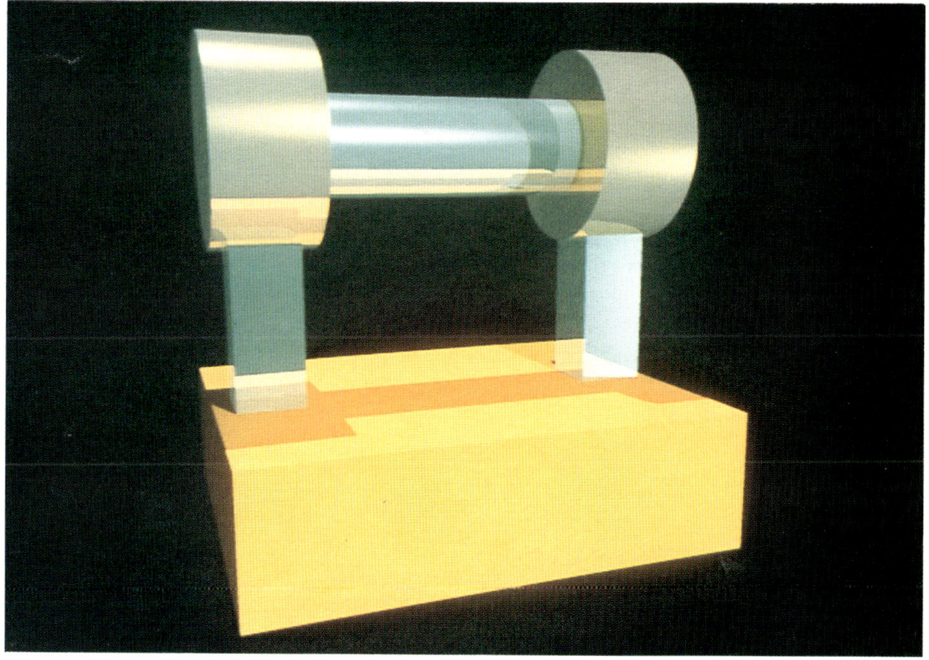

(b) Object created by unioning six primitives

Figure 1. Solid modeling by constructive solid geometry.

References pp. 132–134

(a) Object before shape change

(b) Object after shape change

Figure 2. Shape change with solid models.

appropriate for ensuring solution accuracy since the most powerful mesh control device available for use with these meshing procedures is the manner in which the domain is partitioned into a set of mesh regions [12].

Since it is desirable to allow for general changes in the object's shape and to have the shape optimization process carried out entirely under computer control, it is necessary to use an automatic mesh generation approach that is capable of generating a mesh for any geometry without using intervention.

Overview of Automatic Mesh Generation—There are currently a number of investigators developing fully automatic mesh generation procedures [12–20]. The algorithmic approaches being used include: (1) paring individual elements from the object one at a time [13, 14], (2) recursively subdividing the geometry of the object to the element level [15], (3) volume triangulation of a set of points placed through the domain of the object [16, 17], and (4) spatial decomposition of the object followed by mesh finalization [18–20].

The first two approaches operate using a set of mesh operators that examine the topological description of the object, looking for specific features that are candidates for the application of a meshing operator. Candidates are then tested for geometric validity, and, when valid, the meshing operator is applied, thus reducing the topological complexity of the object. This process is continued until the object has been reduced to a valid finite element mesh. Because these procedures "look at" the geometry to be meshed at each step of the process, they are well suited for the generation of the largest possible elements and are capable of generating the coarsest possible finite element meshes. However, the continual evaluation and examination of the geometry remaining to be meshed tends to make these approaches computationally intensive, with reasonably large computational growth rates as the number of finite elements in the mesh is increased.

The volume triangulation algorithms begin by placing a set of points throughout the domain of interest and then relying on some variation of a Delaunay triangulation to create the finite element mesh. The Delaunay triangulation algorithms have the advantage of being computationally efficient and of ensuring a mesh with certain desirable properties on the shape of the resulting elements. The main disadvantage of these approaches is the development of algorithms to place points throughout the domain of interest in a manner that is automatic and will yield the desired finite element mesh. Although good procedures for this operation have been developed for two-dimensional domains [16], a good fully automatic procedure for three-dimensional domains has yet to be developed.

The procedures based on spatial decomposition [18–20] address the question of the placement of the majority of the nodes through the use of a regularized decomposition scheme such as the variable size grid of an octree representation [21, 22]. The remaining nodes are then placed, based on the intersection of the various geometric features, with the appropriate grid cells. The finite element mesh topology is then generated using these points, information about the grid and information about the geometric features that define the geometry.

References pp. 132–134

Although there are major differences between the various approaches to automatic mesh generation, they share one common feature over the semiautomatic mapped type mesh generators commonly used today. They must carry out an extensive number of geometric calculations and therefore require more computational effort. This is because the automatic mesh generators must "examine" the geometry to determine how to generate the mesh, while the mapped mesh generators are given an implicit definition of the geometry through the manner in which the user has partitioned the geometry into mesh regions and the mapping functions used in each of the regions.

The automatic mesh generation approach best suited for a particular application is a function of the type of finite element mesh desired and the geometric modeling software with which it is to be integrated as well as the analysis software to be used. For example, the generation of fine finite element meshes is most efficiently carried out by a mesh generator based on volume triangulation or spatial decomposition. However, these approaches are not well suited for the generation of very coarse finite element meshes. In those cases the topologically based or recursive mesh generators are more appropriate. Therefore the automatic mesh generation approach most appropriate for use in shape optimization is a function of the software environment in which it is to operate, and the degree and type of mesh control needed.

Adaptive Mesh Control During Shape Optimization—It is possible to use two forms of mesh adaption during shape optimization. The first, which is necessary, is to adapt the mesh to account for the changes in shape from one step to the next. The second form of mesh adaptation is the adaptive improvement of the mesh between the steps of shape change to control the discretization errors.

The procedures used to adapt the mesh to account for changes in shape are dependent on the mesh generation procedure used. One approach that has been used in conjunction with mapped mesh generation [9–11] is to maintain the same mesh topology, but to reposition the interior nodes as dictated by the mesh generator and the updated boundary point positions [3–5]. This approach has the advantage of being computationally efficient and allows for the convenient specification of node movement as needed in the calculation of the design sensitivity information. One disadvantage is that the user is required to partition the original geometry into a set of mesh patches that can be mapped. A more serious drawback is that the amount of shape change allowed will be limited by how far the mesh patches can distort before the element shapes degrade to an unacceptable level.

The use of an automatic mesh generator to develop the element mesh allows for additional flexibilities in accounting for changes in shape. One possibility is simply to regenerate a mesh each time the shape changes. This has the advantage of allowing completely general shape changes, including those that alter the topology of the geometric features that define the object. On the other hand, it has the disadvantage of requiring the sizable computational effort involved in automatic mesh generation and may make the calculation of the sensitivity information needed during the optimization more difficult. Therefore it is desirable to consider alternative forms of mesh modification to account for small shape changes. Various mesh relaxation procedures could account for minor shape changes and should be of similar applicability for the prediction of interior node motion for the calculation of sensitivity information as mapped mesh generation. Forms of local remeshing could account for larger shape changes, with a total remeshing carried out only when needed.

Adaptive Analysis in Shape Optimization—The optimization process uses the results of an analysis to determine the shape changes to be carried out and is therefore dependent on these values. If there are inaccuracies in the analysis results, the shape optimization process is likely to converge to a shape that is not optimal for the intended problem. An example of such an undesirable result is the small oscillations in the boundary caused by erroneous stress calculations in the finite element solution [24]. The only way to avoid this is to ensure that the errors in the numerical solution are controlled to the point at which they do not influence the optimization procedure.

The objective of an adaptive finite element analysis program is to obtain accurate estimates of the discretization error present in a finite element solution and to automatically improve the mesh in areas of unacceptable error until a prespecified degree of accuracy is obtained. The most critical function within an adaptive analysis procedure is obtaining the error estimation proofs used to determine the discretization errors present in a given finite element mesh. A number of procedures, known as a posteriori error estimators, have been developed that employ current solution results to estimate the errors present in a solution [25–29]. The procedures yield accurate estimates of errors in a particular error norm, such as the energy norm, and require little computational effort to calculate. Estimates in more design-oriented norms are also possible [29].

An instructive manner in which to view the process of a posteriori error estimation is to consider a boundary value problem in which a Galerkin based weighted residual finite element formulation is used. The basis functions used in the formulation are selected to ensure an a priori satisfaction of the continuity requirements and essential boundary condition. The errors in such a solution are related to the residual terms that arise due to the dissatisfaction of the governing partial differential equation within the element, the mismatch of the natural boundary conditions between elements (the jump terms) and the mismatch of the natural boundary conditions at the exterior boundary.

The key question in an a posteriori error analysis is to determine how to employ these residual terms to obtain a meaningful estimate of the error in the solution. The first step is to determine the error that is to be measured and the appropriate norm in which to measure this error [25–29]. Energy norms are the most natural for this process and have been considered extensively [25–28].

Given the above stated conditions and the finite element solution on a specific finite element mesh, one approach to a posteriori error estimation is:

References pp. 132–134

- ***Step 1.*** Determine the residuals associated with the sources of dissatisfaction of the governing equations and natural boundary conditions. The current finite element results are the exact solution of the original problem, subject to the applied loads minus the residual loads. Thus the exact solution to the original problem is equal to the current results plus the solution of the problem subjected to the residual loads only. The remaining steps are to obtain an accurate solution for the problem subjected to the residual loads which then represents the information used in error estimation. The current finite element mesh will yield no results when the residual loads are applied, since the mesh was unable to represent that portion of the load to begin with.

- ***Step 2.*** Assuming that most of the error will be picked up by the introduction of the next set of approximation functions, they are introduced for use in error estimation.

- ***Step 3.*** The residual loads are applied to the finite element model with the next set of approximation functions introduced and solved in at least an approximate manner.

- ***Step 4.*** The results obtained for the residual loads acting through the improved modes are used to estimate the error in the current solution.

If the error estimator indicates that the current finite element mesh does not yield the required accuracy, the finite element model must be improved. In most cases, the accuracy of portions of the model is adequate, so it is only necessary to selectively improve those portions of the model where increased accuracy is needed. The terms used to determine those portions of the models which need improvement are referred to as error indicators and are often related to the local contributions to the error estimators.

Once the portions of the model requiring improvement are determined, the finite element mesh in that area must be improved. There are a number of techniques available to improve a finite element mesh, including those presented in references [30–37]: (1) moving node points within a mesh to improve solution accuracy (r refinement), (2) subdividing selected elements into smaller elements of the same type (h refinements), (3) increasing the polynomial order of selected elements (p refinements), (4) various combination of r, h, and p refinements, and (5) defining an entirely new mesh with an improved distribution of elements.

Although certain methods appear to have an advantage over others, some specific choices are superior for specific applications. For example, moving node methods are quite efficient for time dependent problems with moving shocks or fronts [30]. However, in the case of steady elliptic problems, it has recently been proved that exponential rates of convergence are possible with optimal h-p refinements [31, 32].

Two important considerations in the integration of shape optimization and adaptive analysis are the selection of the mesh improvement scheme and how the iterations of adaptive mesh improvement and shape change are integrated together. At a fundamental level, the assurance of the accuracy of a given numerical solution and the definition of the optimal shape of a part are uncoupled problems. This implies that changes in shape should only be carried out after the required degree of solution accuracy is obtained. However, computational efficiencies can be gained by the proper intermixing of the two procedures into a single process, thus allowing changes in shape as the process of adaptive mesh improvement is carried out.

Botkin and Bennett employ h refinements in their shape optimization procedures [2] to control the accuracy of the numerical results. The combination of mesh generator, error estimator and mesh improvement scheme they use is well suited for shape optimization. Other combinations are possible. The selection of adaptive analysis procedures for use in shape optimization must consider the computational efficiencies of the combined process and not just the efficiency of the individual facets.

MODIFIED OCTREE MESH GENERATION FOR THREE-DIMENSIONAL SHAPE OPTIMIZATION

As indicated in the previous two sections, the development of robust three-dimensional shape optimization requires the use of a mesh generator that can automatically generate a mesh for any geometry and can efficiently adapt itself to account for changes in the geometry and to control discretization errors. One approach that appears well suited for this purpose is the modified octree technique [18–20], which is a spatial decomposition procedure loosely based on octree encoding [21, 22]. The two-dimensional version of this approach, the modified quadtree technique, has been used in the automated model generation of problems with geometry changes caused by discrete crack propagation [23].

The Modified Octree Mesh Generator—The modified octree mesh generator makes use of a spatial decomposition procedure to discretize the geometry of an object into a form that can then be converted into a finite element mesh [18–20]. A modified octree discretization of an object is the union of a set of disjoint shapes that are stored in a hierarchic tree. The fundamental shape used in the modified octree discretization is a cube. The size of cube used in any portion of the object is controlled by the mesh gradation information applied to the geometry as explained below.

A simple conceptual device for visualizing the modified octree representation of an object is to envision placing the object into a three-dimensional grid where the grid cells are cubes of varying size and where neighboring cubes, referred to as octants, can be the same size, have side lengths of twice the current octant or a side length equal to half the current octant. Those octants that contain the outer surface of the object, and are thus neither fully inside nor outside the object, have that portion outside the object cut off in some manner.

References pp. 132–134

At the highest levels, the modified octree data structure is the same as in an octree representation [21, 22]. The object to be represented is placed in a cube that totally encloses it. This cube represents the root octant in the tree structure. At the next level in the tree come the eight suboctants created by subdividing the original cube into eight subcubes which have side lengths equal to one-half that of the original. The depth of the tree, i.e., the number of levels used, depends on the smallest octant size used. The final structure of a modified octree contains a root, a number of continuation nodes and terminal nodes. Continuation nodes are those that are subdivided into eight suboctants, while terminal nodes are those at the final level for that portion of the geometry and are at the bottom of that branch of the tree. The three types of terminal nodes possible are (1) complete octants which are entirely inside the object, (2) empty octants which are completely outside the object and (3) cut octants which are partly inside the object. Although it is possible to define the modified octree by starting with the root octant and recursively subdividing octants to the level dictated by the mesh size control information in that portion of the object, this is computationally inefficient and is difficult to implement in a robust manner.

The key to making the modified octree representation an appropriate discretization of the object for use in robust automatic mesh generation is the manner in which one defines those octants partly outside the object—the cut octants. The specification of an acceptable procedure to define the cut octants must consider the requirements placed on the finite element meshes desired. In general it is not critical that the finite element mesh generated match the actual geometry to five significant figures. However, it is important that all relevant geometric features are represented in the resulting finite element model. Since individual finite elements can cover a sizable portion of the domain of the object being represented, the only robust manner to represent all pertinent geometric features desired in the resulting finite element model is to explicitly represent them in the modified octree data structure.

Since the geometric features that define an object are associated with its boundary, the modified octree is defined by first inserting the features that define the boundary of the object [12, 19]. First, each vertex is placed in the appropriate size octant where the octant's size is controlled by the mesh control information associated with the vertex. The discrete edge segments are then inserted, followed by the discretized surface patches. The octant sizes are defined by mesh control information assigned to the edges and surfaces respectively. With the definition of the cut octants complete, the interior of the modified octree is easily complete by means of a tree traversal. For its application to finite element mesh generation, a modified octree is generated such that octants that share an edge or face are forced to have a level difference of no more than one. This is done to help control mesh gradation and element shapes.

The generation of a finite element mesh within the modified octree involves two operations: (1) defining the finite element mesh topology and (2) respositioning the node points to improve the shape of elements and the finite element mesh's

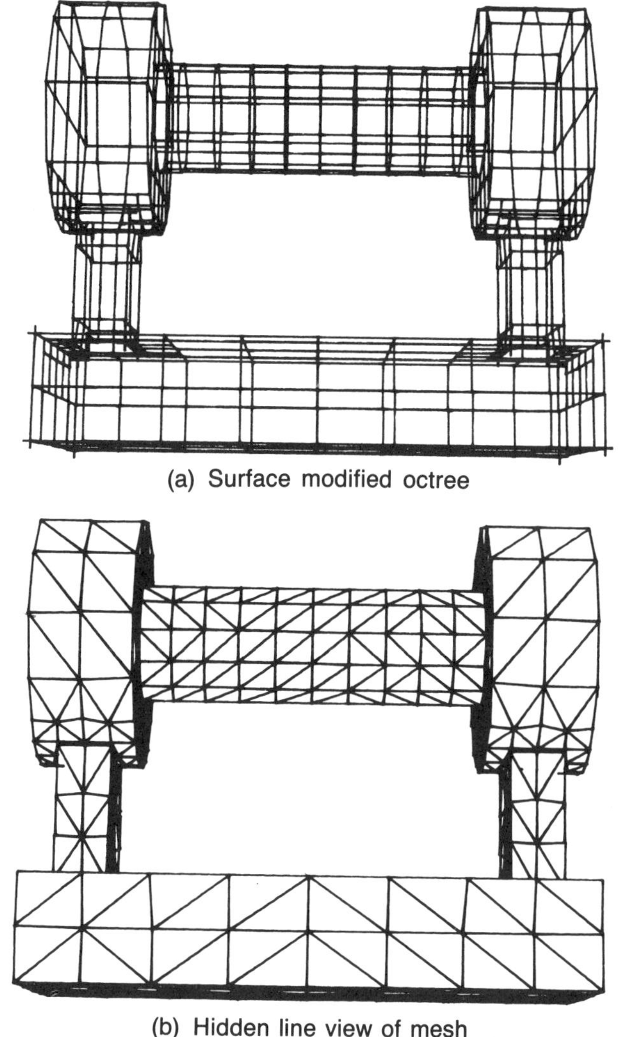

(a) Surface modified octree

(b) Hidden line view of mesh

Figure 3. Modified octree mesh example.

geometric approximation to the original geometry. The mesh topology is defined on an octant-by-octant basis, using a combination of mesh templates and octant triangulation algorithms [19]. Since decisions on how an octant is divided into elements depend on the shape of the octants, the mesh finalization process is carried out in three steps. Nodes are first repositioned based on connection information in the modified octree, the mesh topology is defined and the nodes are repositioned again based this time on connectivities defined by the mesh topology. Figure 3(a) shows the surface of the modified octree and Figure 3(b) shows a hidden line view of the mesh generated for the object of Figure 1(b).

Mesh Control in the Modified Octree Mesh Generator—The size of the octants used at various locations in the modified octree represents the basic mesh control device [12]. Since the sizes can be controlled in a spatially based manner, the octant sizes are the only mesh control device required.

From the viewpoint of generating the finite element model and for adaptive analysis, the mesh control procedures used are associated with the geometric features of the object. Each of the geometric entities that define the object is given a parameter which defines the default octant size used when representing that feature in the modified octree. Control of the mesh gradations dictated by these parameters is inherent in the modified octree since the neighboring octants can differ by no more than one level.

An additional mesh control device, referred to as a refine point, is available. A refine point is a location in space where the octant size value is altered by a specified amount. Although this is a simple device, the spatially based refine points are an important mesh control device for the creation of a priori mesh gradations, and they represent the basis for efficient adaptive mesh control. The determination of the octant size that is appropriate at a given point in space is determined by the geometric parameter values and refine points as explained in [12].

To demonstrate the use of these mesh control devices, one two-dimensional and one three-dimensional example are considered. Figure 4(a) shows a plate with holes and re-entrant corners. Figure 4(a) also shows a set of parameter flags and refine points used to generate the mesh shown in Figure 4(b). The numbers defining the basic mesh size for each edge and vertex as well as the interior are shown in the various shaped boxes in Figure 4(a). The higher the value, the smaller the element size for that feature, with a unit difference in value corresponding to a halving of average element edge length. The refine points are the unboxed numbers with the numerical value of the refine point indicating the number levels of refinement carried out at that location. Figure 5(a) shows a hidden line view of a mesh for a block and cylinder with a finer mesh in the cylinder. Figure 5(b) shows the mesh for the same object with a refine point added along the intersection of the cylinder and block.

The tree-based data structure and spatially based mesh control devices of the modified octree mesh generator make its use for h-type mesh refinement a natural part of the meshing process. The tree-based structure is important for the efficient control of the adaptive analysis.

Storing adaptively refined finite element meshes in a hierarchic tree is common for h-type refinements [33–36]: however, there are some basic differences between the modified octree used for three-dimensional previous approaches and the modified quadtree used for two-dimensional adaptively defined meshes. In previous approaches the tree represents the domain of the mesh that has, at most, two levels below the root for the representation of the original finite element mesh. The modified octree and modified quadtree differ in that they represent a cube or square in which the domain of interest must fit. Typically, the tree is several levels deeper than the mesh trees above. It is a regular tree, with every parent node having eight

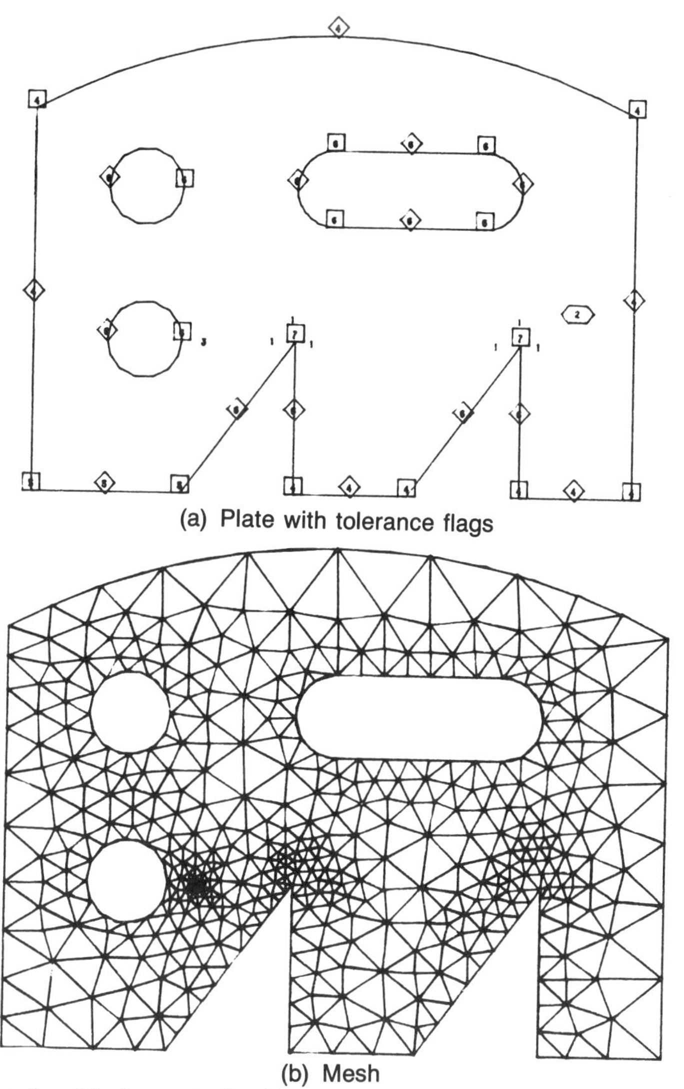

(a) Plate with tolerance flags

(b) Mesh

Figure 4. Mesh control with modified quadtree mesh generator.

children in three dimensions and four children in two dimensions, except for the lowest level of parent nodes, which have the finite elements for children.

The regular structure of the modified octree is well suited for determining the neighbor information (such as shared edges) that is needed during error estimation and other aspects of an adaptive analysis. In addition the modified octree is ideally suited for adaptive refinement at the octant level. When elemental error indicators determine that particular octants must be subdivided, the level of those octants is lowered by the desired amount. The meshing algorithm can then be re-entered in a local manner. It must first be determined whether or not the local refinement has increased the difference in level between the subdivided octant and any of its

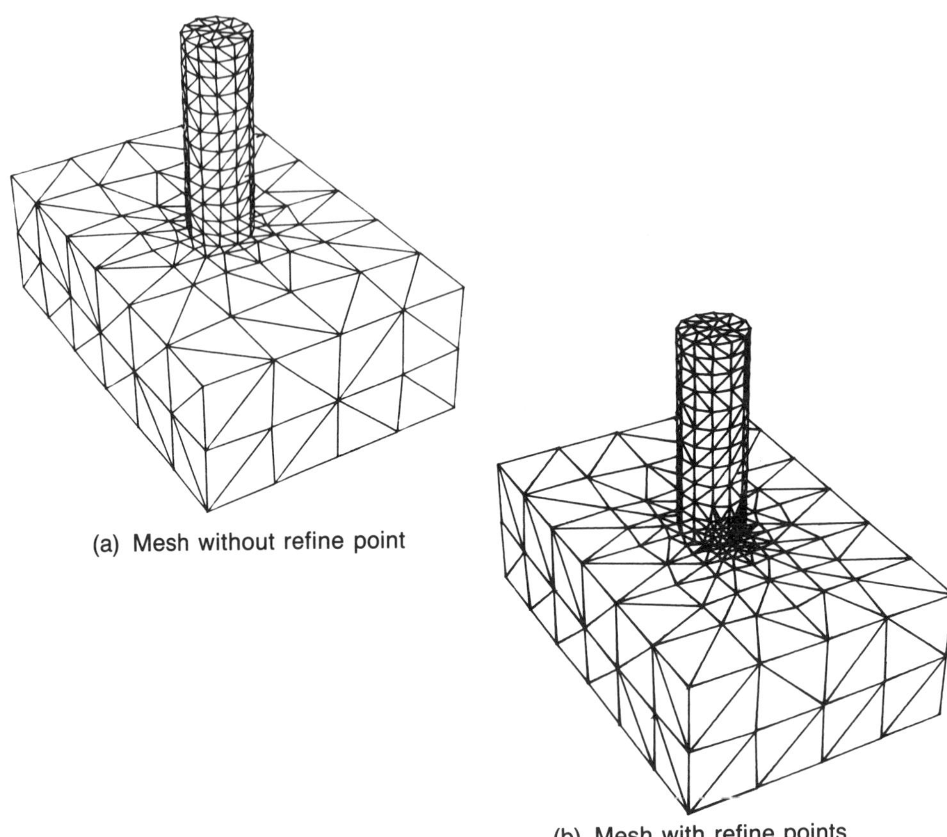

(a) Mesh without refine point

(b) Mesh with refine points

Figure 5. Refine point in a modified octree mesh.

neighbors by more than one. If it has, the appropriate subdivisions are carried out. After all required octants are subdivided, the mesh topology in those and the neighboring octants are regenerated, maintaining control over element aspect ratios. The tree structure can be used to carry out this process efficiently.

Modified Octree for Three-Dimensional Shape Optimization—A minimal requirement of a mesh generator for use in automated three-dimensional shape optimization is the ability to automatically mesh the object as its shape evolves. As indicated earlier, the modified octree mesh generator is capable of carrying out this meshing task and is also capable of efficiently carrying out mesh improvement as dictated by an adaptive analysis procedure.

Since the amount of computational effort required to generate an entirely new mesh each time the object's shape changes is nontrivial, it is desirable to make use of possible local remesh procedures and to remesh the entire geometry only when the element shapes would become unacceptable or there is no direct method to account for the shape changes in the current mesh. The modified octree mesh generator is capable of accounting for shape changes by three methods. If the geometric changes are small, the mesh can be updated by applying a mesh relaxation (mesh smoothing), allowing for the repositioning of nodes on the changed portions of the boundary. If the geometric changes are larger but limited to specific portions of the object, some form of local remeshing is possible. Total remeshing would be invoked in those cases where neither of the other two approaches will provide satisfactory results.

The modified octree generator can be used to determine which form of remeshing is required and then to carry it out. Consider first the shape modifications that do not alter the topology of the geometry. In this case the changes in shape can be reflected in the modified octree structure by updating the representation of the altered geometric entities (vertices, edges and surface patches) in the tree. Based on the octant sizes being used in the affected portions of the object, it is possible to determine if mesh relaxation would be adequate, if local remeshing would be efficient or if a total remesh is most appropriate. Local remeshing could be carried out by locally updating the modified octree and would operate in a manner similar to the local remeshing outlined for adaptive mesh improvements [12] where the affected octants are updated and the mesh is regenerated in those octants.

Shape changes that cause the topology of the geometry to change are more complex to address. It is first necessary to know that there was a topological change. This information must be supplied from the geometric representation and can be detected by tracking the topological changes in the boundary file that is needed by the modified octree mesh generator. In these cases it will not be possible to update the mesh by relaxation. However, local remeshing can be carried out by updating the affected geometric entities and inserting the new ones as defined by the new topology.

INTEGRATION OF THE MODELING FUNCTIONS NEEDED FOR THREE-DIMENSIONAL SHAPE OPTIMIZATION

Recent efforts in shape optimization [2–5] have recognized the need to tie the shape optimization process to a geometric description of the object instead of the finite element model only. With the exception of the work of Botkin and Bennett [2] on plate structures, the procedures developed employ a one-to-one correspondence between the patches used for the geometric description and those used by the mapping-based mesh generator. This limits the amount of shape modification possible and only allows a limited form of geometric description. To date there has not been an attempt to employ automatic mesh generation and solid modeling type of geometric descriptions to three-dimensional shape optimization. The capabilities that must be developed for automated adaptive three-dimensional shape optimization are:

- Shape control parameters tied to the geometric model.

- Mesh generator able to automatically update the finite element model in an

References pp. 132–134

efficient manner to control solution errors and to account for the changes in shape.

- Ability to calculate the required design sensitivity information needed by the optimization procedure.

In addition these capabilities must be fully integrated in an efficient manner.

Integration of Solid Modeling and Automatic Mesh Generation—The integration of an automatic mesh generator with a geometric modeling system requires that the mesh generator obtain the geometric information required to generate a mesh. Only the complete and unique geometric definitions used by solid modelers [6] can satisfy the requirements of an automatic mesh generator. Any solid model representation can address the question as to whether a point is inside or outside an object, and it is possible, although sometimes complex, to construct a boundary representation from any of the solid model representation schemes. Therefore, it is feasible to interface any of the automatic meshing approaches with any solid model representation. However, various combinations of solid modeling and automatic mesh generators can be carried out more readily than others [38].

The complete integration of finite element modeling with geometric modeling also requires the ability to define all the attribute information about the geometric model required for a finite element analysis, and to have that information available for automatic linking to the finite element models generated for the object. The general attribute information needed for a finite element analysis includes the loads, material properties and boundary conditions. Also required to carry out a finite element analysis are mesh gradation information, element types and analysis process control parameters.

For the purpose of generality and modularity, the integration of automatic mesh generation with solid modeling can best be accomplished by means of a group of geometric communication operators [38]. Operators for mesh generation would include geometric interrogation operators and geometric modification operators. Geometric modification operators are used by some automatic mesh generators for the purpose of updating the geometry after a mesh operation has been invoked. The topologically based and recursive subdivision approaches must have these updates carried out before the next mesh operation can be considered.

Automatic Mesh Generation and the Calculation of Sensitivity Parameters—A critical aspect of any shape optimization process is the calculation of the design sensitivity parameters used to predict shape changes. Early procedures for carrying out these calculations employed the change in stiffness matrix as the shape changed and used this in a simple difference equation. This approach has the advantage of being directly usable with an automatic mesh generator even when the geometry is remeshed. However, it has the disadvantage of requiring the evaluation of a new stiffness matrix and does not fit well into the type of modular software structure desired by developers of shape optimization procedures. More recently, variational approaches have been developed [39] that only require a number of adjoint solutions using the current stiffness matrix. These approaches can be cast in either a boundary or domain form [39].

The *boundary form* has the advantage of allowing an evaluation of the sensitivity information which is independent of the change of internal material points, and thus the interior finite element mesh. It has the disadvantage of requiring accurate finite element results at the boundary where they are normally least accurate. The *domain form* requires a prediction of the change of position of internal material points. Experience has indicated that this prediction need not be very accurate, having been successfully predicted on simple geometries by the change in position of nodes produced by involving a mapped mesh generator on both the original and altered geometry. However, it is not obvious how to carry out similar operations on general geometries meshed with automatic mesh generators.

The boundary form of sensitivity parameter calculation would be the easiest approach to integrate with automatic mesh generation procedures. One possibility to improve the accuracy of finite element boundary solutions for stress parameters is the use of better stress smoothing procedures, such as Loubignac's iteration [40], which has been found to produce accurate stress values even when extrapolating to the boundaries.

Since the calculation of domain based sensitivity information only requires a reasonable prediction of the motion of interior points, a number of possible procedures could be used to estimate these motions. As mentioned, remeshing mapped mesh patches will work as long as the element shapes do not degenerate and the topology of the object remains fixed. An approach for automatic mesh generators, which is roughly equivalent to that above, is to maintain the existing mesh topology and apply some form of mesh relaxation with the boundary nodes repositioned due to the shape change. For example, this could be accomplished in the modified octree mesh generator by performing the internal mesh smoothing that is used during the last step of the mesh process.

An alternative approach under consideration by researchers in this area is to apply the boundary motion as prescribed displacements, and to use the displacements of the internal nodes in the calculation of the design sensitivity information. Although computationally more intensive than a simple mesh relaxation, this method yields a good set of values, and substantial computational efficiencies can be gained by the proper use of the existing stiffness matrix.

In cases where the shape changes are large or the topology changes, alternative approaches for use in conjunction with automatic mesh generators must be developed.

CLOSING REMARKS

A number of modeling and analysis capabilities are needed if general three-dimensional shape optimization procedures are to be developed. This paper has concentrated on discussing finite element based modeling procedures which must

References pp. 132–134

be developed and combined to create an automated procedure capable of producing optimum shapes for three-dimensional parts.

A number of the basic modeling procedures required are being developed and are or will shortly be available. In particular, solid modeling procedures are in place and automatic mesh generation procedures are being finalized. However, substantial effort is required to combine these with optimization procedures to produce automated shape optimization programs. As indicated in the previous section, this integration will require the development of major new capabilities. The importance of the proper integration of these modeling procedures with the optimization procedures should not be underestimated. Without this integration the resulting procedures will not represent practical tools to be used by designers.

ACKNOWLEDGEMENTS

The authors would like to express their appreciation to the Engineering Mechanics Department of General Motors Research Laboratories for supporting the development of the modified octree mesh generator. The support of the National Science Foundation under Grant MSM83-05950 for our work on adaptive mesh improvement schemes is also acknowledged.

REFERENCES

1. O. C. Zienkiewicz and J. S. Campbell, Shape optimization and sequential linear programming, Ch. 7 in *Optimum Structural Design* (Edited by R. H. Gallagher and O. C. Zienkiewicz). John Wiley & Sons (1973).
2. M. E. Botkin and J. A. Bennett, Application of adaptive mesh refinement for the optimization of plate structures, in *Accuracy Estimates and Adaptivity in Finite Element Computations* (Edited by I. Babuška, E. R. de Arantes e Oliveria and O. C. Zienkiewicz). John Wiley & Sons (1986), to appear.
3. V. Braibant and C. Fleury, Shape optimal design using b-splines. *Comput. Meth. Appl. Mech. Eng.* **44**, 247–267 (1984).
4. M. H. Imam, Three-dimensional shape optimization, *Int. J. Numer. Meth. Eng.* **18** (5) 635–673 (1982).
5. R. J. Yang and M. E. Botkin, A modular system for three-dimensional shape optimization. *Proc. 27th AIAA Conf. Structures, Structural Dynamics and Materials*, San Antonio, TX, May 1986.
6. A. A. G. Requicha and H. B. Voelcker, Solid modeling: A historical summary and contemporary assessment. *IEEE Comput. Graphics Appl.* **2** (2), 9–24 (1982).
7. M. S. Pickett and J. W. Boyse (Eds.), *Solid Modeling by Computers: From Theory to Applications*. Plenum Press, New York (1984).
8. A. H. Barr, Superquadrics and angle preserving transformations. *IEEE Comput. Graphics Appl.* **1**, 11–22 (1981).
9. W. J. Gordon and C. A. Hall, Construction of curvilinear coordinate systems and applications to mesh generation. *Int. J. Numer. Meth. Eng.* **7**, 461–477 (1973).
10. W. A. Cook, Body oriented (natural) coordinates for generating three-dimensional meshes. *Int. J. Numer. Meth. Eng.* **8**, 27–43 (1974).

11. R. B. Haber, M. S. Shephard, J. F. Abel, R. H. Gallagher and D. P. Greenberg, A generalized two-dimensional, graphical finite element preprocessor utilizing discrete transfinite mappings. *Int. J. Numer. Meth. Eng.* **15**, 1021–1039 (1980).

12. M. S. Shephard, M. A. Yerry and P. L. Baehmann, Automatic mesh generation allowing for efficient a priori and a posteriori mesh refinement. *Comput. Meth. Appl. Mech. Eng.*, to appear.

13. T. C. Woo and T. Thomasma, An algorithm for generating solid elements in objects with holes. *Comput. Struct.* **8** (2), 333–342 (1984).

14. B. Wördenweber, Volume-triangulation. CAD Group Document No. 110, University of Cambridge, Computer Laboratory, Cambridge, England (1980).

15. M. L. C. Sluiter and D. C. Hansen, A general purpose automatic mesh generator for shell and solid finite elements, Vol. 3 in *Computers in Engineering* (Edited by L. E. Hulbert). ASME Book No. G00217 (1982).

16. J. C. Cavendish, D. A. Field and W. H. Frey, An approach to automatic three-dimensional finite element mesh generation. *Int. J. Numer. Meth. Eng.* **21**, 329–348 (1985).

17. V. Ph. Nguyen, Automatic mesh generation with tetrahedronal elements. *Int. J. Numer. Meth. Eng.* **18**, 273–280 (1982).

18. M. A. Yerry and M. S. Shephard, Automatic three-dimensional mesh generation by the modified-octree technique. *Int. J. Numer. Meth. Eng.* **20**, 1965–1990 (1984).

19. M. A. Yerry and M. S. Shephard, Automatic three-dimensional mesh generation for three-dimensional solids. *Comput. Struct.* **20**, 31–39 (1985).

20. M. S. Shephard and M. A. Yerry, Finite element mesh generation for use with solid modeling and adaptive analysis, pp. 53–77 in *Solid Modeling by Computers: From Theory to Application* (Edited by M. S. Pickett and J. W. Boyse). Plenum Press, New York (1984).

21. C. L. Jackins and S. L. Tanimoto, Octrees and their use in representing three-dimensional objects. *Comput. Graphics Image Process.* **14**, 249–270 (1980).

22. D. Meagher, Geometric modeling using octree encoding. *Comput. Graphics Image Process.* **9**, 129–147 (1982).

23. M. S. Shephard, N. A. B. Yehia, G. S. Burd and T. J. Weidner, Automatic crack propagation tracking. *Comput. Struct.* **20**, 211–223 (1985).

24. N. Kikuchi, Adaptive finite element methods for shape optimization of linearly elastic structures, in *The Optimum Shape: Automated Structural Design* (Edited by J. A. Bennett and M. E. Botkin). Plenum Press, New York (1986).

25. I. Babuška, E. R. de Arantes e Oliveria and O. C. Zienkiewicz (Eds.), Accuracy estimates and adaptive refinements in finite element computations. *Int. Assoc. of Comp. Mech.* Palacio Ceia, Lisbon, Portugal (June 1984).

26. I. Babuška and W. C. Rheinboldt, A posteriori error estimates for the finite element method. *Int. J. Numer. Meth. Eng.* **12**, 1597–1615 (1978).

27. I. Babuška and M. R. Dorr, Error estimates for combined h and p versions of the finite element method. *Numer. Math.* **37**, 257–277 (1981).

28. I. Babuška, Feedback, adaptivity, and a posteriori estimates in finite elements: Aims, theory and experience, in *Accuracy Estimates and Adaptive Refinements in Finite Element Computations* (Edited by I. Babuška, E. R. de Arantes e Oliveria and O. C. Zienkiewicz). John Wiley & Sons, (1986) to appear.

29. I. Babuška and A. Miller, The post-processing approach in the finite element method—Part 1: Calculation of displacements, stresses and other higher derivatives of the displacement. *Int. J. Numer. Meth. Eng.* **20**, 1085–1109 (1984).

30. K. Miller and R. Miller, Moving finite elements—Part 1. *SIAM J. Numer. Anal.* **18**, 1019–1032 (1981).

31. I. Babuška, W. Gui and B. A. Szabo, Performance of the p and h-p versions of the finite element method. Tech. Note BN-1027. Inst. for Physical Science and Technology, University of Maryland (Sept. 1984).

32. B. Z. Szabo, Estimation and control of error based on p-convergence, in *Accuracy Estimates and Adaptivity in Finite Element Computations* (Edited by I. Babuška, E. R. de Arantes e Oliveria and O. C. Zienkiewicz). John Wiley & Sons, (1986) to appear.

33. W. C. Rheinboldt and C. K. Mesztenyi, On a data structure for adaptive finite element refinements. *ACM Trans. Math. Software* **6** (2), 166–187 (1980).

34. P. Zave and W. C. Rheinboldt, Design of an adaptive, parallel finite element system. *ACM Trans. Math. Software* **5** (1), 1–17 (1979).

35. R. E. Bank, A. H. Sherman and A. Weisner, Refinement algorithms and data structures for regular local refinement, reprint, Univ. of San Diego (1982).

36. M. C. Rivara, Algorithms for refining triangular grids suitable for adaptive and multigrid techniques. *Int. J. Numer. Meth. Eng.*, to appear.

37. M. S. Shephard, Adaptive finite element analysis and CAD, in *Accuracy Estimates and Adaptivity in Finite Element Computations* (Edited by I. Babuška, E. R. de Arantes e Oliveria and O. C. Zienkiewicz). John Wiley & Sons, (1986) to appear.

38. M. S. Shephard, Finite element modeling within an integrated geometric modeling environment—Part 1: Mesh generation. *Engineering with Computers* **1**, 61–71 (1985).

39. E. J. Haug and K. K. Choi, Material derivative methods for shape design sensitivity analyses, in *The Optimum Shape: Automated Structural Design* (Edited by J. A. Bennett and M. E. Botkin). Plenum Press, New York, (1986).

40. R. D. Cook, Loubignac's iterative method in finite element elastostatics. *Int. J. Numer. Meth. Eng.* **18**, 67–75 (1982).

DISCUSSION

C. Fleury *(University of California–Los Angeles)*

If I have understood correctly, are you suggesting that constructive geometric modeling would be better than pure analytical modeling for three-dimensional structures?

Shephard

That is not what I propose. Looking at what is available now and will be available in the near future, it would be advantageous if we could tie shape optimization to constructive solid geometry, which is the most mature of the modeling capabilities. As geometric modeling systems continue to evolve, they will provide the basic geometric support functions needed by applications such as shape optimization and finite element mesh generation. For example, a number of the constructive solid geometry modeling systems now support the boundary representations which are a needed starting point for many applications.

Fleury

Why it is better to use geometric modeling? For shape optimization, in my opinion, the analytical representation may be advantageous, especially for sensitivity analysis, because you can derive the analysis model almost mathematically from the design model, and from this you can move very easily to sensitivity analysis. I don't see how you will do it with constructive geometric modeling.

Shephard

You can certainly tie your shape changes to an analytic representation—for example, changing the diameter of a cylinder. One advantage of constructive solid modeling is that you are able to maintain manufacturability requirements much more easily, such as maintaining circular holes.

M. S. Shephard

E. Haug *(University of Iowa)*

It seems to me that if we are going to optimize shape, we have to parameterize shape. I would think that if the primitive assembly scheme is the most commonly used and most powerful, we ought to be able to accept classes of designs in which the primitives are characterized by design parameters. Then we can use those parameters to control the shape, enhance the finite element gridding and enhance the sensitivity analysis. Would you comment on that?

Shephard

Most of today's geometric modelers use fairly simple primitives. However a great deal of work is underway throughout the geometric modeling community to support much more powerful primitives where the primitives are defined in terms of free-form surface patches, in which case you have all the geometric flexibility you would like to have. One of the advantages of tying it to a solid model representation, whether it be constructive solid geometry or pure boundary representation, is the ability to update the representation and to ensure that you have a valid topological representation. Topological information can always be extracted from the boundary file, and whenever there is a topological change, that information can be communicated to your optimization procedure. Suppose an edge disappears. You have to know about that because you don't want to be working with design variables that no longer exist.

R. Haber *(University of Illinois–Champaign/Urbana)*

Relating this to the papers presented earlier, if you have to deal with mesh velocity on the interior of a region, there seems to be a problem with this approach because you don't have the advantage of a mapping where you can talk about the motion of an interior node in terms of some boundary parameter. On the other hand, if you were looking at a pure boundary representation—dealing purely with boundary information in your integrals—you might still be able to use this approach.

Shephard

My assumption is that in shape optimization we are looking at overall geometric changes that are large enough that we would not be able to live with the mapped mesh that we started out with. You can deform a mapped mesh quite a bit, but as the shape changes become large, elements start turning inside out. So I'm working under the assumption that at some point we have to remesh. You are certainly right about the sensitivity analysis if you are talking about velocities. If you work with that interior information, you would have to maintain information about a velocity field and interpolate that onto a new mesh. Obviously it would be advantageous just to work with boundary information if the required accuracy can be maintained.

[In a more extended discussion held later by Fleury, Haber and Shephard on the topic of using automatic mesh generators in shape optimization, it became clear that specific developments are needed to support the required sensitivity calculations when fully automatic mesh generation procedures are used. Although no consensus was arrived at as to the best approach, it was agreed that it should be possible to develop such procedures.]

E. Atrek *(Engineering Mechanics Research Corp.)*

How does this automatic mesh generator answer questions on the distortion of the elements and aspect ratio? Those are important questions we have to deal with.

Shephard

Rather than looking at the questions of what happens after we smooth a mesh, let's look at it before the mesh is smooth. We start out with an octant, and we know exactly what that octant's shape is. Certainly if we look at the interior, at worst we can have our nodes at corners and the mid-side, we have a very fixed aspect ratio. We can't be any worse off than that. Smoothing is going to move that only so much. We run into concerns at the boundaries because we are using real intersections and we could get very small segments. Our original version of a mesh generator didn't use real intersections, so we did not have that problem. What we do is to combine the two approaches. When something is small, we throw it out but simply remember that it is there. Controlling aspect ratios can be done quite easily, and actually it should be done in an explicit manner within the mesh generator. We do not have a real problem with that, although there are other problems I could go into in more detail if I had time.

Atrek

So there is no user input required to control the element shapes?

Shephard

No user input is required to control element shapes. By the same token, if the user wants all elements with a 10:1 aspect ratio, he is out of luck with this mesh generator unless it is all in one direction and he can just distort his universe.

ADAPTIVE FINITE ELEMENT METHODS FOR SHAPE OPTIMIZATION OF LINEARLY ELASTIC STRUCTURES

N. KIKUCHI, K. Y. CHUNG, T. TORIGAKI and J. E. TAYLOR

The University of Michigan
Ann Arbor, Michigan

Abstract

In this paper the application of adaptive grid methods with automatic remeshing schemes is discussed to solve shape optimization problems of linearly elastic structures.

After demonstrating that the final optimal shape of a structure strongly depends on the shape of finite elements near the design boundary, we briefly review the discretization error due to finite element approximations. More precisely, we derive the interpolation error to quantify the effect of distortion of the finite elements. Based on the above, an adaptive finite element grid design problem is defined using the idea of structural optimization. A necessary condition is obtained that defines a manner to adapt a given finite element grid.

Since the domain to be discretized in shape optimization problems changes its shape and size very drastically during the iterative process to find the optimal solution, remeshing must be performed at certain design stages in order to maintain the quality of finite elements of undesirable geometrical distortion. However, remeshing must be implemented without interruption of the computing process to obtain the optimal shape. This leads to numerical grid generation and adaptive methods being combined with automatic remeshing schemes.

Several shape optimization problems are solved to demonstrate the capability of the proposed method.

INTRODUCTION

The shape design of linearly elastic structures has been studied for the past decade after extensive development of finite element and optimization methods. In particular, the development of adaptive finite element methods enables us to solve shape design problems without using continuously varying thickness of a plane structure as a design variable. In this paper, the importance of adaptive finite element methods to shape design problems in structural optimization will be discussed.

Zienkiewicz and Campbell [1] applied finite element and penalty methods for shape optimization problems in 1973, and their work was followed by Ramakrishnan and Francavilla [2], Tvergaard [3], Kristensen and Madsen[4], Quéau and Trompette [5], Oda and Yamazaki [6], and others. Tvergaard applied finite difference methods in curvilinear coordinates for the shape optimization of a fillet to minimize the maximum elastic stress for a given load. Quéau and Trompette used straight lines and circles to describe the design boundaries instead of applying polynomials piecewise to represent the arbitrary shape of the boundaries.

Contributions to the analysis and theory of shape optimization have been made by Banichuk]7], Dems [8], Dems and Mróz [9, 10, 11], Choi and Haug [12], Na et al. [13] and others. Dems solved the problem of minimizing the cross-sectional area of a torsion bar by constraining the maximum torsional and bending rigidity. Dems and Mroz provided the first variations of arbitrary stress/strain/displacement functionals in terms of the stress and strain fields of primary and adjoint structures whilc thc boundary shape varies. On the other hand, Choi and Haug introduced the idea of the material derivative in continuum mechanics to derive the design sensitivity. They express design change as design deformation velocity in the dynamic process of continuum deformation. By employing the material derivative of the variational equation and replacing the material derivative of the response by the virtual displacement, they introduce an adjoint equation which leads to design sensitivity from the performance functional. Na et al. applied the idea of structural remodeling, introduced by Olhoff and Taylor [14], to solve shape optimization problems for a tension bar.

The need to apply the adaptive grid design concept to shape optimization problems was recognized in the fillet problem, for defining the optimal shape to minimize the volume of the thin elastic plate, under the constraint that the maximum von Mises stress on the boundary is less than or equal to a given upper bound. The adaptive grid design method introduced by Kikuchi [15] is a combination of automatic remeshing and adaptive finite element methods studied extensively by Babuška and his coworkers [16, 17, 18, 19, 20], Shephard [21, 22], Zienkiewicz et al. [23] and others. Details of adaptive finite element methods and related material can be found in the literature [24, 25, 26]. If the boundary shape of a fillet is defined by the coordinates of nodes, and if an optimality criteria method is used to specify the motion of nodes on the design boundary, then a physically unrealistic optimal shape can be obtained (Figure 1). No matter how

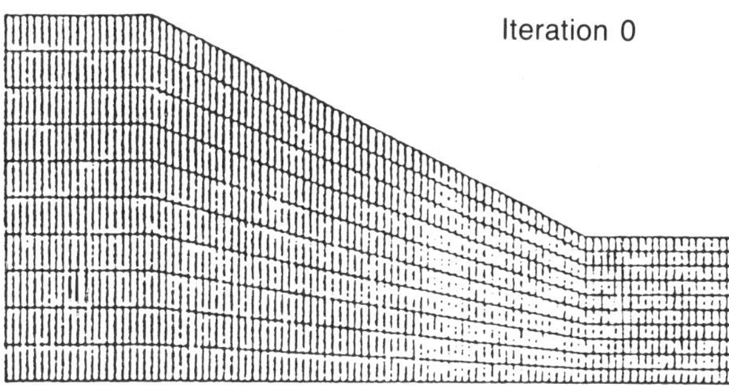

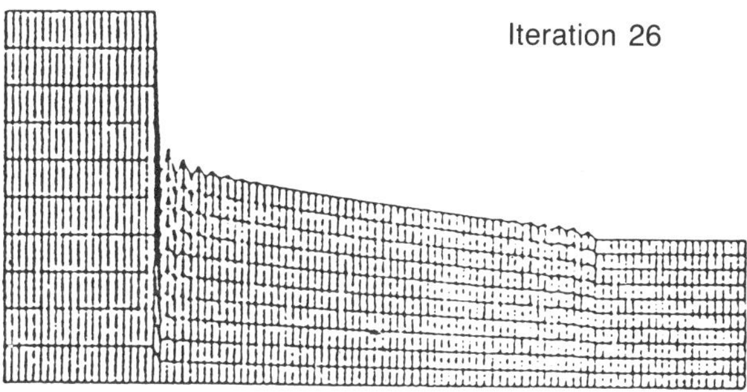

Figure 1. The initial and optimum shapes and finite element grids without remeshing.

refined a grid is used, oscillation of the shape cannot be eliminated, and in fact it tends to be larger as grid refinement progresses. The reason for this could be very simple. Indeed, finite elements near the left edge of the design boundary are severely distorted no matter how refined the grid is because the optimal shape has an infinite slope at the left edge of the design boundary. With these severely distorted grids, stresses must be approximated by finite element methods for shape optimization. It is natural to consider that oscillation of the boundary shape would not be eliminated unless refined and regular-shaped finite elements are automatically assigned near the boundary portion undergoing drastic change. How can we automatically set up refined regular-shaped finite elements in such a boundary portion during an iteration process of optimization?

In this paper, we shall discuss difficulties in shape optimization and some ways to remedy them, as well as a number of applications of the present method to solve illustrative problems.

SHAPE OPTIMIZATION PROBLEMS

Following Chung [27], let us consider a special case of shape optimization problems by minimizing the maximum value of a function of local measure.

References pp. 165–166

In comparison with global measures, local measures have not been used frequently as a criterion for optimal design. This may be in part because of the relative difficulty in controlling local measure throughout the domain by changing only limited portions of the boundaries. Examples of local measures are displacement, von Mises equivalent stress, maximum shear stress, etc. In the formulation of the design problem, the design domain is in general restricted by an isoparametric constraint on the material. A formulation of the design problem is, for example,

$$\min_{\Gamma_d}\{\max_{\mathbf{x}\epsilon\Omega} \mid F(\mathbf{u}, \nabla\mathbf{u}) \mid\} \tag{1}$$

subject to

$$\int_\Omega d\Omega - \bar{A} \leq 0$$

and subject to the equilibrium of a structure where $\bar{A}$ is a specified bound of the area and Γ_d is the design boundary. A way to convert the above formulation into tractable form for a Langrangian function is as follows:

$$\min_{\Gamma_d} \beta \tag{2}$$

subject to

$$\int_\Omega d\Omega - \bar{A} \leq 0 \tag{3}$$

$$\mid F(\mathbf{u}, \nabla\mathbf{u}) \mid - \beta \leq 0 \quad \text{in}\Omega, \tag{4}$$

together with the equilibrium of a structure. The associated Lagrangian L is:

$$\begin{aligned} \mathrm{L}(\mathbf{u}, \mathbf{v}, \Lambda, \lambda, \beta) = \beta + \int_\Omega \{\Lambda_1[\sigma_{ij}(\mathbf{u})\varepsilon_{ij}(\mathbf{v}) - b_i v_i] + \lambda_1[F(\mathbf{u}, \nabla\mathbf{u}) - \beta] \\ - \lambda_2[F(\mathbf{u}, \nabla\mathbf{u}) + \beta] + \Lambda_2\}d\Omega - \int_{\Gamma_t} \Lambda_1 t_i v_i d\Gamma - \Lambda_2\bar{A}. \end{aligned} \tag{5}$$

Here Γ_t is the boundary for the applied traction force $\mathbf{t}, \Gamma_u$ is the fixed boundary, and Γ_0 is the free boundary. Using the procedure of section 4.1 in Chung [27], necessary conditions for the extremality of the Langranian L are:

$$1 - \int_\Omega \lambda_1 d\Omega = 0 \quad \text{in } \Omega, \tag{6}$$

$$\Lambda_i\sigma_{ij}(\mathbf{v})_{,j} = \lambda_1(\partial F/\partial u_i) - [\lambda_1(\partial F/\partial u_{i,j})]_{,j} \quad \text{in } \Omega, \tag{7}$$

$$\mathbf{v} = 0 \quad \text{on } \Gamma_u,$$

$$\sigma_{ij}(\mathbf{v})n_j = \lambda_1(\partial F/\partial u_{i,j})n_j \quad \text{on } \Gamma_0 \text{ and } \Gamma_t,$$

$$\Lambda_1[\sigma_{ij}(\mathbf{u})\varepsilon_{ij}(\mathbf{v}) - b_i v_i] + \lambda_1(F - \beta) = -\Lambda_2 \quad \text{on } \Gamma_d, \tag{8}$$

$$\lambda_1(F - \beta) = 0; \quad \lambda_1 \geq 0, \quad F - \beta \leq 0 \quad \text{in } \Omega. \tag{9}$$

Note that equation (6) represents a normalization of the Lagrange multiplier λ_1, and the optimality condition in equation (8) can be simplified for the case of no body forces by using (9) to obtain

$$\sigma_{ij}(\mathbf{u})\varepsilon_{ij}(\mathbf{v}) = -\Lambda_2/\Lambda_1 = \text{constant} \quad \text{on } \Gamma_d. \tag{10}$$

If Λ_2 is chosen to be 1, the value of the constant in equation (10) is $-1/\Lambda_1$, which is exactly the same optimality condition as for the problem of minimization of weight. This shows an equivalence between the minimization of a maximized objective function with an area constraint and the minimization of area with a behavioral constraint. That is, the values of β in the two design problems are the same for the same values of the area.

Other possible formulations of local measure appear in minimization problems with local measure on the design boundaries Γ_d rather than in the domain Ω. For example:

$$\min_{\Gamma_d} \ (\text{Area}) \tag{11}$$

subject to

$$| F(\mathbf{u}, \nabla\mathbf{u}) | - \beta \leq 0 \quad \text{on } \Gamma_d,$$

and

$$\min_{\Gamma_d}\{\max_{\mathbf{x}\varepsilon\Gamma_d} | F(\mathbf{u}, \nabla\mathbf{u}) |\} \tag{12}$$

subject to

$$\int_\Omega d\Omega - \bar{A} \leq 0.$$

Optimality conditions similar to the above are obtained for these problems, but the adjoint equations and Kuhn-Tucker conditions are different. Equation (12) was used in the work by Ramakrishnan and Francavilla [2] as the approximate formulation of the minimization of stress concentration in domain Ω.

When one looks at the necessary conditions (6) through (9) for optimality in the design problem (1), (6) suggests the existence of a measurable set $\Omega_s \subset \Omega$ such that $\Omega_s = \{\mathbf{x}\varepsilon\Omega : \lambda_1(\mathbf{x}) > 0\}$, since the function F is a function in $L^2(\Omega)$ for a sufficiently smooth stress field. Then, the local measure F must be saturated in Ω_s, that is, $F - \beta = 0$ in Ω_s. A question arises as to whether Ω_s can be the same as Ω for a sufficiently large design boundary Γ_d. As far as our experience is concerned, if Ω_s coincides with Ω, almost all the final designs become somewhat trivial. For example, for the fillet problem shown in Figure 2, the possible solutions would be sets of fibers. This suggests that the design boundary Γ_d should be appropriately restricted. Another possibility is that the maximum value of the function F in the whole domain can never be reduced by changing the shape of boundaries Γ_d if Γ_d is very restricted. For example, if a very flat elliptic hole exists inside a given domain Ω but completely away from the design boundary, no matter how the shape of

References pp. 165–166

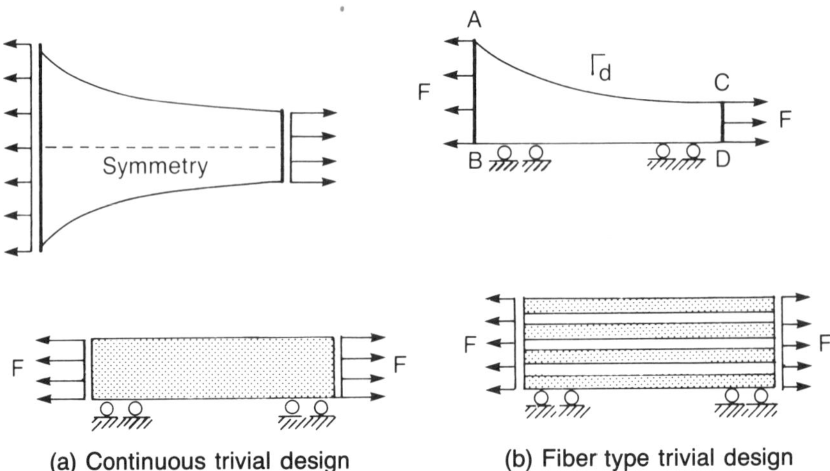

Figure 2. Fillet design problem that yields trivial solutions.

boundaries is changed, the maximum stress is generated around the elliptic hole unless a design change can modify the hole. This implies that for a too restricted design boundary Γ_d the optimization problem (1) would not be so meaningful. It is certain that if the maximum of F in Ω occurs in the neighborhood of the design boundary Γ_d, shape change implies reduction of the maximum value of F. Because of this, problems (11) and (12) are introduced. At this stage of development both (11) and (12) are soluble. But both the theoretical and computational aspects of (1) need more investigation.

COMPUTATIONAL ALGORITHM

In the computational treatment of problems, since analytical solutions are not available for general shape optimization problems in plane linear elasticity, it is necessary to apply approximation methods such as finite element methods to model the state equation. In the present study we shall apply a finite element approximation based on the displacement method for plane problems using four-node quadrilateral isoparametric elements, in which each component of the displacement vector is approximated by a bilinear polynomial. In the displacement method, only the displacement vector is the unknown quantity to be solved for, and the stress tensor is computed a posteriori using a given constitutive equation.

Like many other free boundary problems, shape design problems of linearly elastic structures have been solved using geometric adaptive methods. In most cases, the idea of geometric adaptive methods is stated in the form of a two-step iteration algorithm for the purpose of satisfying the optimality conditions. The first step represents the calculation of some quantities under the assumption that the design boundaries are fixed, and the second step predicts the movement of nodes on the design boundaries in the ratio of differences between the calculated quantities and the given or assumed constants at the nodes of a finite element model.

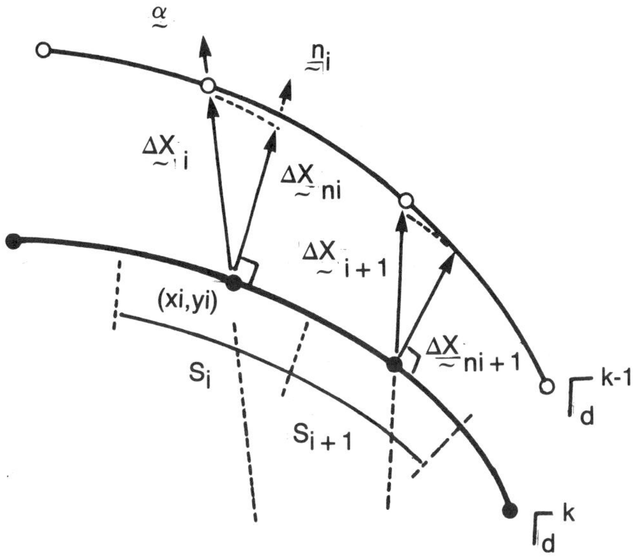

Figure 3. Geometric adaptive method.

Optimality conditions of the design problem (1) are given in (6) through (9), which yield the constant mutual energy on the design boundary Γ_d and the saturation of the design function F in Ω_s at $F - \beta = 0$. If Γ_d is contained in the closure of Ω_s, both the mutual energy and the design function F become constant on the design boundary Γ_d. We shall develop a computational algorithm based on geometric adaptive methods for shape optimization so that design boundaries are moved in the direction of satisfying these conditions (that is, the constant mutual energy and the design function F) until convergence to the final shape is reached within a certain tolerance.

The expression for the new coordinates of nodes on the design boundaries is

$$x_i^{k+1} = x_i^k + \Delta x_i^k \quad i = 1, \ldots, N \tag{13}$$

in the kth iteration, where N is the total number of nodes on the design boundary. Here, Δx_i^k is the movement of nodes in the specified direction and this value can be obtained from the movement in the normal direction:

$$\Delta x_i = \frac{\Delta x_{ni}}{\cos \theta_i} \tag{14}$$

in the kth iteration; θ_1 is the angle between the unit normal vector n_i and the unit direction vector α_i at the ith node in the specified direction (Figure 3). For simplicity, the iteration superscript k is omitted in the following. The normal movement x_{ni} at the ith node in the kth iteration is obtained from

$$\Delta x_{ni} = \frac{\Delta A_i}{S_i} \tag{15}$$

$$S_i = \frac{1}{2}[l_{i-1} + l_i]$$

where ΔA_i is the area allotted at the ith node and l_i is the length of the ith element on the boundary. The allotted area ΔA_i at the ith node is obtained from the ratio between the stresses $F(\sigma)$, calculated in the previous step by the kth iteration, and the prescribed constant β. Note that the value β in min(max F) can be assumed as the average of $F(\sigma)$ values on the nodes along the design boundary.

$$\Delta A_i = \Delta \text{Area} \left[\frac{\frac{F_i - \beta}{\beta} S_i}{\sum_{i=1}^{N} |\frac{F_i - \beta}{\beta}| S_i} \right] \tag{16}$$

where F_i is the value $F(\sigma)$ at node i, which is extrapolated from the Gaussian points using least squares methods, and Δ Area is the area between the design boundary Γ_d^{k+1} and Γ_d^k. The value of Δ Area must be large enough at the beginning and diminish as the iteration proceeds in order to have convergence. For this purpose we define a percent deviation from the optimum as

$$DTP = \left[\frac{1}{N} \sum_{i=1}^{N} (S_i \frac{F_i - \beta}{\beta})^2 \right]^{1/2} \frac{1}{\bar{S}} \tag{17}$$

using the L^2 norm, where $\bar{S}$ is the average length of elements of the design boundary:

$$\bar{S} = \frac{1}{N-1} \left[\sum_{i=1}^{N} S_i \right] \tag{18}$$

The value for ΔArea is given by

$$\Delta \text{Area} = (\text{Total Area of Domain}) \times C \times DTP.$$

From our experience, the value of C is between zero and 1.0, although the proper value must be selected according to the problem and the speed of convergence. That is, the larger the value of C, the faster the convergence speed. However, faster convergence is sometimes accompanied by oscillations of the design boundaries and possibly by oscillations of the deviation as the iterations proceed. On the other hand, a small value of C may result in very slow convergence without oscillations.

The above algorithm possesses the property that if $F_i - \beta \leq 0$ is satisfied at a point $\mathbf{x} \in \Gamma_d^k$ (that is, at node i), then $\mathbf{x}$ moves in the direction so that area A of domain Ω is reduced. This may guarantee that although we have used only the necessary condition, the iteration algorithm automatically yields the minimum.

One disadvantage of this geometric adaptive method is the possibility that the finite element grid may be distorted during the adaptation process. Too much distortion of finite elements certainly yields unnecessary approximation error which may disturb the resulting final shape of design boundaries.

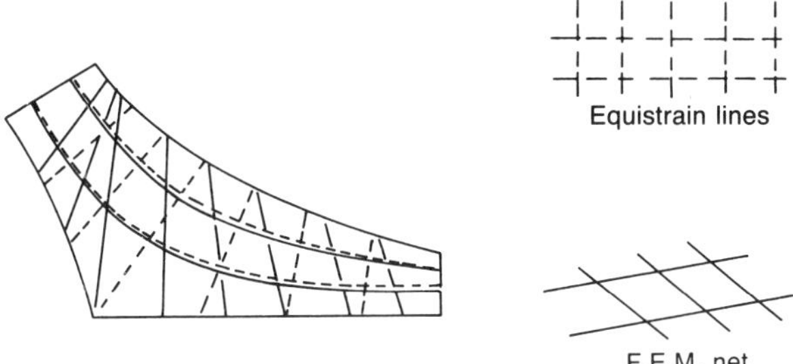

Figure 4. Conflict of the grid direction to the strain field in a design process.

More precisely, if four-node quadrilateral elements are applied to isotropic linearly elastic structures, the components of the strain tensor are approximated by

$$\begin{aligned}\epsilon_\xi &= A_1 + C_1\eta \\ \epsilon_\eta &= A_2 + B_1\xi \\ \gamma_{\xi\eta} &= A_3 + B_2\xi + C_2\eta\end{aligned}$$

in the normalized coordinate system (ξ, η). This means that, for example, the normal strain in the ξ direction is constant in ξ but is linear in η. This may result in a poor approximation in the case that the element distribution is inconsistent with the pattern of the net of equistrain lines as shown in Figure 4.

To quantify the error in the finite element approximation, let us obtain the interpolation error of a function w approximated by four-node finite elements. Suppose that the second derivatives of all the components of the true displacement $\mathbf{u}$ are constant in each finite element. Their values need not be the same in different elements. Define

$$\begin{aligned}A &= \partial^2 w/\partial x^2, \quad B = \partial^2 w/\partial x\partial y, \quad C = \partial^2 w/\partial y^2 \\ \mathbf{x}^T &= \{x_1, x_2, x_3, x_4\}, \quad \mathbf{y}^T = \{y_1, y_2, y_3, y_4\} \\ \mathbf{X}^T &= \{X_1, X_2, X_3, X_4\}, \quad \mathbf{Y}^T = \{Y_1, Y_2, Y_3, Y_4\} \\ \mathbf{L}_s^T &= (1/4)\{-1, 1, 1, -1\} \quad \mathbf{L}_t^T = (1/4)\{-1, -1, 1, 1\} \\ \mathbf{h}^T &= (1/4)\{1, -1, 1, -1\} \\ J_{11} &= (\mathbf{L}_s\mathbf{x}) + (\mathbf{hx})\eta, \quad J_{12} = (\mathbf{L}_s\mathbf{y}) + (\mathbf{hy})\eta \\ J_{21} &= (\mathbf{L}_t\mathbf{x}) + (\mathbf{hx})\xi, \quad J_{22} = (L_t\mathbf{y}) + (\mathbf{hy})\xi \\ J &= J_{11}J_{22} - J_{12}J_{21},\end{aligned} \tag{19}$$

where (x_a, y_a) are the nodal coordinates of the four corner nodes of a quadratic element in the physical coordinate system, and (X_a, Y_a) are four nodes inside the element, say, the four nodes corresponding to the 2×2 Gauss integration points in the master element. It is noted that if an element is a parallelogram the terms

hx and **hy** become indentically zero. Under the assumption stated above, the first derivatives of the difference of the function w and its interpolation w_h by the four-node quadrilateral element can be written as

$$\begin{aligned}
&\partial(w_h - w)/\partial x = \\
&\quad (1/J)[-J_{22}(J_{11}^2 A + 2J_{11}J_{12}B + J_{12}^2 C)\xi + J_{12}(J_{21}^2 A + 2J_{21}J_{22}B + J_{22}^2 C)\eta \\
&\qquad + \{(1-\eta^2)\mathbf{L}_t\mathbf{y}(\mathbf{L}_t - \xi\mathbf{h}) + (1-\xi^2)\mathbf{L}_s\mathbf{y}(-\mathbf{L}_s + \eta\mathbf{h})\} \\
&\qquad\quad \cdot \{\mathbf{hx}(A\mathbf{X} + B\mathbf{Y}) + \mathbf{hy}(B\mathbf{X} + C\mathbf{Y})\} \\
&\qquad + \{\xi(1-\eta^2)(\mathbf{L}_t - \xi\mathbf{h}) + \eta(1-\xi^2)(-\mathbf{L}_s + \eta\mathbf{h})\} \\
&\qquad\quad \cdot \{\mathbf{hx}\,\mathbf{hy}(A\mathbf{X} + B\mathbf{Y}) + \mathbf{hy}\,\mathbf{hy}(B\mathbf{X} + C\mathbf{Y})\}],
\end{aligned}$$

and

$$\begin{aligned}
&\partial(w_h - w)/\partial y = \\
&\quad (1/J)\{-J_{21}(J_{11}^2 A + 2J_{11}J_{12}B + J_{12}^2 C)\xi + J_{11}(J_{21}^2 A + 2J_{21}J_{22}B + J_{22}^2 C)\eta \\
&\qquad + \{(1-\eta^2)\mathbf{L}_t\mathbf{x}(-\mathbf{L}_t - \xi\mathbf{h}) + (1-\xi^2)\mathbf{L}_s\mathbf{x}(\mathbf{L}_s + \eta\mathbf{h})\} \\
&\qquad\quad \cdot \{\mathbf{hx}(A\mathbf{X} + B\mathbf{Y}) + \mathbf{hy}(B\mathbf{X} + C\mathbf{Y})\} \\
&\qquad + \{\xi(1-\eta^2)(-\mathbf{L}_t - \xi\mathbf{h}) + \eta(1-\xi^2)(\mathbf{L}_s + \eta\mathbf{h})\} \\
&\qquad\quad \cdot \{\mathbf{hx}\,\mathbf{hx}(A\mathbf{X} + B\mathbf{Y}) + \mathbf{hx}\,\mathbf{hy}(B\mathbf{X} + C\mathbf{Y})\}]
\end{aligned} \tag{20}$$

where (ξ, η) are the normalized coordinates in the master element. This means that if the second derivatives of the solution are known, the interpolation error can be expressed explicitly in terms of (ξ, η). Thus the error measure defined above can be computed. If an element is a parallelogram, then the interpolation error becomes very simple since $\mathbf{hx} = 0$ and $\mathbf{hy} = 0$, that is, the last two lines are identically zero in each partial derivative. On the other hand, if an element is considerably distorted from a parallelogram, then the terms in the second and third lines in the interpolation error become large in the region where large strain is expected, since $(A\mathbf{X} + B\mathbf{Y})$ and other similar terms are basically strain components in an element. This suggests that regular refined finite elements must be set up in the neighborhood of singular points. Here regularity means that an element is close to a rectangle or parallelogram. Otherwise, the error contribution becomes quite large from the terms in the second and third lines. This means that grids generated by conformal mappings are very appropriate. Similarly, grids generated by the elliptic differential equations method with the orthogonality condition and by the algebraic integer methods are suitable in the sense that errors contributed by grid distortion and high strain (i.e., stress) can be restricted to be small enough.

Figure 5 and Table 1 show the distribution and amount of error, respectively, for a linear elasticity problem. Some finite elements are deliberately distorted in order to see the effect of distortion.

It is clear that undesirable grid distortion generates unnecessary finite element approximation error. One effect of grid distortion can be seen in the example of shape optimization of a triangular plate shown in Figure 6. If shape optimization is performed using the finite element grids shown in Figure 7, different final shapes are

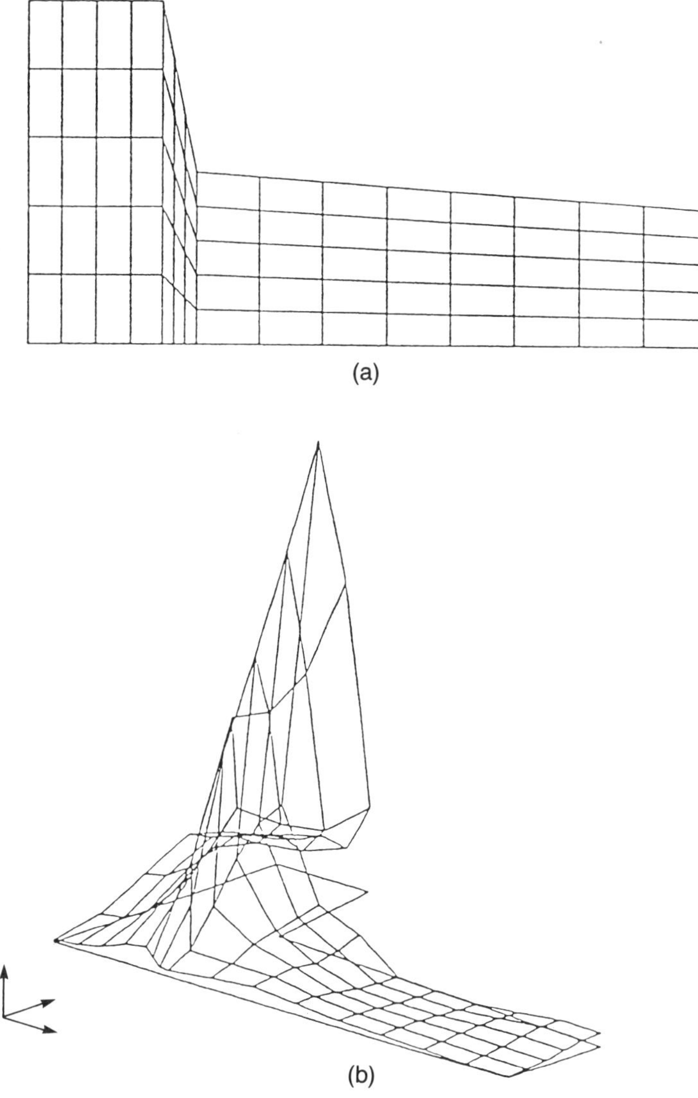

Figure 5. Distribution of the interpolation error estimated by using the form (20).

obtained by using different grids. It is clear that the regular grid shown in Figure 7(c) yields the smoothest structural shape. Therefore, we must develop good grids in the process of shape optimization. One approach is the application of adaptive finite element methods.

References pp. 165–166

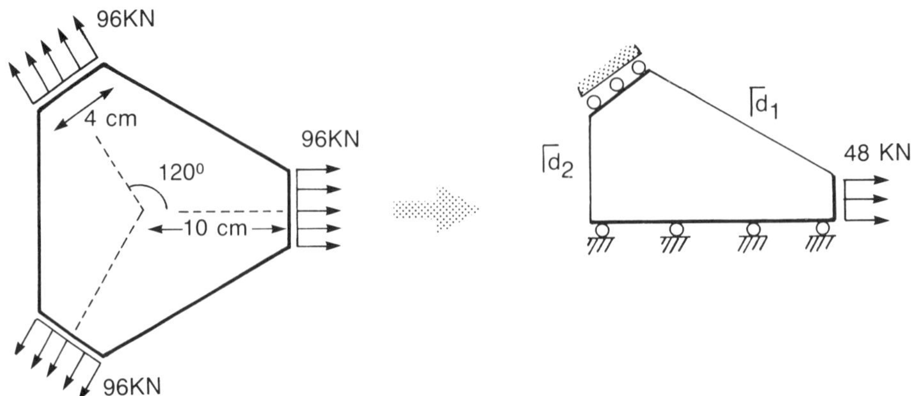

Figure 6. Shape design problem for a triangular plate.

Table 1
Pointwise error value $\partial u/\partial x(\times 10^{-5})$ and integrated error in upper boundary elements

NEL	$\mid(\varepsilon_{xx})_{e,1}\mid$	$\mid(\varepsilon_{xx})_{e,2}\mid$	$\mid(\varepsilon_{xx})_{e,3}\mid$	E_e
5	0.0	0.0	0.0	0.224
10	0.0	0.0	0.0	0.354
15	0.0	0.0	0.0	0.412
20	0.0	0.0	0.0	0.253
25	2.85	0.08	1.01	2.445
30	0.25	0.59	2.81	3.656
35	5.84	0.87	2.96	5.426
40	0.02	0.19	0.29	0.326
45	0.05	0.53	0.70	0.521
50	0.002	0.013	0.02	0.017
55	0.007	0.012	0.11	0.010
60	0.015	0.031	0.17	0.020
65	0.010	0.050	0.13	0.012
70	0.165	0.062	0.255	0.117
75	0.038	0.030	0.605	0.123

$\mid(\varepsilon_{xx})_{e,i}\mid$ = Absolute value of pointwise error of $\partial u/\partial x$ due to the line i in (38)

E_e = Integrated error in each element

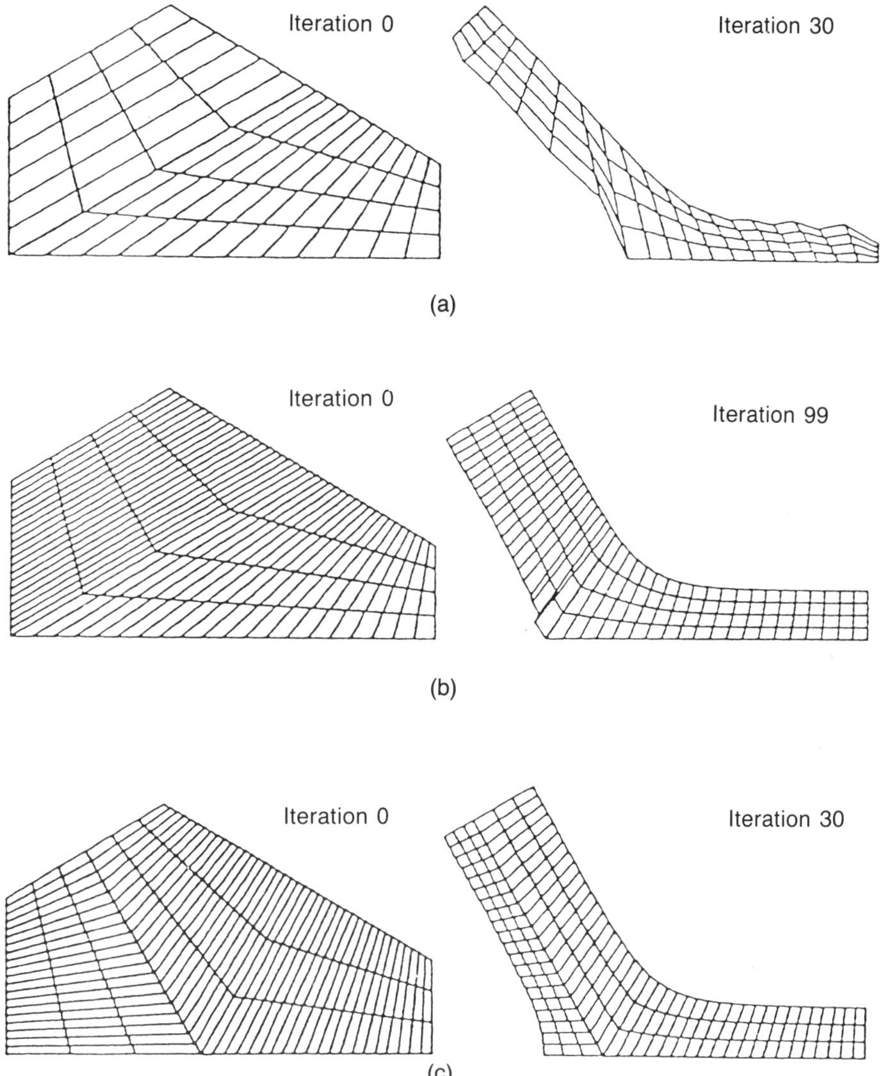

Figure 7. Three different optimal shapes obtained using different finite element grids.

ADAPTIVE FINITE ELEMENT METHODS

As described above, the final shape of design boundaries strongly depends on the finite element grid applied. In particular, if the geometric adaptation for shape optimization is defined by nodal coordinates on the design boundary, and if the moving direction of these nodes is specified as the normal direction, then the finite element grids may be badly distorted and generate unnecessary finite element approximation errors. To avoid this we may represent the design boundary as a set of curves and straight lines which can be represented as simple polynomials using, for example, various kind of splines. Iterative schemes for shape optimization are then introduced to determine the "coefficients" of these polynomials. In this case, if the degree of polynomials is low enough, the flexibility of boundary motion during the optimization process is definitely less than if nodal coordinates were used. It is noted that almost all geometric modeling of structures and machine components are performed by using straight lines, circles, ellipses and quadratic curves. If a combination of geometric modeling and shape optimization is expected, moving all the nodes on the design boundary, would not be the best strategy. The representation of lower order polynomials of a segment of the design boundary, however, does not solve all the difficulties in shape optimization problems. If finite element grids are not modified or reconstructed during the optimization process to be "regular," they are definitely distorted and produce unnecessary errors. As far as interpolation error for four-node quadratic elements is concerned, if the shapes of elements are close to parallelograms, two-thirds of the terms of the interpolation error (20) are automatically zero. Further, if motion of a boundary segment (or more flexible motion of a node on the design boundary) is expected in the normal direction, finite elements near the design boundary should be arranged to be normal to the boundary (see Figure 8) in order to avoid "conflict" between boundary motion for design and finite element grid motion during optimization. Therefore, after a certain amount of remodeling of the initial structure by geometric adaptation for design optimization based on the finite element grid developed at the initial configuration, the finite element grid must be regenerated to maintain its regularity. This requires a capability for automatic remeshing of finite element grids during a shape optimization process without interrupting the design iteration.

However this system is obviously not ready to solve shape optimization problems. As mentioned earlier, the design boundary should not be too large, and certain restrictions must be imposed on the geometric range of possible design change. This, in general, yields a somewhat singular behavior of the design boundary. In other words, the gradient of a function representing the design boundary could be extremely large at the points where the design boundary and the geometrically restricted boundaries intersect. If straight lines, circles and ellipses are used, the radii of circles, for example, could be extremely small at these points. If this singular behavior in the shape generates a nonconvex domain, the stress field will also be singular. It is widely recognized that ordinary finite element methods may not be able to approximate singularities well. To deal with singular behavior and also to control the quality of finite element approximations, adaptive finite element methods have been introduced and extensively studied. It is clear that adaptive finite element methods must be imbedded in the computational algorithm for shape optimization.

Both mathematical and computation aspects of adaptive finite element methods have been developed by Babuška and his coworkers. We shall not discuss details of adaptive finite element methods in this paper since they can be found in the references already cited, but we shall briefly review their use for shape optimization.

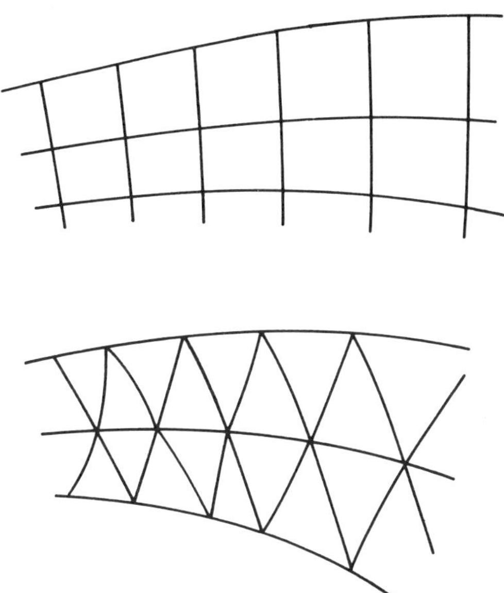

Figure 8. "Regular" finite element grids.

To make a clear distinction from the geometric adaptive method for shape optimization, the adaptive finite element methods will be referred to as *grid adaptive methods* in the present work. The so-called p method, widely applied in adaptive finite element methods, increases the degree of interpolation of polynomials in elements where large finite element errors can be observed. We shall not use this method in shape optimization since it has not been established how boundary segments can be moved according to the values of the design function F evaluated at certain nodal or integration points which are not uniformly distributed as occurs in the p method. We shall apply only the r and h methods in this work.

The r method is based on the optimal design problem for a finite element grid. Defining a finite element approximation error measure E_e in each finite element, we shall consider the problem.

$$\min_{node\,relocation}\{\max_{e=1,\ldots,NE} E_e\} \tag{21}$$

where NE is the total number of elements. Design variables are the coordinates of nodes. A necessary condition for optimality in (21) can be derived as

$$E_e = \text{constant} \quad \text{for } e = 1, \ldots, NE. \tag{22}$$

Thus, we shall determine the location of nodes in a finite element grid so that condition (22) is satisfied. Two questions are involved in this approach: what is the error measure E_e in the above design problem, and how are nodes relocated to satisfy the necessary condition (22)?

Noting that the finite element approximation error is always bounded by the interpolation error in the displacement method for elliptic boundary value problems, we may take the interpolation error in each element Ω_e as an error measure for the grid design problem; that is,

$$E_e = \sqrt{\int_{\Omega e} E_{ijkl}\varepsilon_{kl}(\mathbf{u}-\mathbf{v}_h)\varepsilon_{ij}(\mathbf{u}-\mathbf{v}_h)d\Omega} \tag{23}$$

where $\mathbf{E}$ are the elasticity constants, and $\boldsymbol{\varepsilon}$ is the engineering strain tensor. The first derivatives of the difference between the solution and its interpolation by four-node elements, $\mathbf{u}-\mathbf{v}_h$, can be computed as in (20) under the assumption that the second derivatives of the solution are constant in each finite element. Thus the evaluation of the error measure is possible whenever a method is given to approximate the second derivatives of the solution. These in general can be obtained by applying the least squares method to compute continuous strain tensors at nodes in the finite element model. A more crude error measure may be introduced by defining

$$E_e = \sqrt{\int_{\Omega e} F(\mathbf{u}-\mathbf{v}_h, \nabla(\mathbf{u}-\mathbf{v}_h))^2 d\Omega} \tag{24}$$

using the design function of shape optimization. In this case, we do not have very precise estimates of the finite element approximation error. As far as our computational experience is concerned, even

$$E_e = \sqrt{\int_{\Omega e} (F-F_h)^2 d\Omega} \tag{25}$$

works well for certain problems where F_h is the interpolation of F by four-node finite elements. This error measure does not have any explicit mathematical relation to the finite element error.

The method of relocation of nodal points applied in this work is based on Winslow [28]. The new location of the nth node is defined by

$$\mathbf{x}_n = \sum_e \mathbf{x}_e^c(E_e/A_e)/\sum_e(E_e/A_e) \tag{26}$$

where the summation is taken over the finite elements connecting to the nth node, $\mathbf{x}_e^c$ are the coordinates of the centroid of Ω_e, and A_e is its area. This node relocation scheme does not relocate nodes if the error measures satisfy the necessary condition for a finite element grid consisting of rectangular elements. Furthermore, if finite elements are distorted for reasons other than the solution characteristics, the relocation scheme tends to force grids to be "regular" so that they are close to uniform grids. Another property of equation (26) is that refined finite elements are automatically assigned near the vertices of nonconvex corner points of a polygon since (26) is a very crude difference approximation of the Laplace equation for grid generation. Therefore, (26) is not only useful for enforcing the necessary condition of optimality for the grid design problem, but also for maintaining the regularity of finite element grids with respect to the solution characteristics. After applying the relocation scheme (26), the adapted finite element grid may include some "distorted" elements, but these are generally either elements whose finite element errors are small enough or elements distorted because of the behavior of the solution in order to minimize error as much as possible.

The h method in the present work is restricted to four-node quadratic and three-node finite elements, and is based on the grid design problem similar to equation (21). However, the h method introduced here cannot yield the finite element grid which satisfies the necessary condition in the exact sense. Furthermore, we shall refine grids interatively instead of determining the necessary degree of refinement according to the so-called error indicator that is reduced under a given level, say, for example, 5%. To do this, Babuška and his coworkers introduced several refinements of an appropriate error indicator that provides almost the exact amount of finite element approximation error generated by a given finite element grid. In our case, however, this absolute amount of error is not the issue. The most important thing is the distribution of error and the relative size of error. Once this becomes clear, it is possible to determine which elements must be refined how many times within the allowable number of finite elements. Further, if a quality index which has quite a similar role to that of the error indicator is defined by

$$QI = \left(\sum_{e} E_e^2\right)^{1/2} \Big/ \sqrt{\int_{\Omega} E_{ijk\ell}\varepsilon_{k\ell}(\mathbf{u}_h)\varepsilon_{ij}(\mathbf{u}_h)d\Omega} \tag{27}$$

where $\mathbf{u}_h$ is the finite element solution, it is possible to indicate the upper bound of the finite element approximation error, although it is not as sharp as the one used by Babuška and his coworkers. It is certain that QI goes to zero if the finite element approximation error is getting smaller. Note that refinement of finite elements whose error measures are large will yield a smaller deviation from the average value of error measures than that of the original grid. In this sense, h refinement enforces the necessary condition for optimality of the grid design problem despite the fact that the condition would not be satisfied by this in the exact form. If the condition needs to be satisfied exactly together with substantial reduction of the amount of error, the r and h methods must be applied at the same time.

Using the h method in our work, refinement is performed for finite elements whose error measure is four times larger than the average. If there are no such finite elements, we choose 25%-35% elements of a finite element grid to be refined. These must have the largest error measure in the grid. We repeat this process several times or until $QI = 0.05$ is achieved.

Two examples of the adaptive methods applied in this paper are given in Figure 9 for a gear tooth and in Figure 10 for a shell structure.

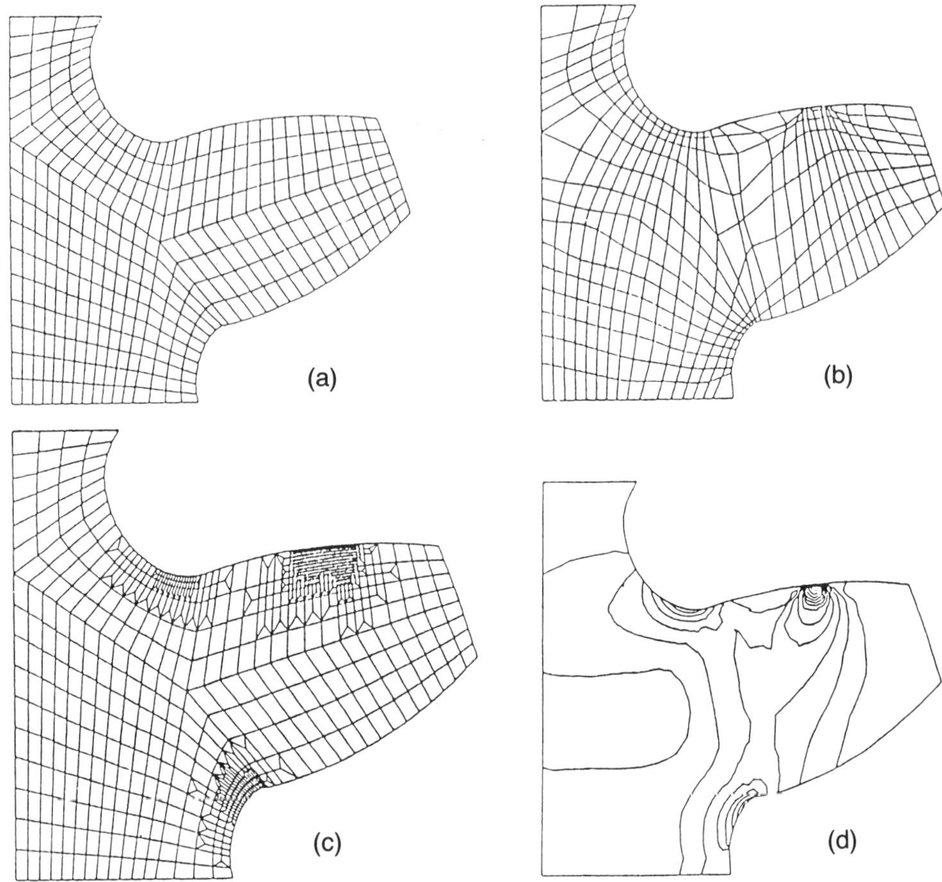

Figure 9. Adaptive finite element methods (*r* and *h* methods).

EXAMPLES OF SHAPE OPTIMIZATION

One fundamental shape design problem is the design of a hole in a square plate where tensile forces are evenly distributed on the horizontal and vertical parallel edges. For an infinite plate the analytical solution is an elliptic hole with a semi-axis ratio (1.5 in this example) of stress components in horizontal and vertical directions at infinity. We shall solve this in order to demonstrate the capability of the geometric adaptive method introduced here. Assume that the design function F in shape optimization is the von Mises equivalent stress and that the design boundary Γ_d is only the hole boundary. Iteration histories of boundary shape and finite element grids are shown for this problem in Figure 11. Large oscillations at the beginning quickly fade away from the exact ellipse. Convergence characteristics are given in Figure 12. It is clear that the maximum value of the von Mises equivalent stress in the whole domain Ω is the same as that on Γ_d. In this case the design problem (12) is equivalent to the original one (1).

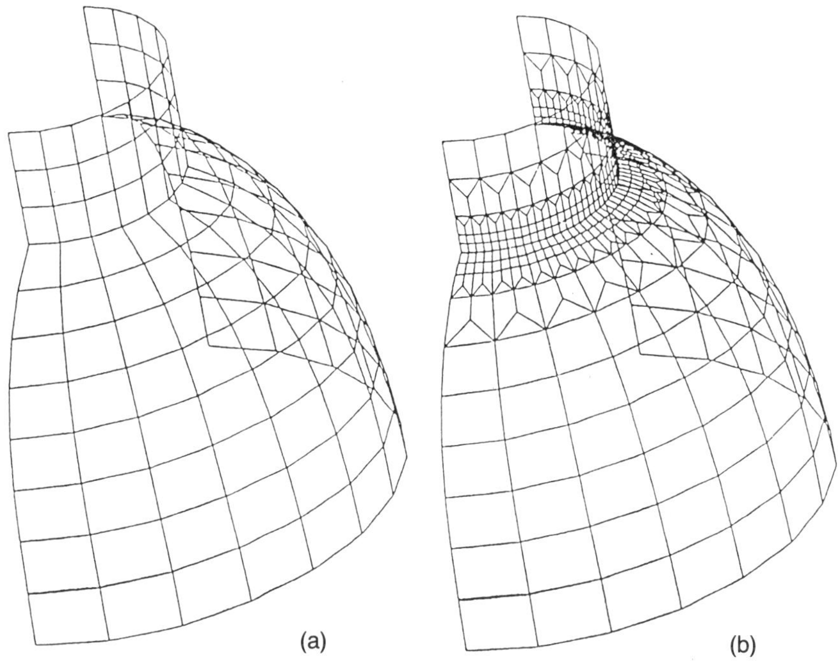

Figure 10. Application of the *h* adaptive method for a shell.

The second example is the shape design of a highway road pole that has been solved by Oda and Yamazaki [29] where a hole was created at the place of minimum thickness to obtain a fully stressed shape, although the fully stressed design was not achieved in their results. Considering only a half portion of the highway road pole, let us first introduce two design boundaries at the center symmetric line and the right-hand side outside boundary. If the bottom line is allowed to move horizontally, the optimal shape is obtained (Figure 13) starting from the initial grid specified in the same figure without applying a remeshing scheme during the geometric adaptive iteration for design change. Also note that the nodes on the design boundary are moved along the grid direction in the initial finite element grid. Thus, after a certain number of iterations and a considerably large design change, the finite element grids are very distorted. The final shape obtained is also bad. Now let us continue the shape design process by applying the least squares method to define a smooth boundary curve in order to set up a finite element grid for the restart of shape design. Further, suppose that we do not want to give up the grid in the final design in Figure 13. More precisely, we do not give up the element connectivities defined at the initial grid. But location of nodes will be modified by applying the r method using the error measures computed at the final design stage. This produces the grid shown in Figure 14, which is assumed to be the second initial grid for shape optimization after "remeshing." If the geometric adaptive method is applied, the optimal shape is obtained as shown in the figure. Note that if the bottom line is fixed, a singular shape design is obtained at the end points of the

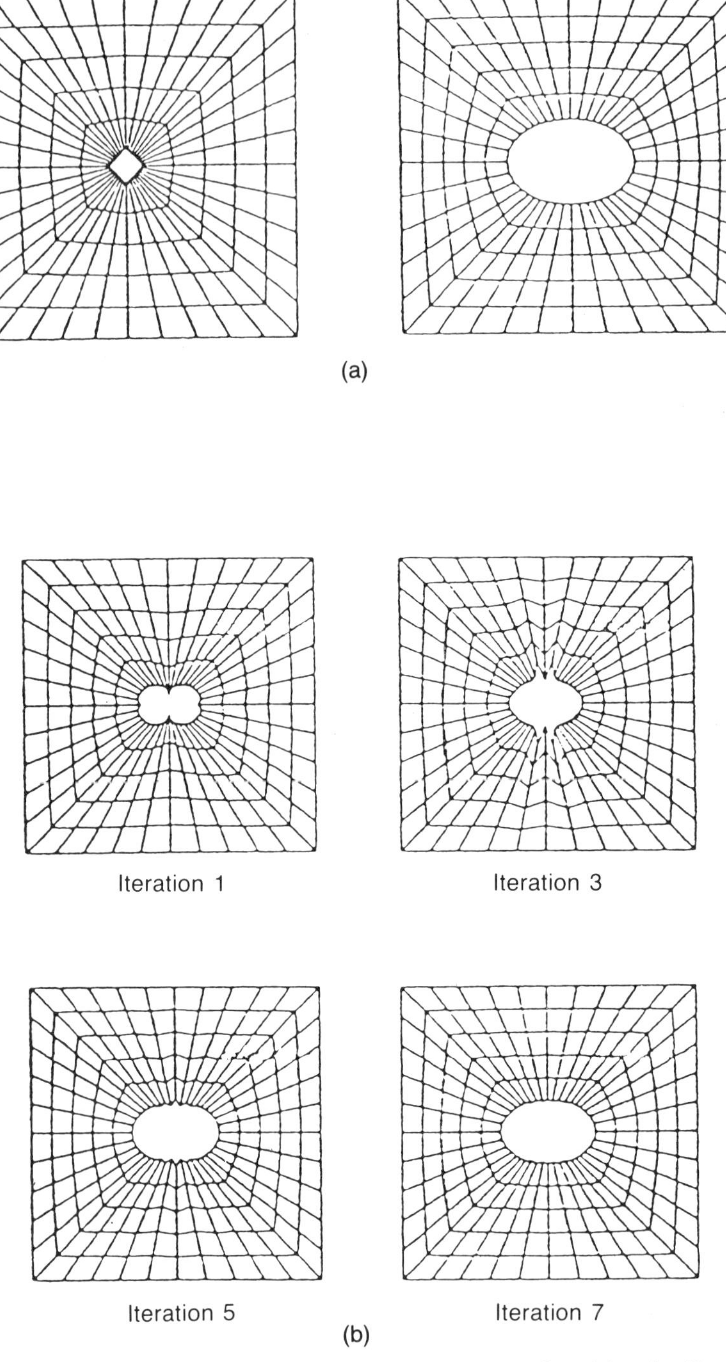

Figure 11. Design histories of a shape design of a thin elastic plate with a hole.

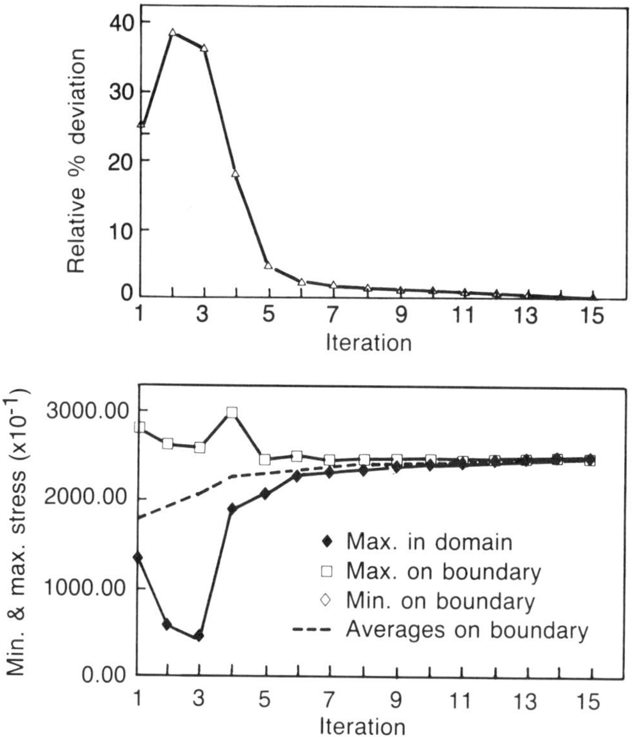

Figure 12. Convergence history of the geometric adaptive method.

bottom line. We cannot expect a hole inside the road pole. Furthermore, if we allow for the possibility that a hole may be generated in the road pole already separated into two parts in the previous design stage, the pole will be further separated into four road poles (Figure 15). Therefore, if we repeat this process infinitely, many poles are generated so that a set of fibers will constitute a road pole—the final optimal structure.

One of the most frequently used examples in shape optimization is the fillet problem because of its practical importance and difficulties of analytic solution. As shown in Figure 1, if remeshing is not performed during shape design, no matter how refined the finite element grids may be, the optimal shape computed will be nonsensical. Thus, following the idea for the road pole problem, let us remesh by applying the least squares method to the design boundary to produce a smooth boundary. After this let us reconstruct finite element grids by applying the r and h methods as shown in Figures 16 and 17 based on the first trial of shape optimization, and using the initial grid in Figure 18. If the geometric adaptive method is again applied for shape optimization for the second initial grids in Figures 16 and 17, the optimal shape of the fillet can be obtained without unreasonable physical oscillation. In this case, note that the value of the maximum von Mises stress in the whole domain is not the same as that on the design boundary, since the right side of

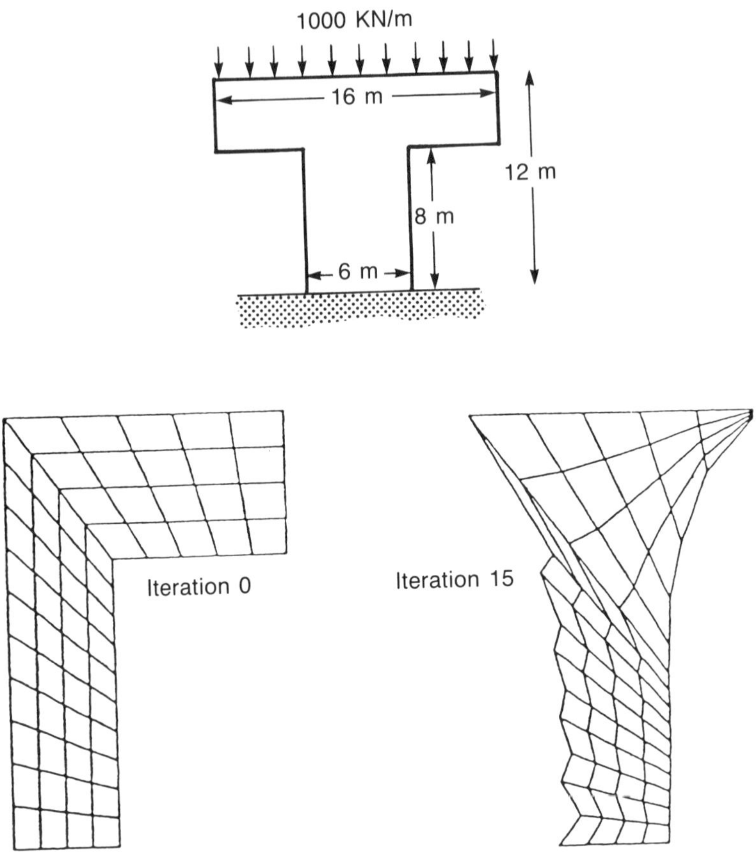

Figure 13. Road pole shape design problem: initial grid and result without applying remeshing.

the design boundary is also restricted. Thus, the maximum value of the von Mises stress appears outside the design boundary. If the right side of the design boundary is released from the design restriction, the maximum of the von Mises stress is on the design boundary. In this case, the optimal shape obtained is shown in Figure 19. Figure 20 compares the finite element result to the photoelastic result obtained by Schnack [30]. It is clear that the optimal shape method presented here results in the same stress fringes as the photoelasticity method.

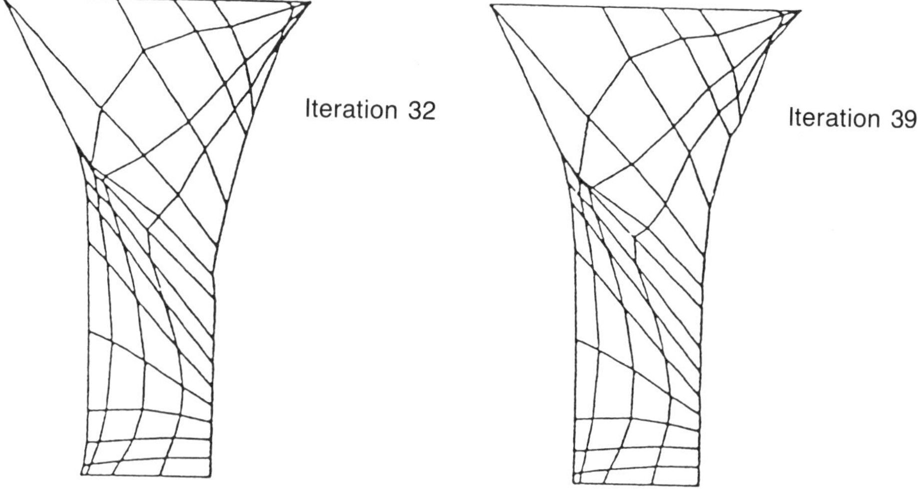

Figure 14. Remeshed second initial grid and the optimum shape of the road pole.

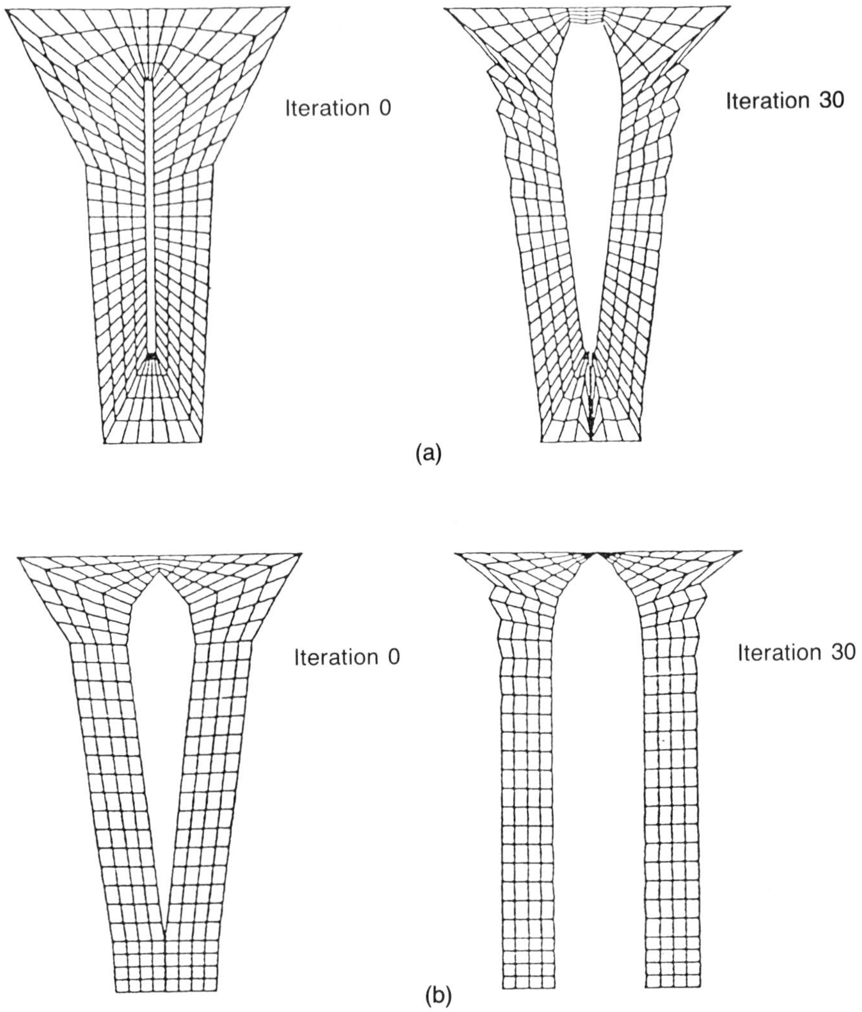

Iteration 0
Iteration 30
(a)
Iteration 0
Iteration 30
(b)

Figure 15. Continuation of the road pole problem with possibility of having a hole inside.

Iteration 0
Iteration 15

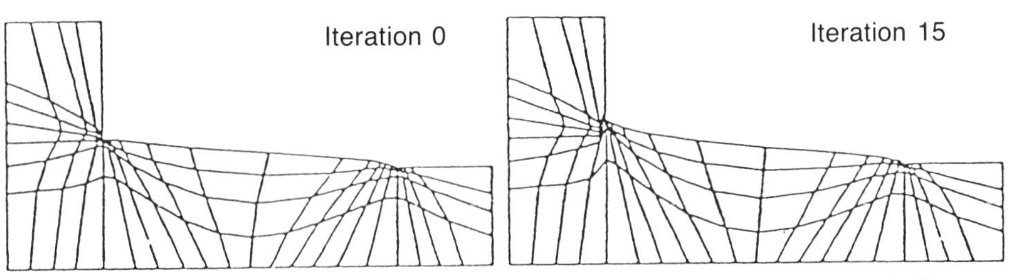

Figure 16. Remeshed second initial grid by the *r* method and the computed optimum shape.

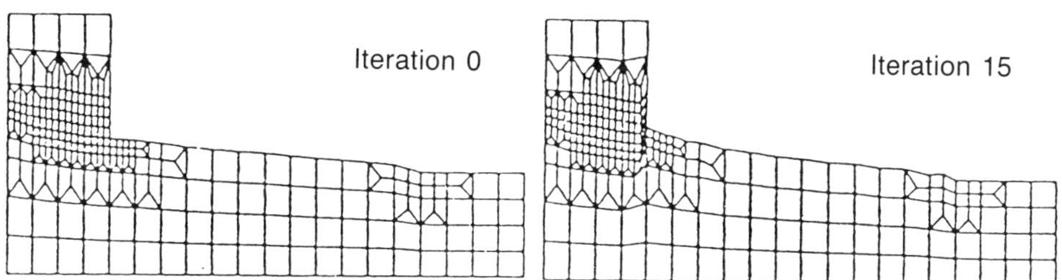

Figure 17. Remeshed second initial grid by the *h* method and the computed optimum shape.

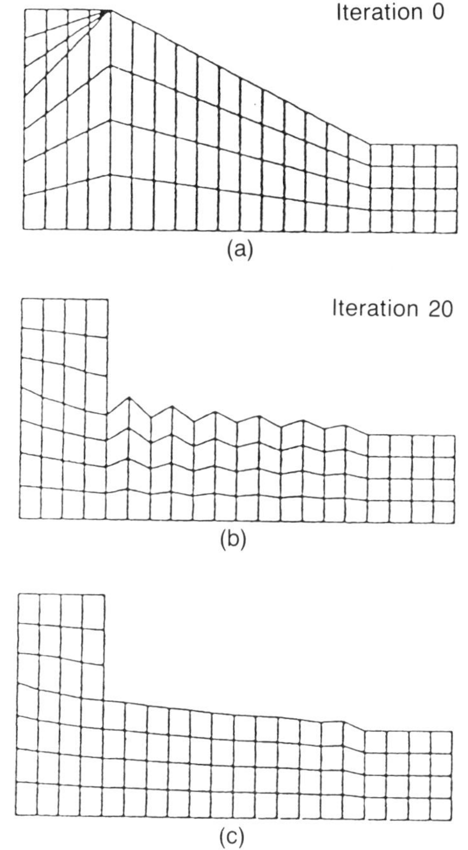

Figure 18. (a) Initial grid, (b) optimum shape without remeshing, and (c) the smoothed shape by the least squares method.

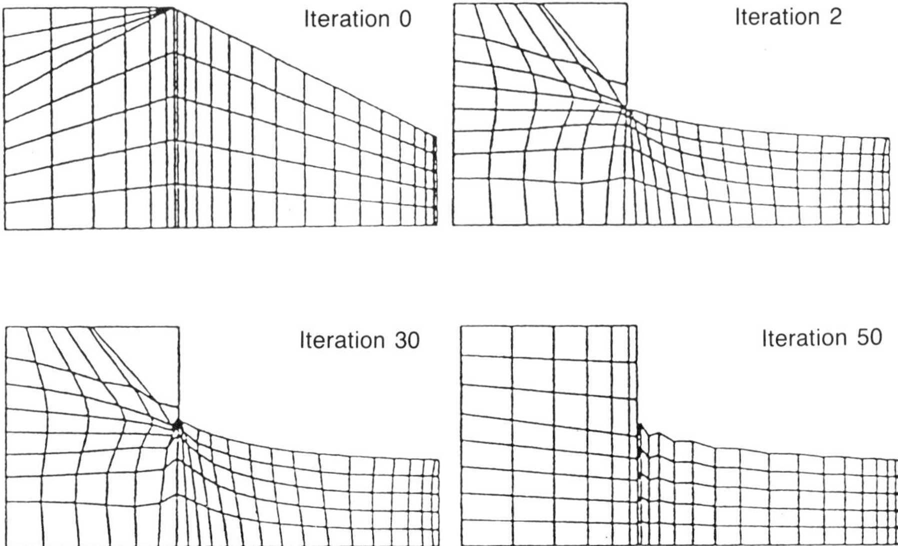

Figure 19. Shape optimization of a fillet without design restriction along the right edge.

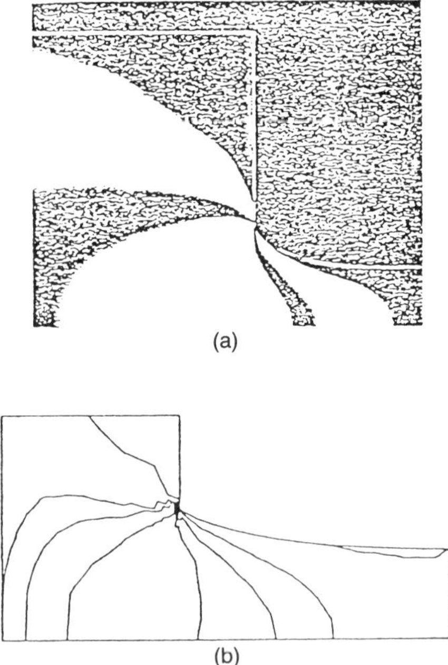

Figure 20. Fringes obtained (a) by photoelasticity and (b) by finite element method.

ACKNOWLEDGMENTS

The authors were partially supported by NASA Lewis Research Center during this work through Grant NAG3-388. This support is gratefully acknowledged.

REFERENCES

1. O. C. Zienkiewicz and J. S. Campbell, Shape optimization and sequential linear programming, Ch. 7 in *Optimal Structural Design* (Edited by R. H. Gallagher and O. C. Zienkiewicz). John Wiley & Sons, London, (1973).
2. C. V. Ramakrishnan and A. Francavilla, Structural shape optimization using penalty functions. *J. Struct. Mech.* **3**(4), 403–422 (1974-1975).
3. V. Tvergaard, On the optimal shape of a fillet in a bar with restrictions. *Proc. IUTUM Symposium on Optimization in Structural Design.* Springer Verlag, Warsaw (1973).
4. E. S. Kristensen and N. F. Madsen, On the optimum shape of fillets in plates subjected to multiple inplane loading cases. *Int. J. Numer. Meth. Eng.* **10**, 1006–1019 (1976).
5. J. P. Quéau and Ph. Trompette Two-dimensional shape optimal design by finite element method. *Int. J. Numer. Meth. Eng.* **15**, 1603–1612 (1980).
6. J. Oda and K. Yamazaki, A procedure to obtain a fully stressed shape of elastic continuum. *Int. J. Numer. Meth. Eng.* **15**, 1095–1105 (1980).
7. N. V. Banichuk, *Problems and Methods of Optimal Structural Design.* (Translated by V. Komkov and E. J. Haug). Plenum Press, New York (1983).
8. K. Dems, Multiparameter shape optimization of elastic bars in torsion. *Int. J. Numer. Meth. Eng.* **15**, 1517–1539 (1980).
9. K. Dems and Z. Mróz, Optimal shape design of multicomposite structures. *J. Struct. Mech.* **8**(3), 309–329 (1980).
10. K. Dems and Z. Mróz, Variational approach by means of adjoint systems to structural optimization and sensitivity analysis-I. *Int. J. Solids, Struct.* **19**(8), 677–692 (1983).
11. K. Dems and Z. Mróz, Variational approach by means of adjoint systems to structural optimization and sensitivity analysis-II. *Int. J. Solids Struct.* **20**(6), 527–552 (1984).
12. K. K. Choi and E. J. Haug, Shape design sensitivity analysis of elastic structures. *J. Struct. Mech.* **11**(2), 231–269 (1983).
13. M. S. Na, N. Kikuchi and J. E. Taylor, Optimal modification of shape for two-dimensional elastic bodies. *J. Struct. Mech.* **11**(1), 111–135 (1983).
14. N. Olhoff and J. E. Taylor, On optimal structural remodeling. *J. Optim. Theory Appl.* **27**(4), 571–582 (1979).
15. N. Kikuchi, Adaptive grid design methods for finite element methods. *Comput. Meth. Appl. Mech. Eng.* (1986), to appear.
16. I. Babuška and M. R. Dorr, Error estimates for the combined h and p version of the finite element method. *Numerische Mathematica*, **25**, 257–277 (1981).
17. I. Babuška and W. C. Rheinboldt, Error estimates for adaptive finite computations. *SIAM J. Numer. Anal.* **15** 736–754 (1978).
18. I. Babuška and W. C. Rheinboldt, Reliable error estimation and mesh adaptation for the finite element method, pp. 67–109 in *Computational Methods in Nonlinear Mechanics* (Edited by J. T. Oden). North-Holland, Amsterdam (1980).

19. I. Babuška and B. Szabo, On the rate of convergence of the finite element method. *Int. J. Numer. Meth. Eng.* **18**, 323–341 (1982).

20. I. Babuška, A. Miller and M. Vogelius, Adaptive methods and error estimation for elliptic problems of structural mechanics, pp 57–73 in *Adaptive Computational Methods of Structural Mechanics* (Edited by I. Babuška, et al.) SIAM, Philadelphia (1983).

21. M. S. Shephard, R. H. Gallagher and J. F. Abel, The synthesis of near optimum finite element meshes with iterative computer graphics. *Int. J. Numer. Meth. Eng.* **15** 1021–1039 (1980).

23. O. C. Zienkiewicz, J. P. Gago and D. W. Kelly. The hierarchical concept in finite element analysis. *Comput. Struct.* **16**, 53–65 (1983).

24. M. S. Shephard (Ed.), *Finite Element Grid Optimization.* ASME Special Publication, PVP-38. American Society of Mechanical Engineers (1979).

25. I. Babuška, J. Chandra, and J. E. Blaherty (Eds.), *Adaptive Computational Methods for Partial Differential Equations.* SIAM, Philadelphia (1983).

26. E. R. de Arantes Oliveira, I. Babuška, O. C. Zienkiewicz and J. P. de S. R. Gago (Eds.), *Proc. of Int. Conf. on Accuracy Estimates and Adaptive Refinements in Finite Element Computation.* Lisbon (1984).

27. K. Y. Chung, Shape optimization and free boundary problems with grid adaptation. Ph.D. Dissertation, University of Michigan, Ann Arbor (1985).

28. A. M. Winslow, Numerical solution of quasilinear Poisson equation in a nonuniform triangular mesh. *J. Computat. Physics*, **2**, 149–172 (1967).

29. J. Oda and K. Yamazaki, On a technique to obtain an optimum strength shape of an axisymmetric body by the finite element methods. *Bull. JSME*, **20** (150), 1524–1532 (1981).

30. E. Schnack. An optimization procedure for stress concentration by finite element technique. *Int. J. Numer. Meth. Eng.* **14**, 115–124 (1979).

DISCUSSION

E. Haug *(University of Iowa)*

It seems to me that if you allow a lot of oscillation and high curvature in the boundary, the optimization algorithm is really working on the inherent or systematic error in the finite element code rather than on the real problem. Do you think it is possible, with the knowledge of error estimates, to parameterize the boundary? And supposing you have ranges of parameters that constrain surface curvatures to be bounded: would it be possible to get uniform error estimates so that you could perform that calculation with confidence that the finite element gridding would be adequate within the full range of shape design variation you have defined?

Kikuchi

Theoretically, anything is possible, but at the moment I really do not know how to implement those things. The expression of the error that I described here can quantify the error very well at each point. Using these data, it may be possible to do the things you suggest, but I do not know how at this moment. It is a good subject to discuss or think over.

Moderator—B. Szabo

If I understood it correctly, you said that the reason for the oscillation is that the error increases as the elements become more distorted. Therefore you advocate that the element shapes should be as close to a parallelogram or to some rectangular shape as possible. I would interpret this as an argument against the r adaptation because in the r adaptation you are moving nodes and therefore distorting the elements.

Kikuchi

Certainly, we can give a certain amount of distortion where we had, for example, the lowest stress. In that portion the distortion is not important. Furthermore, the error will be computed as two terms: 1) the geometric distortion part multiplied by 2) the derivative of the stresses. We have to look at the two sides at once. If the stress or the stress derivative is lower, then the other side can be large. So we should have a regular mesh, as much as possible, for errors to this gradient are high; but otherwise we don't.

N. Kikuchi

Szabo

Are you then advocating the use of very regular linear mapping whenever the gradients are high, but you are free to use other kinds of mapping elsewhere?

Kikuchi

With adaptation, the grading will be almost uniform as a result.

V. Venkayya *(AFWAL/FIBRA)*

Have you made any comparisons of your conclusions to the Michell structures?

Kikuchi

No, that would have taken a larger computer budget than was available. I think that we may be able to go up to the Michell truss, but at this point we would be constrained by manpower and computing resources.

D. J. Wilde *(Stanford University)*

I'd like to talk about these phenomena that you see and Professor Mróz saw. In his case, six stiffeners become two, and in your case one becomes eight. I have seen this in much simpler optimization problems. One example involves three heat exchangers in a row; we used differential methods to obtain values for these exchangers. Then when we applied what we think of as monotonicity analysis, looking at monotonicities rather than derivatives, we found that the optimum—and a much better design—was to have the three exchangers collapse into one. Although I don't know how to do it in a variational sense, I am simply suggesting that looking at the monotonicities of these cases may explain these phenomena.

B. Prasad *(Electronic Data Systems)*

Professor Kikuchi, when you said you had to expend a lot of computational time, does that mean that you are using coordinates of the nodes as design variables?

Kikuchi

Not necessarily. Actually we can end up with the optimal shape after 30 or 40 iterations. At each state we have to re-evaluate the stiffness matrix and solve the system of linear equations. Then we have to obtain the stress computation and apply the least squares method.

Prasad

I am asking whether or not, along the surface, each coordinate is a design variable.

Kikuchi

That is the case.

Prasad

If so, you can use the approach which Bennett and Botkin used, the design line, and then you can reduce the number of the coordinates' points. That way you can constrain the design to follow a certain path and also you do not get into major computational cost problems.

Kikuchi

I believe you may have more problems. If you move a boundary segment, such as a line or a circle, you really have to evaluate more than in the case of a node that is moving; it is more difficult than the node movement on the boundary. I was just lazy and I took a simpler way.

C. Fleury *(University of California - Los Angeles)*

In the elliptical hole, was the problem to minimize the stress concentration or to minimize the volume?

Kikuchi

In that particular case, I believe it was to minimize the area.

Fleury

In your problem, you have to minimize the maximum error and also minimize some particular function such as the area. How do you handle minimizing the objective functions since you have two criteria: to minimize the error and the area at the same time?

J. Taylor *(The University of Michigan)*

I think the answer is simply to do it sequentially.

UNCERTAINTIES IN ENGINEERING DESIGN: MATHEMATICAL THEORY AND NUMERICAL EXPERIENCE

I. BABUŠKA
Institute for Physical Science and Technology
University of Maryland
College Park, Maryland

Abstract

The paper addresses the question of the reliability of engineering computations. It presents a set of paradoxical, unexpected results which shows that the common practice can lead to unreliable results and conclusions. The theory and implementation of the analysis of elasticity problems with stochastic input data (loads, domain, coefficients) are outlined. Numerical examples illustrate the ideas and results.

INTRODUCTION

Shape optimization in structural mechanics came to be in the center of research and applications. Many papers and books dealing with this subject have appeared and special conferences have dealt with these problems. The research is directed toward theoretical questions such as the existence and characterization of the optimal design, bounds for the optimized values, numerical treatment of the optimal design problems, and others. Of the recent and vast literature, we mention only [10], [13], [18] and [21] as examples.

In this paper we will address the problem of the *reliability* of the conclusions based on computational analysis and their relation to the problems of the optimal design. By reliability we mean that the conclusions sufficiently and accurately describe the physical reality.

The problem of optimal design consists, in principle, of the comparison of the solutions of states from the set S of admissible states (for example, solution of the problems from the set of admissible domains) and the selection of the "optimal"

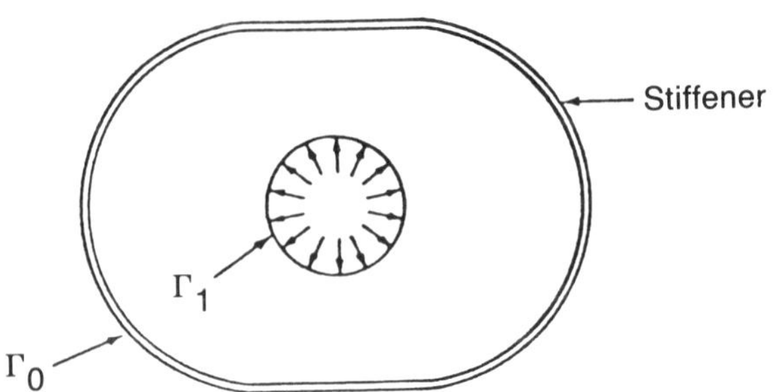

Figure 1. Scheme of a stiffened tube.

state (e.g., the domain) for the engineering design. It is obvious that such a selection can be successful *only if the solution of every state is uniformly reliable with respect to the entire set S.* This requirement creates a serious difficulty because in practice we are used to solving numerically the simplified mathematical formulation of the problem and we *only have experience with a small, limited set of practical problems.* Hence, it is essential to analyze and to be explicitly aware of the assumptions used in the derivation of the model and its numerical treatment as well as the limitations with which we must deal.

It is obvious that we have to focus on the reliability of the analysis of the single states and its uniformity over the entire set S of admissible states. This must always be the starting point of the assesment of the validity of the optimal design. The reliability of the computational treatment of the single states depends on a) the mathematical model, b) the reliability of the input data and c) the reliability of the numerical treatment. These three aspects are of course closely related.

In this paper we will not address the questions in general, but by means of a few concrete engineering examples. We restrict ourselves to relatively simple examples from the engineering point of view in order to present the ideas in the clearest way.

THE PROBLEM OF A TUBE WITH A STIFFENED SURFACE

Let us consider the problem of a tube (plane strain) with a stiffened outer surface. We shall assume that the stiffener is bending free. In general we are interested in the design with respect to the influence of the changes of the outer surface Γ_0. The scheme of the problem is shown in Figure 1.

We formulate the problem as a *linear elasticity* (plane strain) problem. The formulation is the standard one, based on the minimization of the cumulative energy of the tube and the tension energy of the reinforcement on Γ_0.

The Reliability of the Mathematical Model — We will analyze the case when the outer boundary Γ_0 is a regular m polygon and Γ_1 is a concentric circle (see Figure 2). The domain is denoted by Ω_m. The circular domain Ω_0 (see Figure 3) is obviously the limiting case when $m \to \infty$. Assuming that the unit hydrostatic pressure is given on the inner boundary and that the stiffener

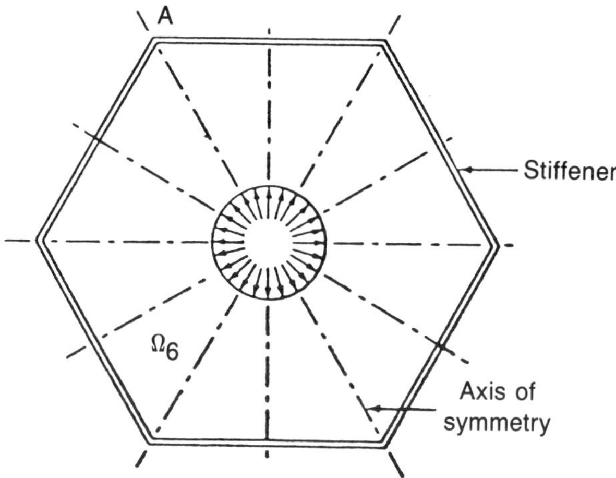

Figure 2. Scheme of a stiffened polygonal tube with m sides.

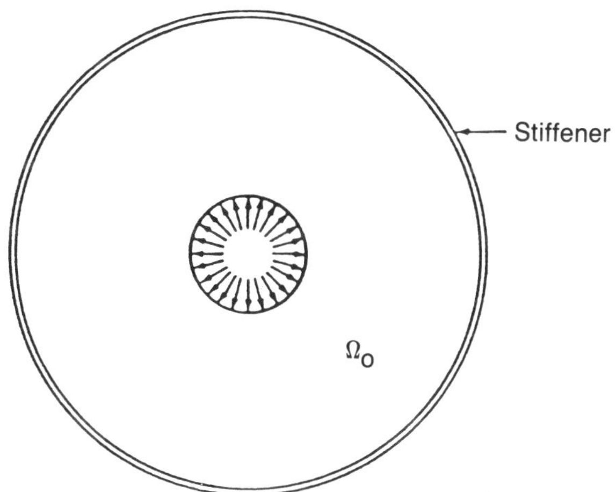

Figure 3. Circular stiffened tube as limit case for $m = \infty$.

is infinitely rigid in tension, the problem of linear elasticity can be formulated on one sector only with the boundary conditions shown in Figure 4. On sides AB and CD the symmetry conditions are prescribed. On side AD the boundary conditions describe the behavior of the stiffened side, and on side BC the tractions are prescribed.

Let us denote by (u_m, v_m) and (u_0, v_0), respectively, the solution (displacement) on Ω_m and Ω_0. From the physical grounds we have to expect that $(u_m, v_m) \rightarrow (u_0, v_0)$ as $m \rightarrow \infty$. If $(u_m, v_m) \not\rightarrow (u_0, v_0)$, we have to have doubts about the reliability of the model.

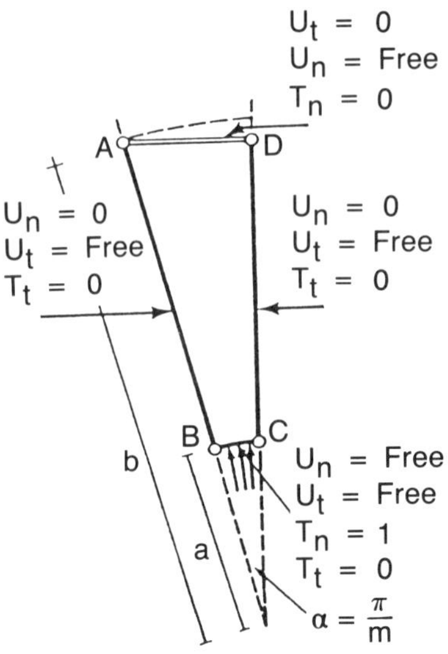

Figure 4. Scheme of the sector analyzed by plane elasticity theory.

We have

THEOREM 1. $\lim_{m\to\infty}(u_m, v_m) = (u_\infty, v_\infty) \neq (u_0, v_0).$ (1)

For the analysis of the dependence of the solution on small changes of the domains we refer, for example, to [1], [3] and [4].

Theorem 1 shows that the *model of linear elasticity is unreliable* at least if the outer boundary is not smooth, and hence it *cannot be used in practice* for optimal design when the admissible domains do not have a sufficiently smooth boundary. We see here that the criterion of uniform reliability with respect to m is clearly violated. The limiting solutions (u_∞, v_∞) and (u_0, v_0) can be found.

THEOREM 2. *The solutions* (u_∞, v_∞) *and* (u_0, v_0) *are radially symmetric. Denoting by* σ_r *and* σ_θ *the stresses in polar coordinates, we get*

$$\sigma_r = \frac{A}{r^2} + B, \qquad \sigma_\theta = -\frac{A}{r^2} + B \tag{2}$$

with

$$A_\infty = a^2, \qquad B_\infty = 0 \tag{3}$$

$$A_0 = \frac{(1-\nu)a^2b^2}{(1-\nu)b^2 + (1+\nu)a^2}, \qquad B_0 = -\frac{(1+\nu)a^2}{(1-\nu)b^2 + (1+\nu)a^2}. \tag{4}$$

Table 1 gives the values of the stresses σ_r and σ_θ on the line CD for the solution (u_∞, v_∞) and (u_0, v_0) when $a = 0.3$, $b = 1.0$ and $\nu = 0.3$. We see clearly that the solutions (u_∞, v_∞) and (u_0, v_0) are essentiallly different.

Table 1
The limit stresses for $m \to \infty$
and stresses for the stiffened circular tube.

r	(u_∞, v_∞)		(u_0, v_0)	
	σ_r	σ_θ	σ_r	σ_θ
0.3	-1.000	1.000	-1.000	0.7135
0.4	-0.5635	0.5625	-0.6251	0.3387
0.5	-0.3600	0.3600	-0.4516	0.1652
0.6	-0.2500	0.2500	-0.3574	0.0710
0.7	-0.1836	0.1836	-0.3005	-0.0142
0.8	-0.1406	0.1406	-0.2637	-0.0227
0.9	-0.1111	0.1111	-0.2384	-0.0480
1.0	-0.0900	0.0900	-0.2208	-0.0609

Before discussing the probable reason for this paradox, let us state a theorem.

THEOREM 3. *Let* (u_m, v_m) *be the solution on* Ω_m *(i.e.,* m*-sided polygon),* $m > 4$. *Then the solution has a singularity in the neighborhood of point* A *(vertex of the polygon)* (see Fig. 2 and 4) *and*

$$\begin{matrix} u_m \\ v_m \end{matrix} = r^{\frac{2}{m-2}} \begin{pmatrix} \phi_m(\theta) \\ \psi_m(\theta) \end{pmatrix} + higher\,order\,terms \tag{5}$$

where (r, θ) *are the polar coordinates with the origin in* A *and* ϕ_m, ψ_m *are smooth functions in* θ.

Theorem 3 shows that the solution has a strong singularity in the neighborhood of A and the strains and stresses there are unbounded. This obviously *violates* the basic assumptions of the linear elasticity model and has unexpected consequences.

We are making the following unproven conjecture: if $(\tilde{u}_m, \tilde{v}_m)$ is the solution of a *nonlinear* problem, then $\lim(\tilde{u}_m, \tilde{v}_m) \to (\tilde{u}_0, \tilde{v}_0)$. This leads to the conclusion that if the set of admissible domains has an unsmooth outer boundary, then *it is necessary to use a nonlinear theory of elasticity* in the optimal design problems. The linear elasticity leads to unreliable results and conclusions.

The Reliability of the Numerical Solution — As we have seen in Theorem 3, the solution has a very strong singularity (note that in the case of a crack the singularity of the solution is $r^{1/2}$) which makes computation very difficult for larger m. The computation we present has been made by the code PROBE which uses the p and h-p versions of the finite element method. See references [23] and [24]. For the theoretical aspects we refer to [6], [7] and [14]. The mesh has to be strongly refined in the area of the singularity if reliable results have to be obtained. See references [14] and [22]. For the p version (i.e., when there is no strong

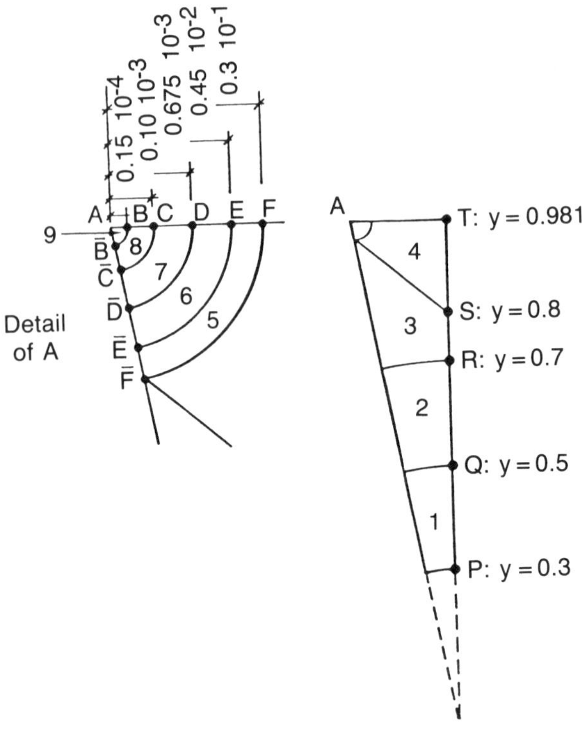

Figure 5 Mesh used for the p version analysis

refinement at A), the energy norm of the error is $\| e \| \approx Cp^{-4/(m-2)}$. See [6] and [9]. For the h version without a properly refined mesh, the situation is still worse. For the properly refined mesh, the rate of convergence in the first phase (when p is not large) is exponential. See [14]. The mesh we used is shown in Figure 5 ($a = 0.3$, $b = 1$, $\nu = 0.3$).

In Table 2 we show the stresses at points P, Q, R, S (see Figure 5) for $m = 8$, 16, 32 for various degrees p of elements. We see that the solution is close to the limiting value of $m = \infty$. Although $\bar{\sigma} = \sigma_{x\infty} + \sigma_{y\infty} = 0$, we see that $\bar{\sigma}$ deteriorates from $m = 16$ to $m = 32$ because the quality of the numerical solution deteriorates as $m \to \infty$. This deterioration is, for example, visible in Table 3 where the computed strain energy for various p and m is given. We see clearly a much larger change in the energy for $m = 32$ than for $m = 8$ when increasing the degree p, which indicates a much larger error for $m = 32$ than for $m = 8$. The strength of the singularity is $r^{2/(m-2)}$ which is so strong that without special care no reasonable accuracy can be achieved for $m = 32$. In our computations for $m = 16$ the error in the energy norm is expected to be 2–4%, and 7–10% for $m = 32$.

Table 4 shows the values of the maximal principal stress at points B and F, and $\bar{B}$ and $\bar{F}$ (Figure 5). We clearly see that the stresses are very large in the neighborhood of the vertices, and as $m \to \infty$ the stresses are increasing because

Table 2

Stresses in selected points for m = 8, 16, 32

POINT	r	p	$m = 8$			$m = 16$			$m = 32$			$m = \infty$
			σ_x	σ_y	τ_{xy}	σ_x	σ_y	τ_{xy}	σ_x	σ_y	τ_{xy}	σ_y
P	0.3	8	+0.9992	−1.000	+1.501 −5	+0.9953	−1.000	−0.579 −4	+0.9634	−1.0000	−0.297 −3	−1.000
		7	+0.9995	−0.9999	+0.110 −3	+0.9960	−0.9998	−0.213 −3	+0.9629	−0.9999	+0.520 −3	EXACT
		6	+0.9990	−0.9999	−0.491 −3	+0.9940	−1.000	−0.552 −3	+0.9600	−0.9999	−0.893 −3	
		5	+1.001	−0.9998	+0.140 −2	+0.9964	−0.9987	+0.134 −2	+0.9600	−0.9987	+0.163 −2	
Q	0.5	8	+0.3564	−0.3567	−0.576 −4	−0.3571	−0.3641	−0.634 −5	+0.3353	−0.3715	+0.108 −3	−0.3600
		7	+0.3569	−0.3566	−0.108 −4	+0.3571	−0.3612	+0.195 −4	+0.3344	−0.3719	+0.108 −4	EXACT
		6	+0.3566	−0.3562	+0.403 −3	+0.3556	−0.3617	+0.185 −3	+0.3317	−0.3734	+0.594 −4	
		5	+0.3548	−0.3568	−0.160 −2	+0.3554	−0.3631	−0.902 −3	+0.3328	−0.3735	+0.396 −4	
R	0.7	8	+0.1578	−0.1558	+0.527 −3	+0.1775	−0.1759	+0.119 −3	+0.1634	−0.1814	−0.483 −2	−0.1836
		7	+0.1598	−0.1643	−0.137 −3	+0.1814	−0.1958	+0.700 −3	+0.1581	−0.2216	+0.679 −2	EXACT
		6	+0.1588	−0.1484	−0.981 −3	+0.1765	−0.1670	−0.218 −2	+0.1672	−0.1705	−0.904 −2	
		5	+0.1550	−0.1756	+0.262 −2	+0.1761	−0.2083	+0.456 −2	+0.1472	−0.2353	+0.117 −1	
S	0.8	8	+0.0891	−0.0888	+0.107 −2	+0.1318	−0.1338	+0.868 −3	+0.1232	−0.1556	−0.128 −3	−0.1406
		7	+0.0884	−0.0879	−0.169 −2	+0.1278	−0.1302	−0.120 −2	+0.1162	−0.1518	−0.449 −3	EXACT
		6	+0.0846	−0.0853	+0.187 −2	+0.1214	−0.1345	+0.234 −2	+0.1017	−0.1655	+0.981 −3	
		5	+0.0934	−0.0893	−0.207 −2	+0.1446	−0.1411	−0.236 −2	+0.1447	−0.1608	−0.204 −2	

the strength of the singularity is increasing. This also indicates the likely reason for the paradox we mentioned above. Not only the mathematical model but also the quality of the numerical solution is very nonuniform with respect to small changes in the boundary (which does not have sufficient smoothness).

Table 3
Values of the strain energy for $4 \leq p \leq 8$ and $m = 8, 16, 32$

p	$m = 8$	$m = 16$	$m = 32$
8	0.229725	0.114518-1	0.559748-2
7	0.229723	0.114491-1	0.559199-2
6	0.229719	0.114456-1	0.558505-2
5	0.229705	0.114402-1	0.557552-2
4	0.229651	0.114298-1	0.556082-2

THE PROBLEM OF PLATES AND SHELLS

In the previous section we addressed the problem of the reliability of the linear elasticity model. Models of plates and shells are two-dimensional although obviously the original problem is three-dimensional. Hence, we will assume that the three-dimensional linear elasticity formulation is reliable and will analyze only the effects of the dimensional reduction from three to two dimensions and the implication for the optimal design.

The Problem of the Simply Supported Plate — For simplicity let us assume that we are concerned with the case *when the Poisson ratio* $\nu = 0$. The plate problem (*with uniform thickness* h) can be formulated in various ways. Let us mention the *projection method* when we assume an *Ansatz* (hypothesis) and use it in the variational principle by minimizing the energy. This approach is sometimes called the Kantorovich method [15]. Denoting u, v and w as the displacement components, we shall consider two *Ansätze* (hypotheses):

1) The K (Kirchhoff) model

$$u(x,y,z) = -z\frac{\partial w}{\partial x}(x,y), \tag{6a}$$

$$v(x,y,z) = -z\frac{\partial w}{\partial y}(x,y), \tag{6b}$$

$$w(x,y,z) = w(x,y). \tag{6c}$$

Using this *Ansatz* in the potential energy principle, we get the usual formulation

$$EI \,\Delta\Delta\, w = f. \tag{7}$$

Table 4
Values of principal stresses for m = 8, 16, 32

POINT	p	Maximal principal stress		
		$m = 8$	$m = 16$	$m = 32$
F	8	−0.3807+0	−0.3758+0	−0.3624+0
	7	−0.3814+0	−0.3767+0	−0.3670+0
	6	−0.3802+0	−0.3716+0	−0.3494+0
E	8	−0.1360+1	−0.1907+1	−0.2043+1
	7	−0.1358+1	−0.1916+1	−0.2076+1
	6	−0.1355+1	−0.1879+1	−0.1990+1
D	8	−0.4805+1	−0.9498+1	−0.1150+2
	7	−0.4810+1	−0.9502+1	−0.1153+2
	6	−0.4748+1	−0.9394+1	−0.1119+2
C	8	−0.1709+2	−0.4695+2	−0.6393+2
	7	−0.1709+2	−0.4683+2	−0.6383+2
	6	−0.1769+2	−0.4585+2	−0.6176+2
B	8	−0.5296+2	−0.1764+3	−0.2722+3
	7	−0.6709+2	−0.2762+3	−0.4340+3
	6	−0.4972+2	−0.1442+3	−0.1931+3
F	8	−0.3851+0	−0.4044+0	−0.5062+2
	7	−0.3855+0	−0.4048+0	−0.4998+0
	6	−0.3821+0	−0.3947+0	−0.4767+0
E	8	−0.1356+1	−0.2111+1	−0.3358+1
	7	−0.1358+1	−0.2135+1	−0.3427+1
	6	−0.1335+1	−0.2083+1	−0.3351+1
D	8	−0.4818+1	−0.1130+2	−0.2136+2
	7	−0.4825+1	−0.1147+2	−0.2184+2
	6	−0.4748+1	−0.1121+2	−0.2136+2
C	8	−0.1740+2	−0.6362+2	−0.1400+3
	7	−0.1747+2	−0.6459+2	−0.1435+3
	6	−0.1722+2	−0.6349+2	−0.1408+3
B	8	−0.6361+2	−0.3590+3	−0.9021+3
	7	−0.6481+2	−0.3824+3	−0.9606+3
	6	−0.6437+2	−0.3636+3	−0.8924+3

The simple support is obtained by minimization of the energy with the *only* constraint being $w = 0$ on the boundary Γ of the domain.

2) The R-M (Reissner-Mindlin) model

$$u(x,y,z) = -z\phi(x,y) \tag{8a}$$

$$v(x,y,z) = -z\psi(x,y) \tag{8b}$$

$$w(x,y,z) = w(x,y). \tag{8c}$$

Utilizing equation (8) in the expression for three-dimensional potential energy and imposing the *only* constraint $w = 0$ at Γ, we obtain a system of three differential equations of second order in contrast to one equation of fourth order in the K model.

The dimensional reduction has been analyzed asymptotically when $h \to 0$ and the solution is smooth. For example, see [11] [12] [20]. In this asymptotic frame we *cannot* distinguish between the K and the R-M models. Physically the R-M model takes into account the shear stresses while the K model neglects them. Let us once more assume that Ω_m is the regular m polygon inscribed in the circle of radius a and that Ω_0 is the circle with radius a, and let us consider the problem of a uniformly loaded (by load p) and simply supported plate.

Denote by w_m and w_0, and by $(\phi_m, \psi_m, \bar{w}_m)$ and $(\phi_0, \psi_0, \bar{w}_0)$, respectively, the solutions of the K and R-M models on Ω_m and Ω_0. Then we have

THEOREM 4. $$w_m \to w_\infty \neq w_o \tag{9a}$$

$$(\phi_m, \psi_m, \bar{w}_m) \to (\phi_\infty, \psi_\infty, \bar{w}_\infty) = (\phi_0, \psi_0, \bar{w}_0). \tag{9b}$$

See [3] [5]. We can compute the limiting solution w_∞ and $(\phi_\infty, \psi_\infty, \bar{w}_\infty)$ analytically. For the K model we have

$$w_\infty(0,0) = -\frac{3}{64}\, a^4\, \frac{p}{EI} \tag{10a}$$

$$w_0(0,0) = -\frac{5}{64}\, a^4\, \frac{p}{EI} \tag{10b}$$

and for the R-M model we have

$$w_\infty(0,0) = w_0(0,0) = -\frac{5}{64}\, a^4\, \frac{p}{EI} - \frac{pa^2}{EF} \tag{10c}$$

where $I = h^3/12$ and $F = h$ are the moment of inertia and the thickness, respectively, and E is the modulus of elasticity.

Theorem 4 and equations (10a)–(10c) show that the effects of the shear stress in the neighborhood of the corners are essential. Although we discussed only the problem of the polygon plate, the analysis (see [3] and [5]) covers much more general solutions and clearly points to the following conclusion: *the optimal design of a plate has to be based on the R-M and not on the K model.* We will not discuss here the reliability of the numerical treatment. A similar but more complicated situation occurs in the case of the shells.

The Problem of the Plate with a Variable Thickness — Let us consider a plate with variable thickness. If the thickness is very slowly varying with respect to the average thickness of the plate, then the derivation (dimensional reduction) can be made in the same way as for the constant thickness. However, if the thickness is varying rapidly, then the classical derivation is not valid. Recently a theory has been developed (see [16] and [17]) that shows an important relation between the thickness and thickness variation which strongly influences the reliability of the

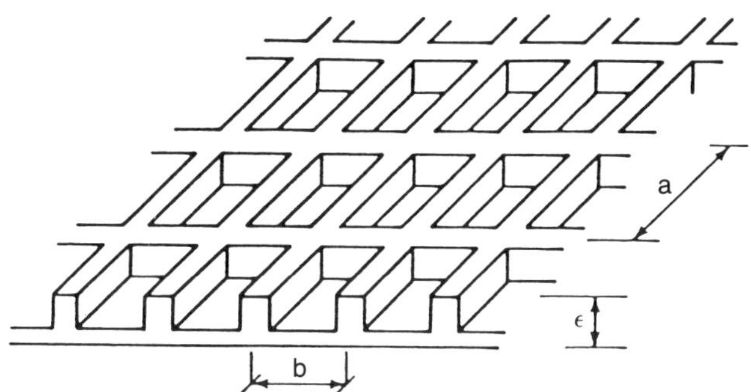

Figure 6. Scheme of a stiffened plate.

mathematical model. We will show it in the most simple setting. Consider the stiffened plate shown in Figure 6. The main idea of the classical plate derivation is to consider the limiting process $\varepsilon \to 0$ and apply the results for $\varepsilon > 0$.

We can assume that $a = C_1 \varepsilon^{\lambda_1}$, $b = C_2 \varepsilon^{\lambda_2}$ and consider the limiting process $\varepsilon \to 0$. In [17], $\lambda_1 = \lambda_2 = \lambda$ is assumed and it is shown that we get a different model for $\lambda < 1$, $\lambda = 1$ and for $\lambda > 1$.

In the case of $\lambda < 1$ the stiffeners are far apart when $\varepsilon \to 0$; in the case $\lambda > 1$ they are close together. In all three cases the dimensional reduction leads to the plate formulation with effective coefficients depending on the value of λ. This example shows that optimal design based on one model, say, $\lambda < 1$ for fixed but small thickness, can lead to a design in which the model is no longer valid and reliable. Using a proper model for this design and redoing the optimal design once more, we can again exceed the reliability range of the model. Hence we have to consider here *the simultaneous design optimization and the model selection.* For important aspects of this problem directly related to the optimal design, we refer to [17].

THE PROBLEM OF A SUPPORTED STRUCTURE

Let us consider the optimal design of a supported structure (see Figure 7). The problem is how to model the support at point B. To show the difficulty, let us consider the problem shown in Figure 8 and solve the linear elasticity (plane stress $\nu = 0.3$) problem. The standard procedure in finite element modeling is to make the constraint $v = 0$ at the node located at the support. This modeling is *incorrect* because the solution strongly depends on the finite element mesh.

Let M be the moment at side AA' and M_N the moment computed by the finite element method. Assume that the size h of the maximal element $h_N = \max h \to 0$ as $N \to \infty$. Then we have

THEOREM 5. $\lim_{N \to \infty} M_N = M_0$ (11)

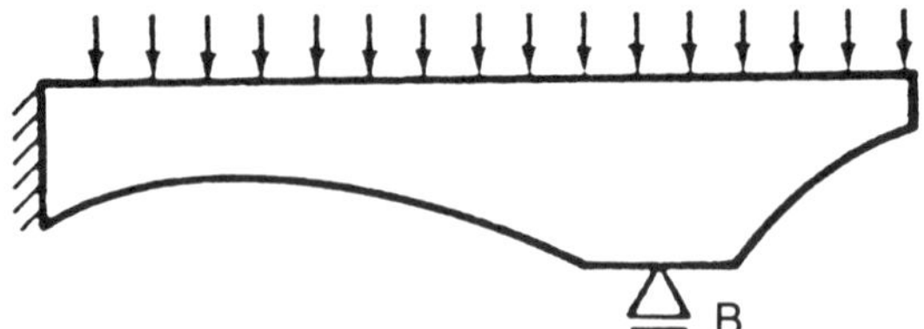

Figure 7. Scheme of a supported structure.

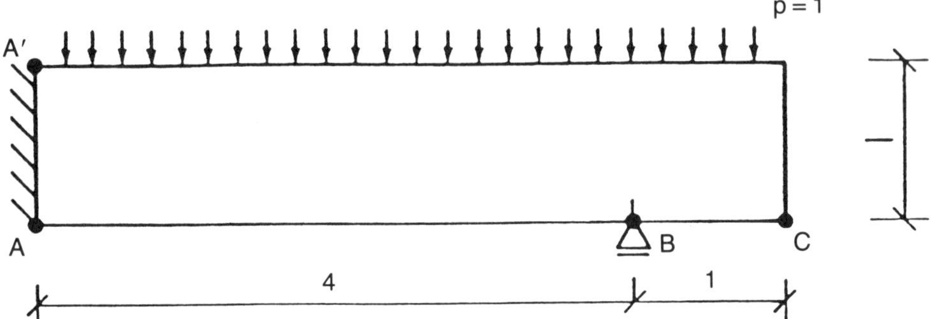

Figure 8. Scheme of the supported beam under consideration.

where M_0 *is the moment when there is no support (and hence* M_0 *can be analytically computed).*

Theorem 5 shows that by selecting different meshes we can get completely different results; thus optimal design will strongly depend on the mesh used. In fact, the situation is still more complicated because $M_N \to M_0$ slowly, and we have no means to establish how reliable the solution is.

Before discussing this effect, let us show the computation by the code PROBE. The mesh used is shown in Figure 9. There is refinement in the neighborhood of AA' and especially strong refinements in the neighborhood of B (see Figure 10). We used two meshes — A_4 with smallest ring of radius a_4, and A_5 with radius a_5. Table 5 shows the moments on AA' and Table 6 shows the displacement v at point C. Although the moments and the displacement are significantly smaller than that of the unsupported beam, the mesh dependence is obvious. Note that the difference between the values obtained by the mesh A_4 and A_5 is nearly independent of p. The reason for the effects we have shown is that the support is not correctly modeled. The reaction is a *point force* which leads to infinite energy and an *infinite displacement* at the point of the reaction. The infinite displacement at the reaction point can be seen from the analytical solution on the half plane with a concentrated load; hence the reaction has to be zero and we obtain the solution of an unsupported beam.

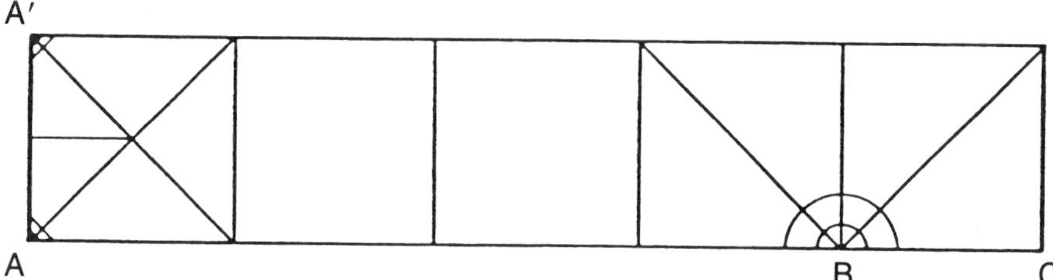

Figure 9. Mesh used for the *p* version analysis.

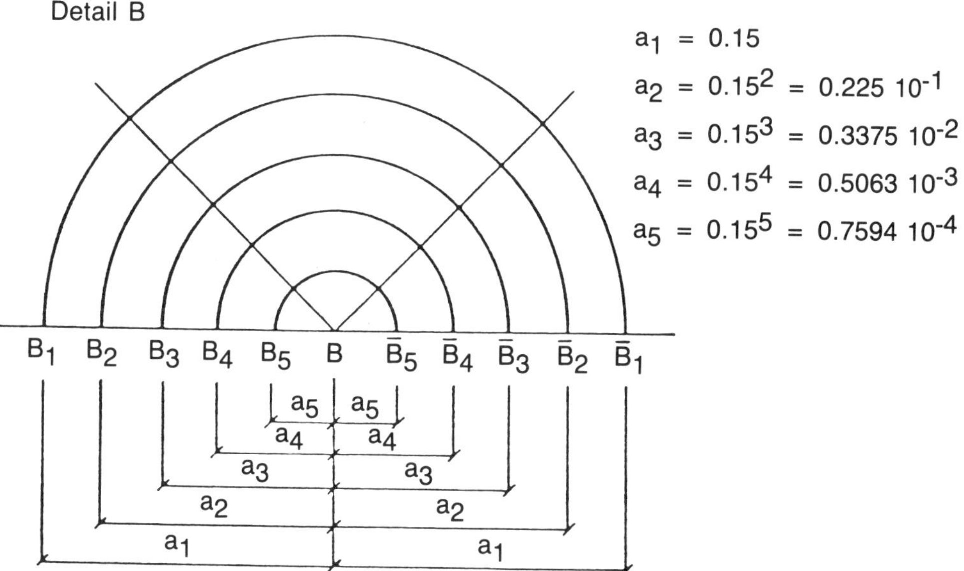

Figure 10. Detail of the mesh in the neighborhood of the support.

Table 5
The values of the bending moment in $A\bar{A}$ with dependence on the mesh selection

p	Mesh A_4	Mesh A_5
3	1.875	1.918
4	1.909	1.954
5	1.921	1.965
6	1.931	1.976
7	1.939	1.984
8	1.946	1.991

Table 6
The values of the displacement in C with dependence on the mesh selection

p	Mesh A_4	Mesh A_5
1	− 7.92	−10.68
2	−10.94	−14.91
3	−13.14	−17.31
4	−14.56	−18.77
5	−15.64	−19.85
6	−16.37	−20.58
7	−17.38	−21.21
8	−17.56	−21.76

In Table 7 we show the displacements at points B_i and $\bar{B}_i$ computed by meshes A_4 and A_5. Realizing that the distance between B and B_5 is $0.75\ 10^{-4}$ and the constraint at B is $v = 0$, the numerical results show the effect mentioned above. This clearly indicates that the mathematical model of the supported beam is unreliable because it does not distinguish between a supported and an unsupported beam. Hence, *a more sophisticated model of the support is needed.* Nevertheless, we will not discuss here the question of a reliable model. [Usually it is claimed that a concrete but not strongly refined mesh models the support. It is obvious that without a reference to the proper formulation of the support, the claim has no firm meaning.]

THE PROBLEM OF STOCHASTIC INPUT DATA

The basic input data describing the elasticity problems are the domain, the material properties and the loads. Assume now that the data are stochastic functions. For example, the boundary of the domain can be described by a stochastic function which expresses the uncertainty of fabrication. Then the solution is also a stochastic function. In addition, the failure criterion which can be the basis for the optimal design is always stochastic. Hence, we have combined both stochastic characters to get the desired information. Because of the uncertainty of the input data, the dispersion of the results can be significant. Recently we developed a theory of the solution with stochastic input data (see [2] and [19] and forthcoming papers) and its numerical treatment by the finite element method. The implementation is based on the PROBE code mentioned earlier.

The Case of the Stochastic Load — Let us consider a container of the form shown in Figure 11. Side AB is loaded by a horizontal stochastic function

Table 7
The values of the displacement in the neighborhood of the support

POINT	p	Mesh A_4	Mesh A_5	POINT	Mesh A_4	Mesh A_5
B_5	8		− 9.189	$\bar{B}_5$		− 9.188
	7		− 8.777			− 8.777
	6		− 8.313			− 8.312
	5		− 7.764			− 7.763
B_4	8	− 9.231	−12.36	$\bar{B}_4$	− 9.226	−12.36
	7	− 8.818	−11.95		− 9.831	−11.95
	6	− 8.352	−11.49		− 7.795	−11.48
	5	− 7.801	−10.94		− 7.123	−10.94
B_3	8	−12.43	−15.54	$\bar{B}_3$	−12.39	−15.51
	7	−12.02	−15.14		−11.98	−15.11
	6	−11.55	−14.67		−11.57	−14.64
	5	−11.00	−14.13		−10.97	−14.10
B_2	8	−15.72	−18.80	$\bar{B}_2$	−15.48	−18.61
	7	−15.31	−18.40		−15.07	−18.20
	6	−14.85	−17.94		−14.05	−17.73
	5	−14.31	−17.40		−13.37	−17.19
B_1	8	−19.60	−22.52	$\bar{B}_1$	−18.01	−21.27
	7	−19.22	−22.14		−17.58	−20.84
	6	−18.78	−21.17		−16.52	−20.36
	5	−18.27	−21.20		−15.80	−19.79

$X_x = \Lambda(y,\omega)$, $0 < y < H$ and we will assume that $\overline{X_x(y,\omega)} = 1$ where we denoted the mean by $\overline{X_x(y,\omega)}$. The correlation function is assumed to be

$$K(y_1, y_2) = 0.1^2\, e^{-a|y_1 - y_2|}. \tag{12}$$

A simulated sample of the load from a given probability field ($\alpha = 0.03781$) is shown in Figure 12. We will assume that the linear elasticity provides reliable results for all loads under consideration. The solution of our model problem is a stochastic function with the mean being the deterministic solution for the mean load.

References pp. 192-193

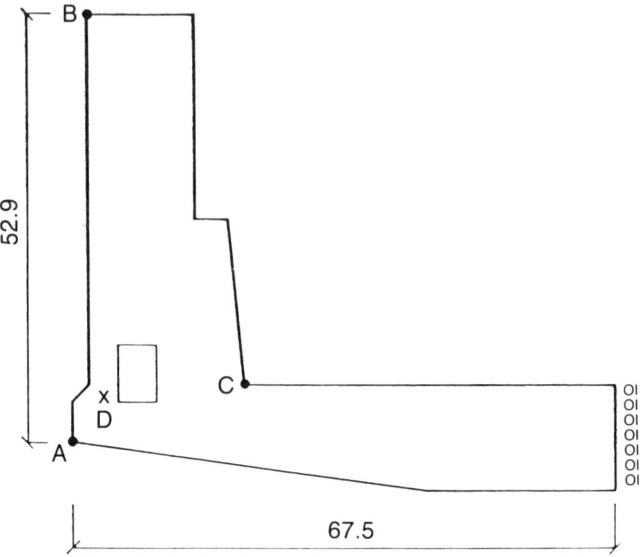

Figure 11. Scheme of a container.

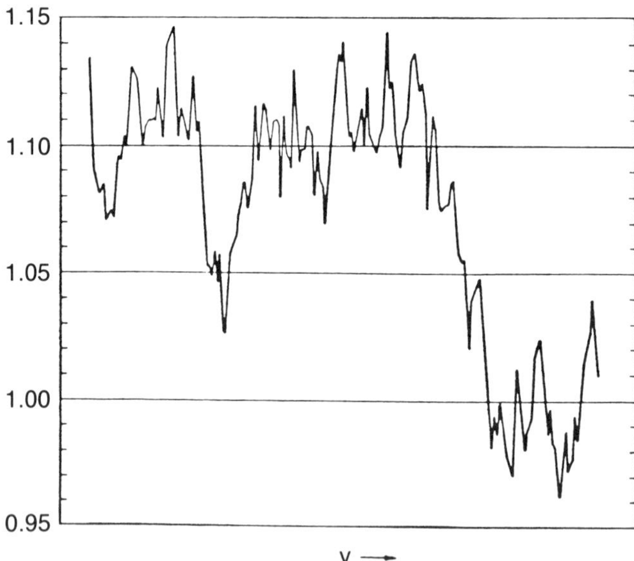

Figure 12. Simulated sample of the stochastic load.

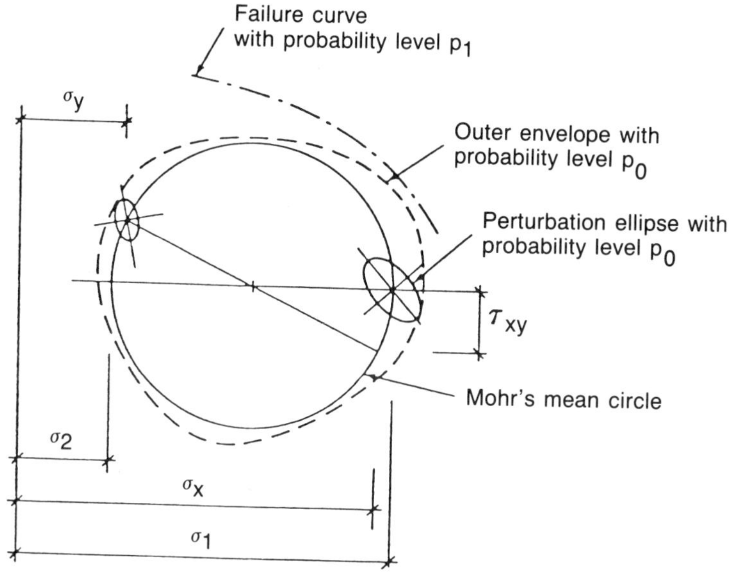

Figure 13. Mohr's circle and its stochastic form.

Our aim is to determine the variance and covariance of the values of interest. Concerning the failure criterion we will consider: a) the criterion of stress intensity factor F at point C (Figure 11), and b) the failure criterion based on the envelope of the Mohr circles at point D (Figure 11).

Knowing the stress intensity factor as a random variable characterized by its mean and standard deviation, we can establish the probability level of failure when the material probabilistic characterization of admissible stress intensity factor is given.

Criterion b is more complicated. Here we need the correlation of the components of the stress tensor which allows us to compute not only the mean of the Mohr circles but also its perturbation at every point of the circle which for a given probability level has an elliptic character. The envelope of these ellipses is compared with the admissible failure curve in Figure 13.

The concrete computation of our model problem used the PROBE program and the mesh shown in Figure 14. The refinement in the neighborhood of the reentrant corners is not shown. In Table 8 we show the mean values and the standard deviation $\mathrm{Sd}(F)$ of the stress intensity factor F at point C with dependence on p. For the technique used in PROBE for the computation of the stress intensity factor, see reference [8].

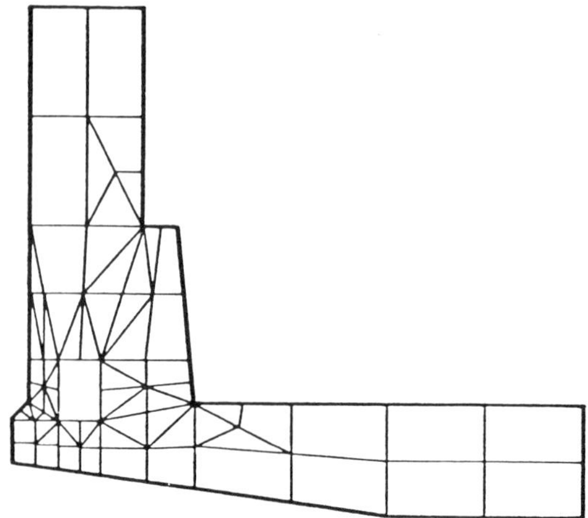

Figure 14. Mesh used for the *p* version analysis.

Table 8
The mean value and standard deviation of the stress intensity factor.

p	F	Sd(F)
1	−46.5958	3.71714
2	−51.7433	3.92931
3	−49.3796	3.94039
4	−49.0721	3.91575

The Mohr circles for the 90% probability level are shown in Figure 15.

The Problem of Stochastic Boundary — The problem of a stochastic boundary is more complicated but it can be transformed to the case of stochastic load. For simplicity of exposition, let us consider the problem of a symmetric, cracked panel (plane strain, $\nu = 0.3$) shown in Figure 16, and assume that the deterministic traction T at the boundary is such that the exact stress tensor is given by the following:

$$\begin{aligned} \sigma_x &= (2\pi r)^{-1/2} \cos\frac{\theta}{2}(1 - \sin\frac{\theta}{2}\sin\frac{3\theta}{2}) \\ \sigma_y &= (2\pi r)^{-1/2} \cos\frac{\theta}{2}(1 + \sin\frac{\theta}{2}\sin\frac{3\theta}{2}) \\ \tau_{xy} &= (2\pi r)^{-1/2} \sin\frac{\theta}{2}\cos\frac{\theta}{2}\cos\frac{3\theta}{2}. \end{aligned} \tag{13}$$

These functions are symmetric mode functions of the stress intensity factor. Let us assume that side A has a stochastic perturbation so that the boundary is given

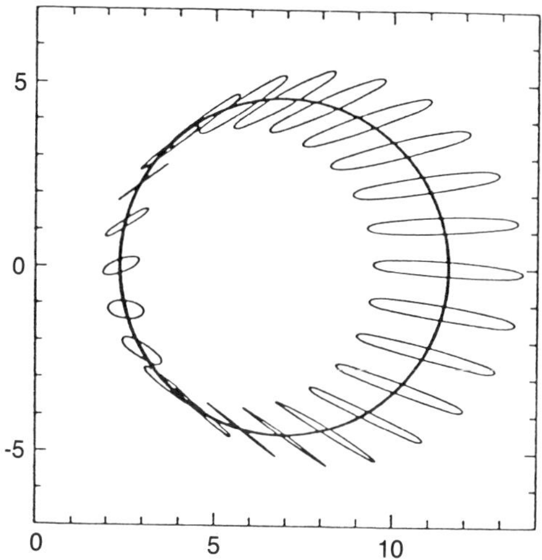

Figure 15. Stochastic Mohr circle for 90% probability level.

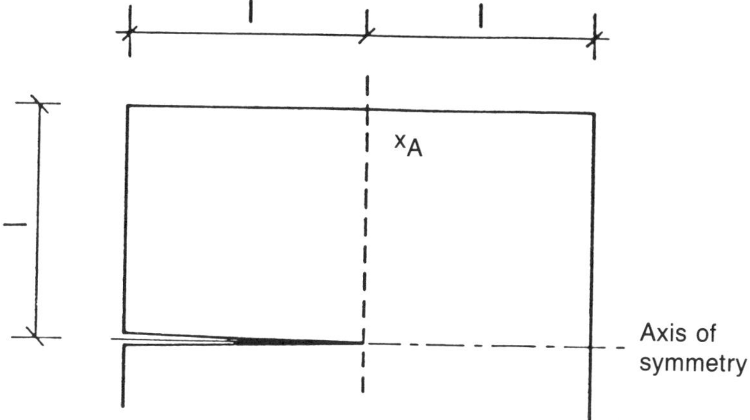

Figure 16. Scheme of a cracked panel with stochastic boundary.

by the function $y = 1 + \Lambda(x,\omega)$, $-1 < x < 1$, $\Lambda(\pm 1,\omega) = 0$, where $\Lambda(x,\omega)$ is the stochastic function with the correlation function $K_1(x_1, x_2)$.

We use in our model problem

$$K_1(x_1,x_2) = f(\frac{|x_1 - x_2|}{2}) - f(\frac{|x_1 + x_2|}{2} + 1), \qquad |x_1| \leq 1, \quad i = 1,2, \qquad (14)$$

where

$$f(\xi) = \frac{8 - 15\xi^2(2-\xi)^2}{720}.$$

A simulated sample of the perturbance is shown in Figure 17. Our aim is to find the stress intensity factor F and its standard deviation $\mathrm{Sd}(F)$ caused by the random

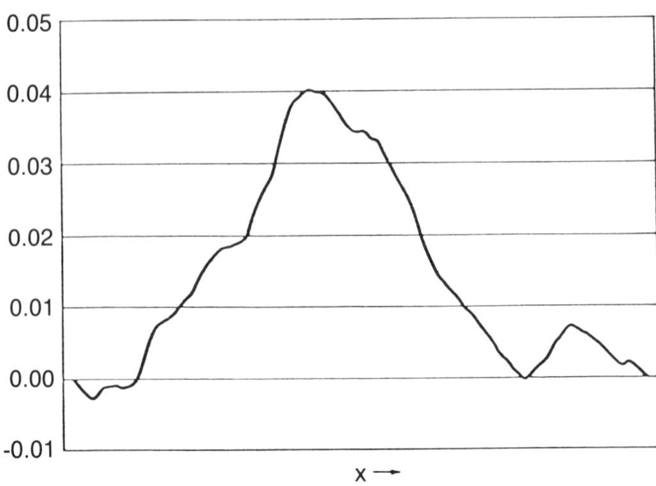

Figure 17. Simulated sample of the stochastic boundary.

boundary, and to find the stresses and their variances and covariances at point (0.1, 0.9).

Before addressing this problem, we have to know how the traction will change when the domain is changing so that the equilibrium is always guaranteed. To this end we will assume that functions $\tilde{\sigma}_x(x,y)$, $\tilde{\tau}_{x,y}(x,y)$ and $\tilde{\sigma}_y(x,y)$ are defined in the neighborhood of side AB such that

$$\frac{\partial \tilde{\sigma}_x}{\partial x} + \frac{\partial \tilde{\tau}_{xy}}{\partial y} = 0$$
$$\frac{\partial \tilde{\tau}_{xy}}{\partial x} + \frac{\partial \tilde{\sigma}_y}{\partial y} = 0 \tag{15}$$

and $\tilde{\sigma}_y = \sigma_y$ and $\tilde{\tau}_{xy} = \tau_{xy}$ on AB where (τ_{xy}, σ_y) is the given traction vector at AB. If point D now lies on the perturbed boundary, then the traction vector $\mathbf{T}$ is

$$\mathbf{T} = \begin{bmatrix} \tilde{\sigma}_x, & \tilde{\tau}_{xy} \\ \tilde{\tau}_{xy}, & \tilde{\sigma}_y \end{bmatrix} \begin{bmatrix} n_1 \\ n_2 \end{bmatrix} \tag{16}$$

where (n_1, n_2) is the outer normal to the perturbed boundary. This guarantees the equilibrium condition for every perturbation.

Assume that the magnitude of the perturbance is λ. Then we have

THEOREM 6. *The solution of (up to higher order terms in λ) the perturbed problem is the solution of the original domain with the modified load* $\mathbf{T}$

$$\mathbf{T} = \mathbf{T}_0 - \left[\Lambda(x,\omega)\frac{\partial}{\partial y}\begin{bmatrix} -\tilde{\tau}_{xy} + \tau_{xy} \\ -\tilde{\sigma}_y + \sigma_y \end{bmatrix} + \Lambda'(x,\omega)\begin{bmatrix} \tilde{\sigma}_x + \sigma_x \\ -\tilde{\tau}_{xy} + \tau_{xy} \end{bmatrix}\right] \tag{17}$$

where $\mathbf{T}_0 = (\tau_{xy}, \sigma_y)^{\mathrm{T}}$ *is the traction on* AB.

Theorem 6 immediately gives us the possibility to solve the problem in the same vein as in the previous section.

We have used in our model problem

$$\begin{aligned}\tilde{\sigma}_x(x,y) &= \sigma_x(0,1)\\ \tilde{\tau}_{xy}(x,y) &= \tau_{xy}(x,1)\\ \tilde{\sigma}_y(x,y) &= \sigma_y(x,1) - (y-1)\frac{\partial \tau_{xy}}{\partial x}(x,1).\end{aligned}$$

Using the PROBE program for $p = 8$, we obtained:

a) The stress intensity factor F with mean value $F = 0.99830$ (exact value $F = 1$) and standard deviation $\mathrm{Sd}(F) = 2.54(-4)$.

b) The stress at point $A = (0.1, 0.9)$ with mean value $\bar{\sigma}_x = 0.1426$, $\bar{\sigma}_y = 0.4821$, $\bar{\tau}_{xy} = 0.1206$; standard deviation $\mathrm{Sd}(\sigma_x) = 0.48418(-2)$, $\mathrm{Sd}(\sigma_y) = 0.35178(-2)$, $\mathrm{Sd}(\tau_{xy}) = 0.2088(-2)$; covariance $c(\sigma_x, \sigma_y) = 0.1697(-4)$, $c(\sigma_x, \tau_{xy}) = 0.9493(-5)$, $c(\sigma_y, \tau_{xy}) = 0.7019(-5)$; and normalized covariance $\rho(\sigma_x, \sigma_y) = 0.9966$, $\rho(\sigma_x, \tau_{xy}) = 0.9389$, $\rho(\sigma_y, \tau_{xy}) = 0.9555$.

We see that the variance of F is much smaller than the variance of the stress at point A. If the failure criterion is based on the stress intensity factor F, then in our case it is practically uninfluenced by the uncertainty of the boundary. If the failure criterion is based on the Mohr circle in A, then it is much more sensitive to the uncertainty of the boundary. This shows very clearly that the same uncertainties can lead to the uncertainties of different magnitude in the failure criterion parameters.

Let us mention that in equation (17) we need the derivatives of the stresses of the deterministic solution which is computed by the finite element method. This, of course, requires special care and the computation can be made by the postprocessing technique (see [8]).

The selection of functions $\tilde{\sigma}_x, \tilde{\sigma}_y, \tilde{\tau}_{xy}$ does not usually cause any problems. Many times we have a traction-free surface and then, of course, $\tilde{\sigma}_x = \tilde{\sigma}_y = \tilde{\tau}_{xy} = 0$ is the proper choice. We have assumed that the tractions are not stochastic. We can also treat the combined case when both the domain and the traction are stochastic.

Here we have shown only illustrative examples of relatively simple structures. The theory and implementation principles were developed for the general case. It is also possible to compute higher correlation functions and obtain, e.g., the skewness of the distribution of the stress intensity factors, etc. In the case of stochastic material coefficients, we can proceed similarly and by an iterative technique obtain the desired data for small variations of the material coefficients. In general the optimal design should take into account the stochastic character of the input data.

References pp. 192-193

CONCLUSIONS

Solving the problems of optimal design and engineering problems in general, one has to take into account various aspects of the mathematical model and its numerical treatment for getting reliable results. Detailed a priori mathematical analysis is of utmost importance to obtain reliable conclusions.

ACKNOWLEDGEMENT

This work was partially supported by the Office of Naval Research under Contract N00014-85-K-0169.

REFERENCES

1. I. Babuška, The continuity of the solutions of elasticity problems on small deformation of the region. *ZAMM* **39**, 411–412 (1959) (in German).
2. I. Babuška, On randomized solution of Laplace's equation. *Casopis Pest. Mat.* **86**, 269–276 (1961).
3. I. Babuška, The stability of the domain of definition with respect to basic problems of the theory of partial differential equations especially with respect to the theory of elasticity I, II. *Czechoslovak Math. J.* **11**, 76–105, 165–203 (1961) (in Russian).
4. I. Babuška, The theory of small change in the domain of definition in the theory of partial differential equations and its applications. *Proc. Conf. EQUADIFF*, pp. 13–26. Prague (1962).
5. I. Babuška, The stability of domains and the questions of the formulation of plate problems. *Apl. Mat.* **7**, 463–467 (1962) (in German).
6. I. Babuška, B. A. Szabo and I. N. Katz, The p-version of the finite element method. *SIAM J. Numer. Anal.* **18**, 515–545 (1981).
7. I. Babuška and B. A. Szabo, On the rates of convergence of the finite element method. *Int. J. Numer. Meth. Eng.* **18**, 323–341 (1982).
8. I. Babuška and A. Miller, The post-processing approach in the finite element method, Part I, II, III. *Int. J. Numer. Meth. Eng.* **20**, 1085–1109, 1111–1129, 2311–2325 (1984).
9. I. Babuška and M. Suri, The optimal convergence rate of the p-version of the finite element method. Tech. Note BN 1045, Institute for Physical Science and Technology, University of Maryland (Oct. 1985).
10. N. V. Banichuk, *Problems and Methods of Optimal Structural Design.* Plenum Press, New York (1983).
11. P. G. Ciarlet and P. Destuynder, A justification of the two dimensional linear plate model. *J. Mécanique* **18**, 315–344 (1979).
12. P. G. Ciarlet and P. Rabier, *Les Equations de von Karmán.* Lecture Notes in Mathematics No. 826. Springer-Verlag, Berlin (1980).
13. R. H. Gallagher, E. Atrek, K. Ragsdell and D. C. Zienkiewicz (Eds.). *Optimum Structural Design.* John Wiley & Sons, New York (1983).
14. B. Guo and I. Babuška, The h-p version of the finite element method—I: The basic approximation results; II: General results and applications. *Computat. Mech.* (1986), to appear.

15. L. V. Kantorovich and V. I. Krylov, *Approximate Methods of Higher Analysis.* Noordhoff Gronigan, The Netherlands (1958).

16. R. V. Kohn and M. Vogelius, A new model for thin plates with rapidly varying thickness. *Int. J. Solids Struct.* **20**, 333–350 (1984).

17. R. V. Kohn and M. Vogelius, Thin plates with rapidly varying thickness and their relation to structural optimization. Institute for Mathematics and its Applications, University of Minnesota, IMA Preprint Ser. 155 (June 1985).

18. R. V. Kohn and G. Strang, Optimal design and relaxation of variational problems *Comm. Pure Appl. Math.*, to appear.

19. S. Larsen, Numerical analysis of elliptic partial differential equations with stochastic input data. Ph.D. Dissertation, University of Maryland (1985).

20. D. Morgenstern, Herleitung der Plattentheorie aus der dreidimensionalen Elasticitätstheorie. *Arch. Ration. Mech. Anal.* **4**, 145–152 (1959).

21. D. Pironneau, *Optimal Shape Design for Elliptic Systems.* Springer-Verlag, New York (1984).

22. B. A. Szabo, Mesh design for the p-version of the finite element method. Center for Computational Mechanics, Washington University, Rep. No. WU/CCM-85/2. St. Louis, MO (1985).

23. B. A. Szabo, Implementation of a finite element software system with h- and p-extension capabilities. *Proc. 8th Invitational UFEM Symposium: Finite Element Software Systems* (Edited by H. Kardestuncer). University of Connecticut (1985).

24. B. A. Szabo, *PROBE: Theoretical Manual.* Noetic Technologies Corporation, St. Louis, MO (1985).

DISCUSSION

R. Haftka *(Virginia Polytechnic Institute and State University)*

The example you used sounds familiar because we always seem to be optimizing and getting an answer which is not valid, even when we just did an analysis that seems to indicate it is OK. Several years ago we ran into a similar problem with the plate stiffening where people optimized plate thicknesses and then got results which did not seem to be optimum. Similar problems occurred with bimodal solutions of a design for buckling. So your point is well taken. Of course we can construct examples of this happening, but we mainly need to be able to recognize them when they actually occur. How do you recognize these problems?

Babuška

The computed results have to indicate that something is wrong. This indication can be accomplished by incorporating a set of error indicators and estimators into the program. In the case of the point support that I mentioned, an error estimator indicates that the error of the finite element solution is still also large if the mesh is (for example adaptively) very refined. The problem of the bonded tube has a different origin. It is caused by the fact that the assumptions of linear elasticity are violated in the corner of the tube. Nevertheless, here also an error indicator could warn that the effects of nonlinear behavior cannot be neglected.

In my opinion, any program should have a set of various error indicators which will give a necessary warning, and I believe that this can be done. In the case of the error of the finite element solution, as in the case of the point support, the estimator built into our program clearly indicates the problem. Nevertheless, a lot of work needs to be done.

I. Babuška

V. B. Venkayya *(Wright-Patterson AFB)*

You have a physical model and a mathematical model. You made some assumptions and transformed the physical model into a mathematical model. Something could go wrong in this model or even in a simpler model to thoroughly botch it up. This happens in a number of cases even when there is nothing wrong with optimization. For that matter, even if we take a stochastic model, we make many assumptions about distributions and confidence levels. In this case we should be aware of the physical model in the mathematical model. Does the stochastic model answer the problem?

Babuška

In mathematics one must deal with a mathematical model. One cannot solve a physical problem. The question now becomes how to decide whether the model used is applicable in a concrete situation; certainly it is possible to optimize in a mathematical framework. But this does not mean that the solution is applicable in a given and concrete engineering setting. For example, in the case of a bonded tube, the assumption has been made that linear elasticity is applicable, i.e., that stress and strain are small. Nevertheless, this assumption is violated in the corners.

(Is this a deadly sin? I have shown that this is the case.) One has to try to answer mathematically the question about the effects of the violation of mathematical assumptions. In any case, one has to be aware of all assumptions which have been made. Of course, this also applies to the stochastic model, which is also based on a set of assumptions.

L. A. Schmit *(University of California–Los Angeles)*

I'd like to try to tie some questions together. Everything might be related to the question of design. We need to be able to do analysis if we are going to automate design, but analysis is also necessary if we only try and test and do not automate the design. Your examples remind me that we must be alert to problems that involve sharp corners, cracks, load introduction, introduction of discrete stiffening in members and so forth. To handle those problems we may need many levels of analysis. Do you see any potential to resolve at least some of these difficulties using global-local types of finite element methods?

Babuška

Error indicators and estimators are tools that can lead to the solution you mentioned. We have already developed some error estimators that indicate whether the finite element solution is acceptable. Nevertheless, we are far from the point where we could be fully satisfied.

E. Haug *(University of Iowa)*

You seem to have three criteria, which I don't recall exactly, but they may boil down to completeness or compactness of the design space. Let's suppose I wanted to formulate my shape optimal design problem with a finite number of design variables. If I did this in such a way that I have a uniform bound on curvature of the domain, which assures adequate smoothness of the boundary, it seems to me that I have a closed and bounded set in design space and I can even apply optimality criteria with some confidence. Do you see any potential for defining reasonably broad classes of criteria, such as those I mentioned, that would help us in reducing problems to reliably solvable form?

Babuška

Restricting the class of problems under consideration will resolve some difficulties, but it could create others. For example, consider the plate thickness optimization which I mentioned. Restriction of the admissible variation of the thickness will guarantee uniform validity of the model, but the optimal solution will strongly depend on this (artificial) class and could be from a solution in another class, for example, plates with stiffeners. So the major problem is to have a model that is uniformly valid over a sufficiently large class. For example, in this context the

Reissner-Mindlin model has to be preferred to the Kirchhoff model, which I elaborated on in my talk. Of course, to resolve these problems within reasonable cost constraints is not easy.

Moderator—B. Szabo

Professor Haug, did your question refer to the fact that the polygon itself with the stiffener or stiffening tape formed a closed set, or did it refer to a general set?

Haug

I asked about a closed and bounded set in the design space. It has always seemed to me that once we realize these pitfalls can arise, we can guard against loss of existence. In some sense, God gave us existence of solutions of strongly elliptic boundary value problems, but God never gave us existence of solutions of optimal design problems.

Szabo

He reserved that for himself.

Babuška

I would say that different admissible sets constitute different problems.

R. Kohn *(Courant Institute of Mathematics)*

To the last observation I can't resist adding that by placing a constraint on the set and the possible designs (such as the curvature of the domain boundary) we do get existence of an optimal design. But you have to worry about whether this possibly artificial constraint is active at the optimum.

Szabo

Professor Babuška, could you explain how much computational labor is involved in stochastic analysis? Most of us are familiar with the Monte Carlo methods, but that is not what you advocate. Comparing the computational effort to a nonstochastic analysis, how much effort would you spend using the methods you described?

Babuška

I did not refer to the Monto Carlo method. Use of this method is out of the question in this context because of its ineffectivity. We also did some numerical experimentation with this method to confirm our theoretical conclusions about the Monte Carlo method. Our approach, which I referred to, directly computes the correlation function which satisfies some partial differential equations. I would estimate that in many practical cases the cost of computation is on the order of dealing with, say, 30 load cases.

Szabo

Instead of one, there are 30 back substitutions?

Babuška

Yes, 20 to 50 back substitutions and computations of stresses, postprocessing, etc. for these 20 to 50 load cases.

BOUNDARY ELEMENTS IN SHAPE OPTIMAL DESIGN OF STRUCTURES

C. A. MOTA SOARES

Center for Mechanics and Materials
Technical University of Lisbon
Lisbon, Portugal

K. K. CHOI

Center for Computer Aided Design
The University of Iowa
Iowa City, Iowa

Abstract

The shape optimal design of shafts and two-dimensional elastic structural components is formulated using boundary elements. The design objective is to maximize torsional rigidity of the shaft or to minimize compliance of the structure, subject to an area constraint. Also a model based on minimum area and stress constraints is developed, in which the real and adjoint structures are identical but have different loading conditions. All degrees of freedom of the models are at the boundary, and there is no need for calculating displacements and stresses in the domain. Formulations based on constant, linear and quadratic boundary elements are developed. A method for accurately calculating the stresses at the boundary is presented, which improves considerably the design sensitivity information. A technique for an automatic mesh refinement of the boundary element models is also developed. The corresponding nonlinear programming problems are solved by Pshenichny's linearization method. The models are applied to shape optimal design of several shafts and elastic structural components. The advantages and disadvantages of the boundary element method over the finite element techniques for shape optimal design structures are discussed with reference to applications. A literature survey of the development of the boundary element method for shape optimal design is presented.

INTRODUCTION

The finite element method has been extensively used in structural optimization during the last decade, including successful application to shape optimal design of shafts and elastic structural components. In contrast, the boundary element method has only recently been applied to shape optimal design of structures.

Application of the finite element method for shape optimal design of structural components has been successfully demonstrated, but it has some disadvantages. It is often necessary to redefine new finite element meshes as the geometry of the structure changes. Inaccurate evaluations of stresses at the boundary can be responsible for the calculation of very inaccurate design sensitivity analysis, thus leading to a large number of optimization iterations or even unrealistic designs.

These difficulties with the finite element formulation can be partially overcome by using the boundary element method to discretize the structure. Results of the boundary element analysis of elasticity problems are more accurate than the corresponding solutions of the finite element models, and they are expected to yield improved design sensitivity information. Consequently, a smaller number of iterations is needed to find the optimum shape. In the last years about 20 papers have been published in the development of the boundary element method for shape optimal design of engineering systems. A literature survey and a review of the state of the art is presented.

The boundary element method is less versatile for structural analysis than the finite element technique. Its applicability to shape optimal design of structures is at present limited to elasticity problems, subject to static constraints. However, with continuing development of the boundary element method, the range of shape optimal design problems that can be efficiently solved is expected to increase in the near future.

The shape optimal design of shafts and two-dimensional elastic structural components is formulated using boundary elements. The optimal design objective is to maximize torsional rigidity of the shaft or to minimize compliance of the structural component, subject to a fixed amount of material. Also, a model based on minimum area and stress constraints is developed, in which the real and adjoint structures are identical but have different loading conditions. All the degrees of freedom of the boundary element models are at the boundary of the structural system and there is no need for internal cells. Displacements and stresses are only calculated at the boundary. The boundary element models are based on constant, linear and quadratic elements.

A method for accurately calculating the stresses at the boundary is presented which considerably improves the design sensitivity information. We develop a technique for automatic refinement of the boundary discretization based on the continuity of the tangential boundary stresses for unloaded smooth surfaces.

The shape optimization nonlinear programming problem is solved by Pshenichny's linearization method. The models are applied to the shape optimal design of several shafts and elastic structural components. The advantages and disadvantages of the boundary element method compared to the finite element technique for shape optimal design are discussed with reference to applications.

LITERATURE SURVEY

An extensive body of literature has been published on numerical methods for optimization of structures whose shapes are defined by cross-section and thickness variables. Only limited literature has appeared in the area of shape optimal design. Recently Pironneau [1] and Haug, Choi and Komkov [2] have published books dedicated to this subject. The finite element method has been applied extensively to shape optimal design of structures since 1973 [3], while it is only in the last few years that the boundary element method has been used in this field.

Mota Soares, Rodrigues, Oliveira Faria and Haug [4–8] developed models for the shape optimal design of solid and hollow shafts, based on constant, linear and quadratic boundary elements and nonlinear programming techniques. The design objective is to choose a shaft with a given area, which has maximal torsional stiffness. These models are much more efficient and robust than the corresponding finite element discretizations, since the sensitivity information is more accurate and there is no need to calculate the state variable in the domain.

A similar model for the shape optimal design of shafts, based on the boundary element method, has also been developed by Burczynski and Adamczyk [9]. Optimality conditions are generated and the Newton–Raphson method is used to solve a set of nonlinear algebraic equations. The examples show that the number of analyses required is smaller than a corresponding finite element discretization.

Models for the shape optimal design of bidimensional elasticity problems based on the boundary element method and linear programming technique have been developed by Zochowski and Mizukami [10]. The design objective is to minimize the area, subject to displacement and geometrical constraints. The adjoint structure generated is not identical to the real structure. The boundary element model is compared with equivalent finite element models, and results show that the boundary element technique is more accurate but less efficient in computational time than the finite element method.

Mota Soares, Rodrigues and Choi [11–12] have developed models for the shape optimal design of bidimensional structures based on linear and quadratic boundary elements and nonlinear programming techniques. The design objective is to minimize compliance, subject to a constant area. The adjoint and real structures are identical and subjected to the same loading conditions. Applications show that the boundary element model is more accurate and efficient than the corresponding finite element model. For general shapes, the technique used to calculated the stresses at the boundary was not very accurate. This problem has been overcome by Leal, who also has developed an automatic technique for mesh refinement [13]. This adaptive scheme improves the discretization and the accuracy of the boundary

References pp. 226–228

stresses, and the technique is based on the continuity of the boundary stresses for smooth unloaded surfaces.

Shape optimal design models for two- and three-dimensional elasticity problems, based on the boundary element method, have recently been developed by Burczynski and Adamczyk [14–16]. The design objective is to maximize stiffness, subjected to constant volume. The optimality conditions are derived for an optimal boundary. An iterative process, based on finite differences and the Newton-Raphson method, is used to solve a set of nonlinear algebraic equations, making it possible to determine the unknown optimal shape.

Eizadian and Trompette [17–18] have also developed a model for shape optimal design of two-dimensional structures, based on the boundary element method and nonlinear programming techniques. The design objective is to minimize the tangential stress subjected to geometrical constraints. The geometry is defined by linear and circular elements. Substructures are used to represent the fixed and moving boundaries. The multiplier method is employed to solve the nonlinear programming problem. Several applications are presented, including the shape design of a connecting rod and a rotor. Numerical instabilities are reported.

The boundary element method has also been applied to the shape optimal design of heat transfer problems. Futagami [19–21] presents a model for steady state and transient optimal heat conduction control based on linear and dynamic programming. A combined boundary element and finite element model is also developed. The applications show that boundary elements constitute a powerful technique for these types of problems.

Barone and Caulk [22] optimize the position, size and surface temperature of circular holes inside a two-dimensional heat conductor to produce a minimum variation in surface temperature over a portion of the outer boundary. In this problem, which arises in thermal design of molds and dies, the internal geometry of the heat conductor depends on the design variables. Since the objective function depends only on the boundary temperatures, there is no need to determine temperature in the interior. Also, it is not required to regenerate a boundary mesh every time the boundary is changed. The model is applied to the thermal design of compression molds.

Boundary elements have been used by Meric [23–25] to analyze the optimal heating of solids. The design objective is to achieve a desired temperature profile along a segment of a solid boundary with a minimum amount of boundary heat flux. Adjoint equations and the necessary optimality conditions are derived. The conjugate gradient method is used to solve the mathematical programming problem. Numerical results show the efficiency and the accuracy of the boundary element model.

The boundary element method has also been applied to the shape optimization of airfoils and wings by Pironneau [1]. The author argues that the boundary element method is more advantageous than the finite element or finite difference techniques when the solution of the partial differential equations is needed only at

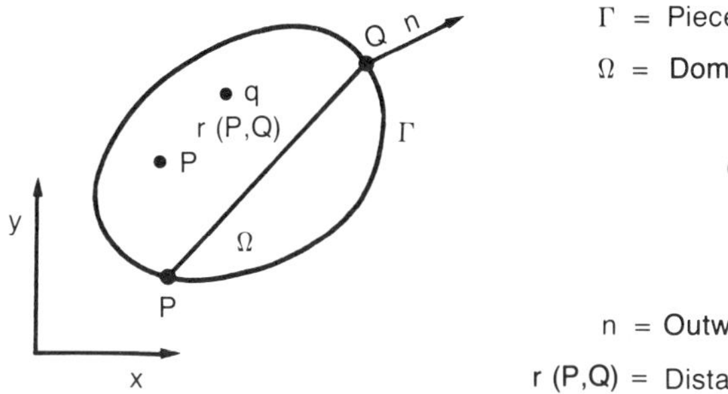

Γ = Piecewise smooth boundary

Ω = Domain

$\bar{\Omega} = \Omega \cup \Gamma$

$Q \in \Gamma$

$q \in \Omega$

$P \in \bar{\Omega}$

n = Outwards normal to Γ

r (P,Q) = Distance between P and Q

x,y = Coordinate systems

Figure 1. Definitions.

the boundary, and consequently its range of applicability is limited.

BOUNDARY ELEMENT METHOD FOR SHAFT TORSION

The boundary element method applied to the torsion of shafts is based on Green's formula, which allows the formulation of certain boundary value problems as integral equations, involving the solution of the state variable and its normal derivative only on the boundary. Consider a shaft defined in Figure 1.

Let z be the stress function of the torsion problem. This stress function must satisfy the Poisson state equation:

$$\begin{aligned} \frac{\partial^2 z}{\partial x^2} + \frac{\partial^2 z}{\partial y^2} = \nabla^2 z = -2 \text{ in } \Omega \\ z = 0 \text{ on } \Gamma. \end{aligned} \tag{1}$$

This equation can be transformed by introducing a new variable,

$$u = z + \frac{1}{2}x^2 + \frac{1}{2}y^2, \tag{2}$$

into a Laplace state equation:

$$\begin{aligned} \nabla^2 u = 0 \text{ in } \Omega \\ u = \frac{1}{2}x^2 + \frac{1}{2}y^2 \text{ on } \Gamma. \end{aligned} \tag{3}$$

Using Green's identity,

$$\int_\Omega (v\nabla^2 u - u\nabla^2 v)\, d\Omega = \int_\Gamma \left(v\frac{\partial u}{\partial n} - u\frac{\partial v}{\partial n} \right) d\Gamma \tag{4}$$

where u and v are solutions of the Laplace equation (3); and assuming that

$$v = \ell nr, \tag{5}$$

equation (4) becomes

$$\int_\Omega u\nabla^2 \ell nr\, d\Omega = \int_\Omega \ell nr \nabla^2 u\, d\Omega - \int_\Gamma \ell nr \frac{\partial u}{\partial n}\, d\Gamma + \int_\Gamma u \frac{\partial}{\partial n} \ell nr\, d\Gamma. \tag{6}$$

For any sufficiently smooth function $u(x,y)$ defined in $\bar{\Omega}$, equation (6) becomes

$$c(P)u(P) = \int_\Omega \nabla^2 u(q) \ell nr(P,q)\, d\Omega + \int_\Gamma u(Q) \frac{\partial}{\partial n} \ell nr(P,Q)\, d\Gamma - \int_\Gamma \ell nr(P,Q) \frac{\partial u}{\partial n}(Q)\, d\Gamma \tag{7}$$

where

$$c(p) = \int_\Gamma \frac{\partial}{\partial n} \ell nr(P,Q) d\Gamma. \tag{8}$$

Since

$$\frac{\partial}{\partial n} \ell nr(P,Q) = \frac{d\theta}{d\Gamma}(P,Q) \tag{9}$$

where

$$\theta(P,Q) = \tan^{-1} \frac{y(Q) - y(P)}{x(Q) - x(P)}, \tag{10}$$

it follows that

$$c(P) = \begin{matrix} 2\pi & \text{for } P \text{ inside } \Omega \\ 0 & \text{for } P \text{ outside } \bar{\Omega} \end{matrix} \tag{11}$$

and if Γ has an unique tangent at P, then

$$c(P) = \pi \text{ in } \Gamma. \tag{12}$$

For the Laplace equation, $\nabla^2 u(P) = 0$, and consequently equation (7) becomes

$$c(P)u(P) - \int_\Gamma u(Q) \frac{\partial}{\partial n} \ell nr(P,Q)\, d\Gamma = -\int_\Gamma \ell nr(P,Q) \frac{\partial u}{\partial n}(Q)\, d\Gamma. \tag{13}$$

This equation can be approximated numerically by dividing the boundary into N segments Γ_j; on each segment u and $\partial u/\partial n$ are assumed to be constant. Writing equation (13) at the middle point of each segment, an N equation can be obtained of the form

$$cu_i - \sum_{j=1}^{N} u_j \int_{\Gamma_j} \frac{\partial}{\partial n} \ell nr\, d\Gamma = -\sum_{j=1}^{N} \frac{\partial u_j}{\partial n} \int_{\Gamma_j} \ell nr\, d\Gamma, \qquad i = 1, 2, \ldots, N \tag{14}$$

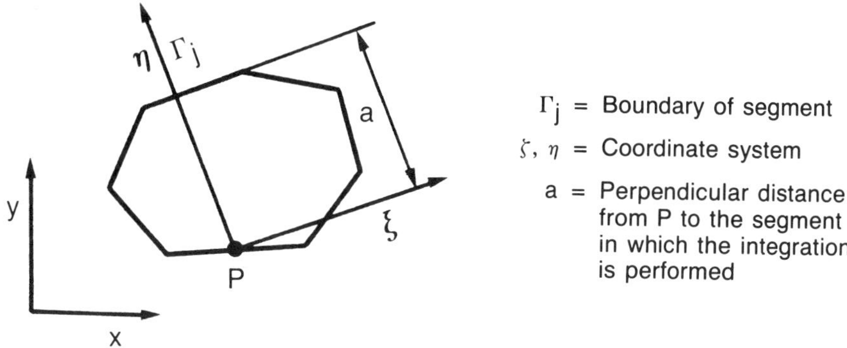

Figure 2. Transformation of the coordinate system.

where u_i and $\partial u_i/\partial n$ are nodal values. This equation can be written as

$$\mathbf{H} \quad \mathbf{u} = \mathbf{G} \quad \mathbf{p} \tag{15}$$

where

$$\mathbf{u} = \begin{Bmatrix} u_1 \\ \vdots \\ u_N \end{Bmatrix} \quad , \quad \mathbf{p} = \begin{Bmatrix} \partial u_1/\partial n \\ \vdots \\ \partial u_N/\partial n \end{Bmatrix} \tag{16}$$

and $\mathbf{H}$ and $\mathbf{G}$ are full unsymmetric matrices.

Imposing the boundary conditions, equation (15) becomes

$$\mathbf{G} \quad \mathbf{p} = \mathbf{f} \tag{17}$$

where

$$\mathbf{f} = \mathbf{H} \quad \mathbf{u}. \tag{18}$$

The solution of equation (17) gives the values of $\partial u/\partial n$ at the boundary nodes. It should be noted that for the solution of the Laplace equation it is only necessary to discretize the boundary. Although the boundary element equations are based on full and unsymmetric matrices, the number of degrees of freedom are small when compared with finite element models.

For the constant element, the integrals of equation (14) can be evaluated in closed form, transforming it to a system (ξ, n) centered at point P and with ξ parallel to the segment in which the integration is performed (see Figure 2).

The integrals of equation (14) are of the type

$$\int_{\Gamma_j} \ell n(\xi^2 + a^2)\, d\xi \quad ; \quad \int_{\Gamma_j} \frac{a}{\xi^2 + a^2}\, d\xi. \tag{19}$$

For the linear boundary element, a linear variation of u and $\partial u/\partial n$ is assumed in each segment:

$$\begin{aligned} u &= N_1 u_1 + N_2 u_2 \\ \frac{\partial u}{\partial n} &= N_1 \frac{\partial u_1}{\partial n} + N_2 \frac{\partial u_2}{\partial n} \end{aligned} \tag{20}$$

where N_i are the unidimensional linear shape functions, and u_i and $\partial u_i/\partial n$ are the nodal values of u and $\partial u/\partial n$ at the extremes of the element. Thus, equation (13) becomes

$$c_i u_i - \sum_{j=1}^{N} \int_{\Gamma_j} u \frac{\partial}{\partial n} \ell nr \, d\Gamma = - \sum_{j=1}^{N} \int_{\Gamma_j} \ell nr \frac{\partial u}{\partial n} \, d\Gamma, \qquad i = 1, 2 \ldots N \tag{21}$$

and

$$c_i u_i - \sum_{j=1}^{N} \int_{\Gamma_j} [N_1 N_2] \frac{\partial}{\partial n} \ell nr \, d\Gamma \begin{Bmatrix} u_1 \\ u_2 \end{Bmatrix} = - \sum_{j=1}^{N} \int_{\Gamma_j} [N_1 N_2] \ell nr \, d\Gamma \begin{Bmatrix} \partial u_1/\partial n \\ \partial u_2/\partial n \end{Bmatrix} \tag{22}$$

where $i = 1, 2, \ldots, N$, and c_i is the constant c for node i. The integrals of this equation can be evaluated using numerical and analytical integration. Equation (22) can be written in the same form as equation (15).

When the boundary is not smooth, equation (12) is not valid and the diagonal values of matrix $\mathbf{H}$ are calculated analytically or from rigid body considerations [26]:

$$\mathbf{H} \ \mathbf{I} = \mathbf{O}, \quad \mathbf{I} = \begin{Bmatrix} 1 \\ \cdot \\ \cdot \\ 1 \end{Bmatrix}, \quad \mathbf{O} = \begin{Bmatrix} 0 \\ \cdot \\ \cdot \\ 0 \end{Bmatrix} \tag{23}$$

The boundary element method can also be applied to the solution of the Poisson equation (1). Following Fairweather, Rizzo, Shippy and Wu [27] the domain integral of equation (7) can be transformed to a boundary integral. Thus, the boundary element model for the solution of the Poisson equation does not need internal cells; all the calculations are at the boundary. The boundary integral equation for the torsion of shafts, in terms of the stress function, is:

$$c(P)z(P) - \int_{\Gamma} z \frac{\partial}{\partial n} \ell nr \, d\Gamma = - \int_{\Gamma} \ell nr \frac{\partial z}{\partial n} \, d\Gamma - \frac{1}{2} \int_{\Gamma} \frac{\partial}{\partial n} (r^2 (\ell nr - 1)) \, d\Gamma. \tag{24}$$

This equation can be used for the development of boundary elements. Full details of the boundary element method are presented in the books of Banerjee and Butterfield [26] and Brebbia, Telles and Wrobel [28].

BOUNDARY ELEMENT METHOD IN TWO-DIMENSIONAL ELASTICITY

The boundary element method for elasticity is based on Somigliana's identity [26]. Using the notation in Figure 3, this boundary integral equation is given by

$$C_{ij}(P) u_i(P) + \int_{\Gamma} u_j(Q) T_{ij}(P, Q) \, d\Gamma = \int_{\Gamma} t_j(Q) U_{ij}(P, Q) \, d\Gamma + \int_{\Omega} b_j(q) U_{ij}(P, q) \, d\Omega \tag{25}$$

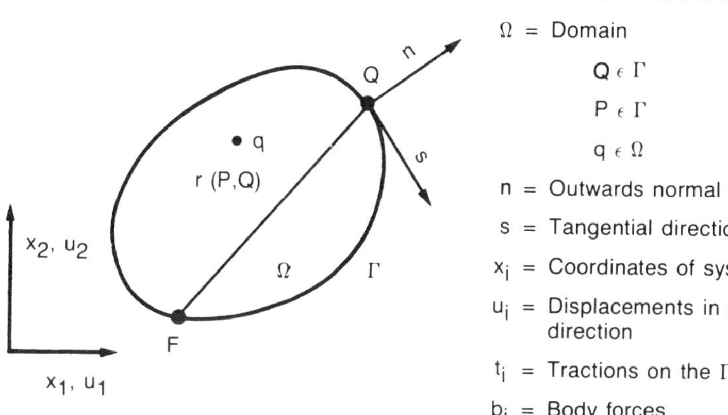

Figure 3. Nomenclature.

where U_{ij} and T_{ij} are the fundamental Kelvin solutions for displacements and tractions, due to a unit concentrated force in an elastic infinite space; C_{ij} is the coefficient that depends on the geometry of the boundary at point P.

When body forces are not present, equation (25) is only dependent on the boundary displacements and tractions. In this case, there is no need for internal cells in the domain. The boundary can be divided into N segments, or elements, with surfaces $\Gamma_K, K = 1, \ldots, N$. Within each element, the geometry, displacement, and traction fields can be assumed to be linear or quadratic, as shown in Figure 4.

Any variable within an element is assumed to be given by

$$\theta(\xi) = \sum_{m=1}^{\ell} N_m(\xi)\theta^m \tag{26}$$

where $N_m(\xi)$ are the shape functions in local nondimensional coordinates $(-1 \leq \xi \leq 1)$. For the linear $(\ell = 2)$ or quadratic element $(\ell = 3)$, θ^m represents the nodal values of the variable.

For problems without body forces, equation (25) becomes

$$\begin{aligned} C_{ij}(P_n)u_i(P_n) + \sum_{k=1}^{N} \int_{\Gamma_k} N_m(\xi)T_{ij}(P_n, Q)J(\xi)\, d\xi u_j^m \\ = \sum_{k=1}^{N} \int_{\Gamma_k} N_m(\xi)T_{ij}(P_n, Q)J(\xi)\, d\xi t_j^m \end{aligned} \tag{27}$$

where u_j^m is the value of u_j at local node m, t_j^m is the value of t_j at local node m, and J is the Jacobian of the transformation of coordinates. Note that P_n refers to a particular node. For all nodes, equation (27) can be expressed in matrix form as

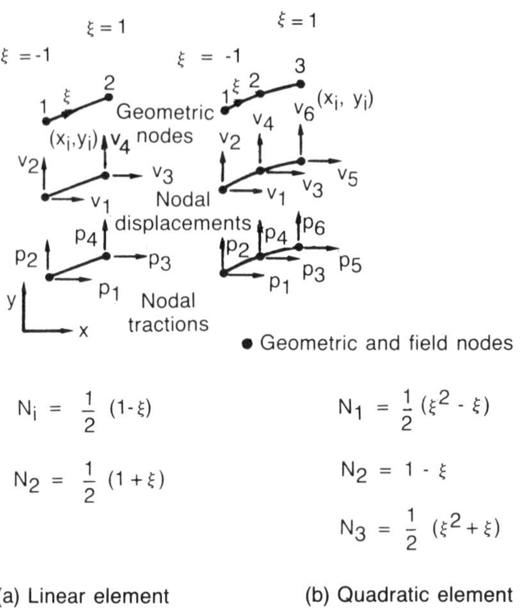

Figure 4. Linear and quadratic boundary elements for two-dimensional elasticity.

$$\mathbf{C}\quad\mathbf{u}+\hat{\mathbf{H}}\quad\mathbf{u}=\mathbf{G}\quad\mathbf{t} \tag{28}$$

or

$$\mathbf{H}\quad\mathbf{u}=\mathbf{G}\quad\mathbf{t} \tag{29}$$

where $\mathbf{u}$ and $\mathbf{t}$ are the boundary nodal displacements and tractions. The elements of matrices $\hat{\mathbf{H}}$ and $\mathbf{G}$ can be obtained from the integrals

$$\hat{H}_{ij}^{nm}=\int_{\Gamma_k} N_m(\xi)T_{ij}(P_n,Q(\xi))J(\xi)\,d\xi \tag{30}$$

and

$$G_{ij}^{nm}=\int_{\Gamma_k} N_m(\xi)U_{ij}(P_n,Q(\xi))J(\xi)\,d\xi. \tag{31}$$

The strong singular integral of equation (30) and the corresponding coefficients C_{ij} can be evaluated by rigid body considerations [26]. The weak singular integrals of equation (31) lead to integrals of the type

$$\int_{-1}^{1} \ell n\frac{1}{r} f(\xi)d\xi \tag{32}$$

which can be transformed to

$$\int_{-1}^{1} \ell n\frac{1}{r} f(\xi)\,d\xi=\int_{-1}^{1} \ell n\left(\frac{1+\xi}{2r}\right) f(\xi)\,d\xi+2\int_{0}^{1} \ell n\frac{1}{\varsigma} f(\varsigma)\,d\varsigma \tag{33}$$

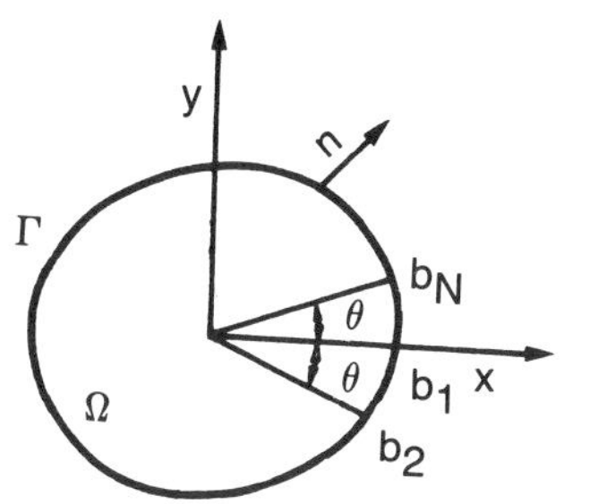

Γ = Boundary
Ω = Domain
x,y = Coordinate system
n = Normal to the boundary
b_i = Design variables
N = Number of design variables
$\theta = 2\pi/N$
A = Given area of shaft

Figure 5. Design variables of solid shaft.

where

$$\varsigma = \frac{1}{2}(1 + \xi). \tag{34}$$

The first integral on the right side of equation (33) is evaluated using standard Gaussian quadrature with four integration points, while the second integral is calculated numerically by formulas given by Banerjee and Butterfield [26].

OPTIMIZING SOLID SHAFT GEOMETRY

The design objective is to choose the shape of a solid shaft with a given cross-sectional area and subject to constraints on the design variables which will have maximal torsional stiffness. Full details of the theory are presented by Haug, Choi and Komkov [2].

With reference to Figure 5, let z be the stress function of the torsion problem. The torsional rigidity is given by the negative of

$$\psi_0 = -\int_\Omega 2z \, d\Omega \tag{35}$$

which, using Green's identity, can be transformed to

$$\psi_0 = -\int_\Omega (x^2 + y^2) \, d\Omega + \int_\Gamma u \frac{\partial u}{\partial n} \, d\Gamma. \tag{36}$$

For all simply connected domains, the problem is expressed as finding Ω to minimize ψ_0, subject to the constraint

$$\psi_1 = \int_\Omega d\Omega - A = 0 \tag{37}$$

and to prescribed constraints on the design variables.

It has been shown by Haug, Choi and Komkov [2], using total derivatives and variational calculus, that the variations of the objective and constraint functionals

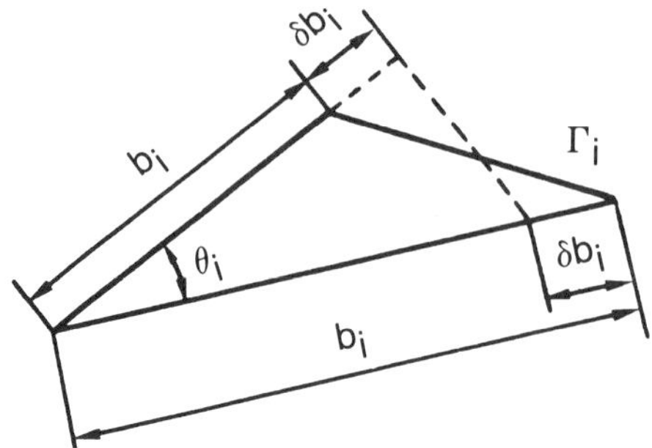

Figure 6. Perturbation of the boundary of an element.

in terms of the design variations are given by

$$\delta\psi_0 = -\int_\Gamma \left(\frac{\partial z}{\partial n}\right)^2 v_n \, d\Gamma = -\int_\Gamma \left(\frac{\partial u}{\partial n} - xn_x - yn_y\right)^2 v_n \, d\Gamma \tag{38}$$

and

$$\delta\psi_1 = \int_\Gamma v_n \, d\Gamma \tag{39}$$

where n_x and n_y are the direction cosines of the boundary, and v_n is the "normal perturbation" of the boundary.

With reference to Figure 6 and assuming that the boundary is divided into constant or linear boundary elements, the first order approximation of the sectorial area change due to a perturbation in the design variables is given by

$$\int_{\Gamma_i} v_n \, d\Gamma = \frac{1}{2} \sin\theta (b_i \delta b_j + b_j \delta b_i). \tag{40}$$

It should be observed that equations (36–39) are boundary integrals of the state variable and domain integrals of the geometry of the shaft. These are the necessary equations for the solution of the nonlinear programming problem by the Pshenichny's linearization method [29]. Also, to solve the Laplace and Poisson torsion equations by the boundary element method, it is only necessary to evaluate the state variable at the boundary. Thus for the shape optimal design of solid shafts, all the calculations of the state variable are at the boundary.

OPTIMIZING HOLLOW SHAFT GEOMETRY

The design objective is to maximize the rigidity of a hollow shaft with a known hole and a given cross-sectional area, subject to some other constraints.

Referring to Figure 7, the state equation in terms of the stress function is

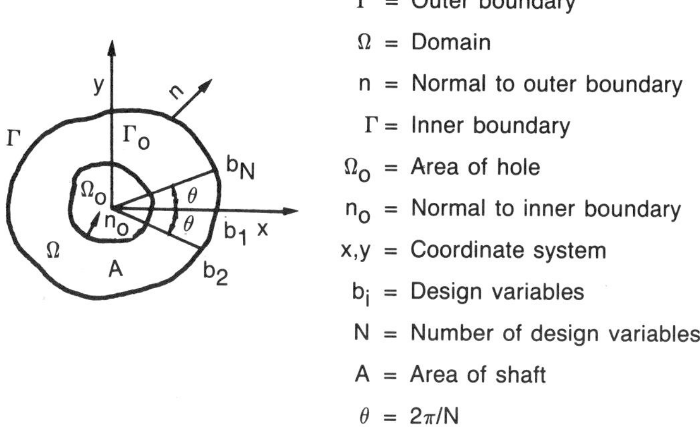

Figure 7. Design variables of hollow shafts.

$$\begin{aligned} \nabla^2 z &= -2 \text{ in } \Omega \\ z &= 0 \text{ on } \Gamma \\ z &= z_0 \text{ on } \Gamma_0 \\ \int_{\Gamma_0} \frac{\partial z}{\partial n}\, d\Gamma &= 2\Omega_0 \end{aligned} \tag{41}$$

where z_0 is a constant to be determined. This Poisson equation can be transformed into a Laplace equation by using equation (2). The torsional rigidity is given by the negative of

$$\psi_0 = -\int_\Omega 2z\, d\Omega - 2z_0\Omega_0 \tag{42}$$

which, using Green's theorem, can be transformed to

$$\psi_0 = -\int_\Omega (x^2 + y^2)\, d\Omega + \int_\Gamma u \frac{\partial u}{\partial n}\, d\Gamma + \int_{\Gamma_0} u \frac{\partial u}{\partial n}\, d\Gamma. \tag{43}$$

The area constraint for the hollow shaft is identical to equation (37). Following Gelfand and Fomin [30], it can be shown that the first variation of the objective function of the hollow shaft is identical to the first variation of the solid shaft (38). It should be noted that equation (43) is a boundary integral of the state variable and a domain integral of the geometry of the shaft. Thus for the shape optimal design of hollow shafts, it is only necessary to evaluate the state variable at the boundaries.

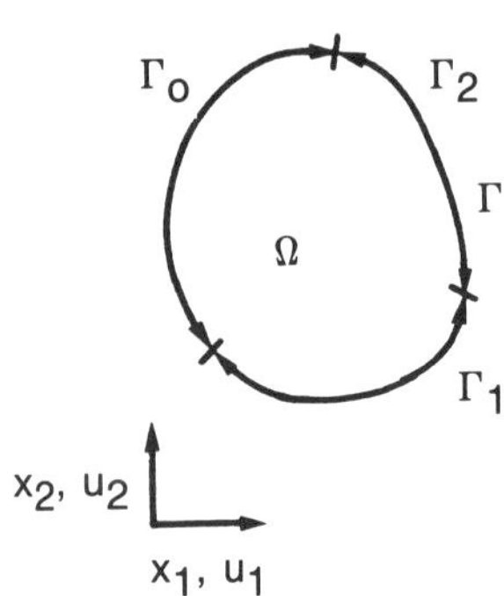

Γ_0 = Boundary where displacements are zero

Γ_1 = Boundary where tractions are prescribed at t_i^0

Γ_2 = Design boundary (unloaded surface)

Ω = Domain; $\Gamma = \Gamma_0 \cup \Gamma_1 \cup \Gamma_2$

u_i = Displacements

t_i = Tractions

x_i = Coordinates

Figure 8. Definition of domain.

SHAPE OPTIMAL DESIGN OF STRUCTURES BASED ON MINIMUM COMPLIANCE

The design objective is to find the shape of an unloaded boundary of a specified structure which has a given area, is subject to constraints in the design variables, and which has minimum compliance.

Consider an elastic body that is rigidly supported on a boundary Γ_0 and loaded by tractions on boundary Γ_1 (see Figure 8). Also let the design boundary Γ_2 be free from loading. It is assumed that there are no body forces. The nomenclature used is shown in Figure 8. The objective function is the compliance of the structure, which is given by

$$\psi_0 = \frac{1}{2} \int_\Gamma t_i u_i \, d\Gamma. \tag{44}$$

The optimization problem is expressed as finding Γ_2 to minimize ψ_0, subject to the area constraint

$$\psi_1 = \int_\Omega d\Omega - A \leq 0 \tag{45}$$

where A is the given area of the domain.

To solve this nonlinear programming problem numerically, it is necessary to evaluate the first variation of the objective and constraint functionals. A general formulation for sensitivity analysis of volume and boundary functionals is presented in [2]. For unloaded design boundaries of linear materials, the first variation of the objective functional of equation (44) becomes

$$\delta\psi_0 = -\int_{\Gamma_2} U v_n \, d\Gamma \tag{46}$$

where U is the strain energy density and v_n is the normal perturbation of the boundary. The variation of the constraint functional of equation (45) is

$$\delta\psi_1 = \int_{\Gamma_2} v_n \, d\Gamma. \tag{47}$$

The boundary can be divided into N linear and/or quadratic boundary elements. It is assumed that the design boundary is represented by M geometrical linear elements. With reference to Figure 6, the first-order approximation of area change due to a small perturbation in the design variables is given by

$$\int_{\Gamma_i} v_n \, d\Gamma = \frac{1}{2} \sin\theta_i (b_i \, \delta b_j + b_j \, \delta b_i). \tag{48}$$

The variation in compliance given by equation (46) can be accurately and efficiently calculated by the boundary element method. For linear elements, the strain energy density is constant on each element. Thus, equation (46) becomes

$$\delta\psi_0 = -\sum_{m=1}^{M} U_m \int_{\Gamma_m} v_n \, d\Gamma \tag{49}$$

where U_m is the strain energy density of a boundary element. This energy can be evaluated from the boundary tangential stress. For an unloaded design boundary, the only stress component that exists is the tangential stress.

For the shape optimal design, all equations that are necessary (44–47) to implement the Pshenichny linearization method [29] of nonlinear programming are boundary functionals of the displacement, tractions and stresses, and domain functionals of the geometry. Consequently, shape optimal design of structures, based on minimum compliance, can be efficiently and accurately solved using the boundary element method to discretize the structure. Also, there is no need to calculate displacement and stress in the domain.

SHAPE OPTIMAL DESIGN OF STRUCTURES BASED ON STRESS CONSTRAINTS

The design objective is to find the shape of an unloaded boundary for the minimum area, subject to constraints on the stresses and design variables. The notation of Figure 8 is used.

The objective function

$$\psi_0 = \int_{\Omega} d\Omega \tag{50}$$

should be a minimum. The variation of this area functional is given by equation (47).

The stress should be less than the allowable stress σ_a anywhere in the domain or boundary. For plane stress or strain problems with smooth boundaries and without body forces, it can be proved [31] that the maximum von Mises stress is always at the boundary. The von Mises yield stress constraint functional, averaged over a small region Ω_k (defined in Figure 9) can be represented by

References pp. 226–228

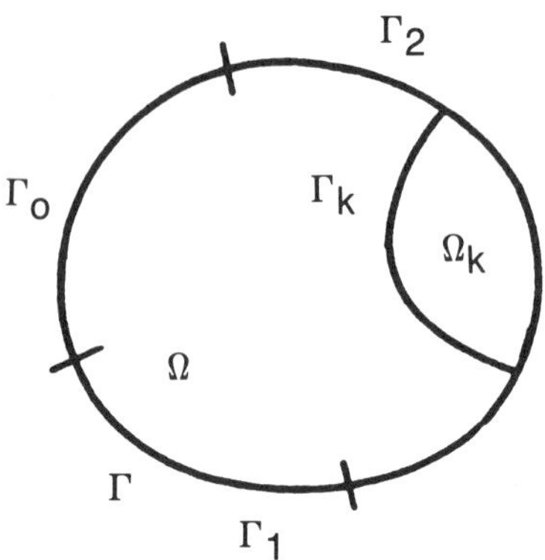

Figure 9. Domain of the stress constraint.

$$\psi_k = \int_\Omega \phi m_k \, d\Omega = \frac{\int_{\Omega_k} \phi \, d\Omega}{\int_{\Omega_k} d\Omega} \tag{51}$$

where

$$\phi = \frac{\sigma_y - \sigma_a}{\sigma_a}, \tag{52}$$

$$\sigma_y = \sqrt{\sigma_{11}^2 + \sigma_{22}^2 + 3\sigma_{12}^2 + \sigma_{11}\sigma_{22}} \tag{53}$$

and σ_{ij} are the components of the stress tensor. In equation (51) m_k is a characteristic function defined as

$$\overline{m}_k = \frac{1}{\int_{\Omega_k} d\Omega} \text{ in } \Omega_k$$

and

$$\overline{m}_k = 0 \text{ in } \Omega \backslash \Omega_k. \tag{54}$$

Following Haug, Choi and Komkov [2], the variation of equation (51) is given by

$$\delta\psi_k = -\int_{\Gamma_2} \sigma_{ij}\varepsilon_{ij}(\lambda) v_n \, d\Gamma + \overline{m}_k \int_{\Gamma_k} (\phi - \psi_k) v_n \, d\Gamma \tag{55}$$

where $\varepsilon_{ij}(\lambda)$ are the components of the strain tensor of the adjoint structure. The adjoint structure is identical to the real structure, but with different loading conditions. For each constraint, the adjoint problem is defined by the equilibrium equation

$$\sigma_{ij,j}(\lambda) + F_i^* = 0$$

where λ is the adjoint variable and F_i^* are the adjoint body forces

$$F_i^* = -\sum_{j=1}^{2} \left(\sum_{k,\ell=1}^{2} \frac{\partial \phi}{\partial \sigma_{k\ell}} D_{k\ell ij} m_p \right)_{,j} \tag{56}$$

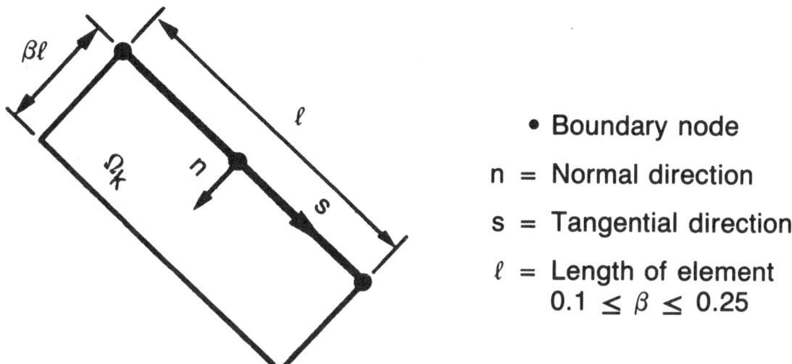

Figure 10. Domain of the stress constraint for the boundary element model.

and $D_{k\ell ij}$ are the stress/strain relations. The boundary conditions of the adjoint structures are

$$\begin{aligned} \lambda_i &= 0 \text{ on } \Gamma_0 \\ \sigma_{ij} n_j &= 0 \text{ on } \Gamma_1 \, U \, \Gamma_2 \backslash \Gamma_k \\ \sigma_{ij} n_j &= t_i^* \text{ on } \Gamma_k \end{aligned} \tag{57}$$

where t_i^* are the adjoint tractions,

$$t_i^* = \sum_{j=1}^{2} \left(\sum_{k,\ell=1}^{2} \frac{\partial \phi}{\partial \sigma_{k\ell}} D_{k\ell ij} m_p \right) n_j \tag{58}$$

and n_j are the normal components to the boundary.

It should be noted that the sensitivity equation (55) depends on the stress at the boundary of the real structure and on the strains at the boundary of the adjoint structures. Also, there is no need of calculating displacements and stresses of the real and adjoint structures in the domain. Thus, the boundary element method should be efficient and accurate in the shape optimal design of structures based on stress constraints.

It is assumed that the design boundary is represented by quadratic elements with straight geometries. For the boundary element model, the small area Ω_k is defined in Figure 10. Also within Ω_k the tangential stress σ_{ss} and strain, the shear σ_{ns} and normal σ_{nn} stresses are linear in the s direction but constant in the n direction. The shear and normal stresses are only dependent on the boundary tractions. These assumptions are only accurate if the parameter β in the boundary element model is small, preferably much less than 0.25.

With these approximations, the distributed adjoint body forces are zero. Also, the concentrated adjoint body forces are identical to the adjoint boundary tractions. Figure 11 illustrates the loading conditions of the adjoint structure.

References pp. 226–228

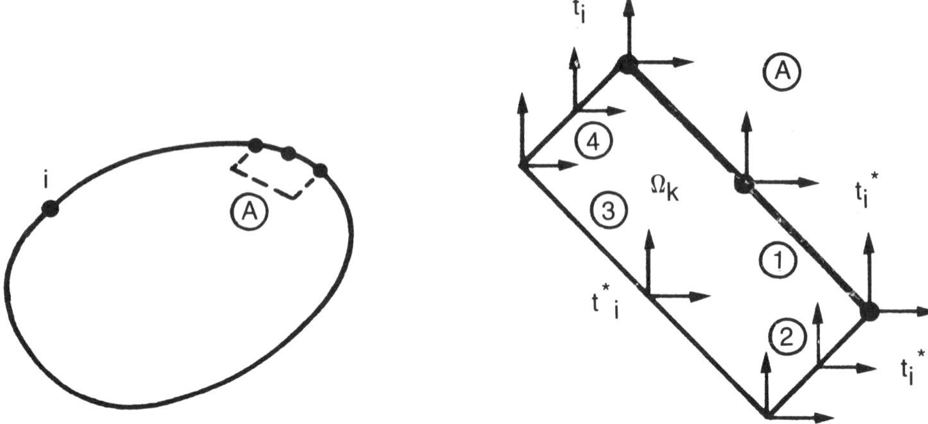

Figure 11. Adjoint loading for stress constraint.

The application of the boundary element method to calculate the stresses and strains of the adjoint structure is almost standard. However, we should consider the integration of the pseudotractions in surfaces 2, 3 and 4 of Figure 11. Because the integrals are almost singular when $i \in \Omega_k$, it is necessary to integrate the pseudoadjoint tractions with nine Gaussian points. For this reason the parameter β in the boundary element model should be larger than 0.1.

BOUNDARY STRESSES AND AUTOMATIC MESH REFINEMENT

The accuracy of the boundary stresses and strains is crucial for the shape optimal design of elastic structures. It is well known that the stresses at the boundary are more accurately calculated by the boundary element method than by the finite element technique. Unlike the stresses at interior points, the stresses at the boundary are not calculated directly by the boundary element method. Banerjee and Butterfield [26] present a technique for calculating boundary stresses which has been found to be not very accurate as design changes [13]. Another technique has been developed by Hartmann [32] to calculate accurate boundary stresses [13].

The boundary element method accurately calculates the displacement u_i and the tractions t_i at the boundary. Also the derivative of the displacement with respect to tangential coordinate $s(\partial u_i/\partial s)$ can be calculated accurately. The stresses at the boundary must also obey the Hooke and Cauchy laws. Thus the following equations can be derived:

$$\sigma_{ij} = \frac{2\mu\nu}{1-2\nu}\delta_{ij}\varepsilon_{kk} + 2\mu\varepsilon_{ij} \tag{59}$$

$$t_i = \sigma_{ij}n_j \tag{60}$$

$$\frac{\partial u_i}{\partial s} = \frac{\partial u_i}{\partial x_j}\frac{\partial x_j}{\partial s} \tag{61}$$

where μ is the rigidity modulus, ν is the Poisson ratio, δ_{ij} is the Kronecker delta, and n_j are the normal components of the boundary.

For two dimensional elasticity these equations form a system of seven algebraic equations:

$$\begin{bmatrix} 1 & 0 & 0 & a & 0 & 0 & -\lambda \\ n_1 & 0 & n_2 & 0 & 0 & 0 & 0 \\ 0 & n_2 & n_1 & 0 & 0 & 0 & 0 \\ 0 & 1 & 0 & -\lambda & 0 & 0 & a \\ 0 & 0 & 0 & -n_2 & 0 & n_1 & 0 \\ 0 & 0 & 1 & 0 & -\mu & -\mu & 0 \\ 0 & 0 & 0 & 0 & -n_2 & 0 & n_1 \end{bmatrix} \begin{Bmatrix} \sigma_{11} \\ \sigma_{22} \\ \sigma_{12} \\ u_{1,1} \\ u_{2,1} \\ u_{1,2} \\ u_{2,2} \end{Bmatrix} = \begin{Bmatrix} 0 \\ t_1 \\ t_2 \\ 0 \\ u_{1,s} \\ 0 \\ u_{2,s} \end{Bmatrix} \tag{62}$$

where

$$a = -\lambda - 2\mu; \quad \lambda = \frac{2\mu\nu}{1-2\nu}. \tag{63}$$

The solution of equation (62) at any boundary point gives the boundary stresses. Since the right-hand side of equation (62) is calculated accurately by the boundary element method, the boundary stresses are accurate. It should be noted that equation (62) is only valid for smooth boundaries.

Although the boundary element discretization is able to adapt itself to a new configuration without major distortion of the boundary elements, it is convenient to have an automatic generator of boundary element meshes. Alarcon, Avia and Revester [33] and Rencis and Mullen [34] have developed techniques for automatic boundary element mesh generation of potential problems. It is concluded that the efficiency and accuracy of the boundary element method is increased with an automatic mesh generator.

For elasticity problems, Leal [13] has developed a technique based on the continuity of the tangential stresses for unloaded continuous boundaries. With reference to Figure 12, the tangential stress is calculated at the nodal points. For an acceptable boundary element mesh, σ_{i2} and σ_{j1} should be almost identical for each nodal point. When the discrepancy between σ_{i2} and σ_{j1} is more than an allowable error, the adjacent elements are divided. It is not necessary to divide elements where the boundary stresses are low compared to the maximum stress. The number of divisions is limited to avoid numerical integration problems. Leal's results [13] show that this simple technique can be used in practice.

APPLICATIONS

The models developed are applied here to the shape optimal design of several shafts. In all applications, the boundary element model is based on the boundary integral formulation (13) of the Laplace equation (3).

The constant element model is applied to the shape optimal design of a shaft presented by Choi, Haug, Hou and Sohoni [35]. The design objective is to choose

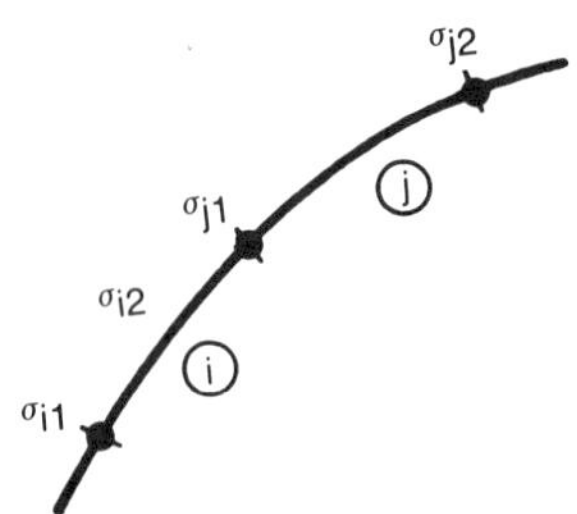

Figure 12. Nodal tangential stresses.

Table 1

Numerical results for optimal design of shaft

Iteration number	b_1	b_2	b_3	b_4	b_5	Torsional rigidity	Area
1	2.0000	2.0392	2.1648	2.1865	2.2109	31.0226	14.0906
2	2.0000	2.0392	2.1648	2.1669	2.2218	30.8578	14.0451
3	2.0000	2.0392	2.1648	2.1639	2.2212	30.8162	14.0338
4	2.0000	2.0392	2.1648	2.1593	2.2206	30.7538	14.0169
5	2.0000	2.0392	2.1648	2.1581	2.2206	30.7382	14.0126
6	2.0000	2.0392	2.1648	2.1544	2.2203	30.6912	14.0000
Finite elements* (30 iterations)	2.0000	2.0392	2.1648	2.1675	2.1953	30.4541	14.0021

Initial design: circle with 2.2 radius
b_i = design variable at 11.25 $(i-1)$ degrees
*See reference [35]

the shape of a shaft with a cross-section of 14.0 which must fit a square housing of 16.0 and has maximal torsional stiffness. This problem has been solved using finite elements and the model is shown in Figure 13(a). The discretization used for the boundary element formulation is shown in Figure 13(b).

Table 1 shows results obtained and compares them with a finite element solution. It can be concluded that the boundary element results converge faster than the finite element solution because the nodal values of the normal derivative of the state variable at the boundary are more accurate.

The constant boundary element model is also applied to the same problem but with different initial shapes. In all the applications the model converges to the correct shape in a few iterations without redefining the boundary element mesh (see Figures 14 and 15).

The constant element model for hollow shafts is applied to the shape optimal

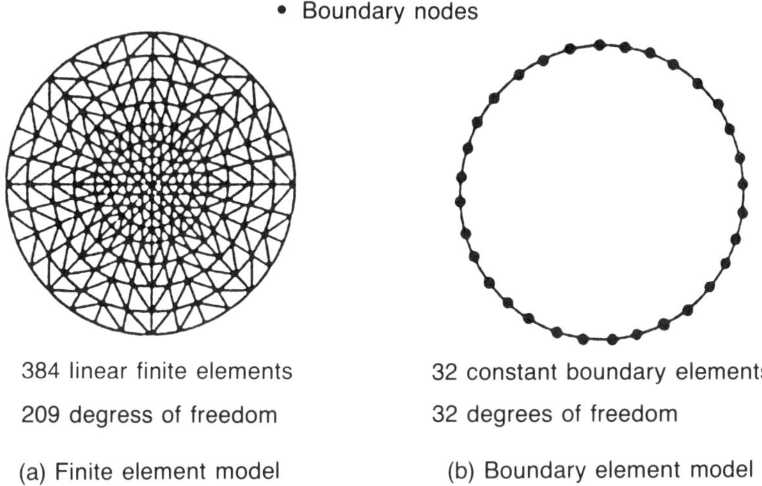

Figure 13. Finite element and boundary element models.

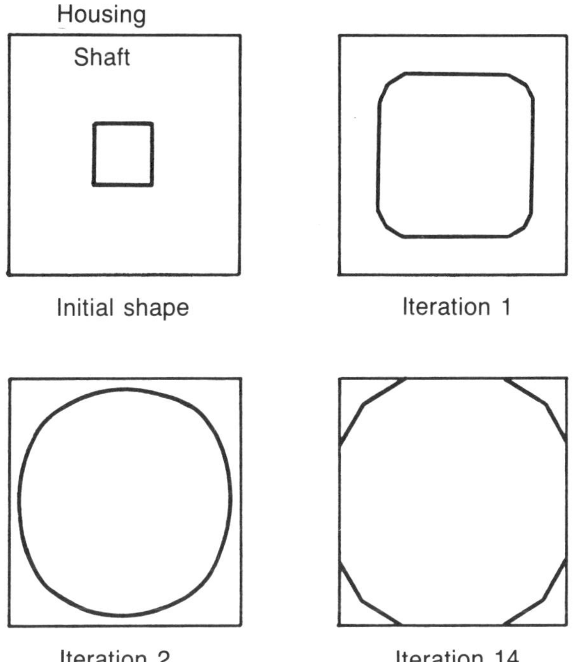

Figure 14. Modification of the geometry of shaft with iteration process: square initial shape.

design of several shafts. The discretization used is presented in Figure 16. All the shafts have an elliptical, square or round hole, and an initial circular or elliptical outside boundary. Several cross-sectional areas of the shaft are considered. In all the

References pp. 226–228

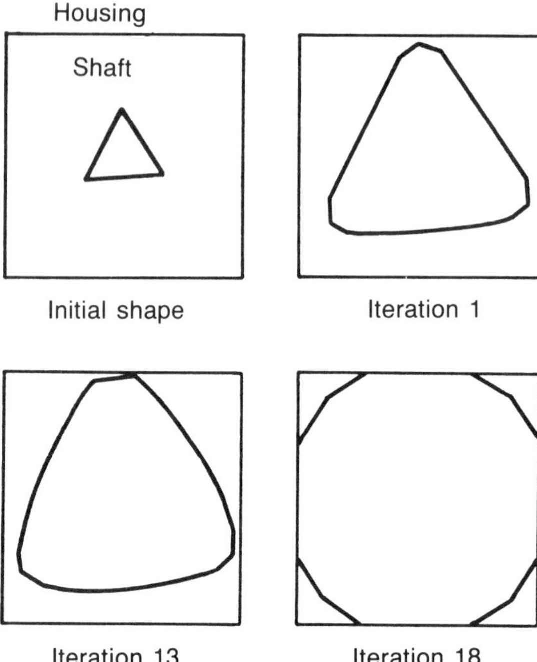

Figure 15. Modification of the geometry of shaft with iteration process: triangular initial shape.

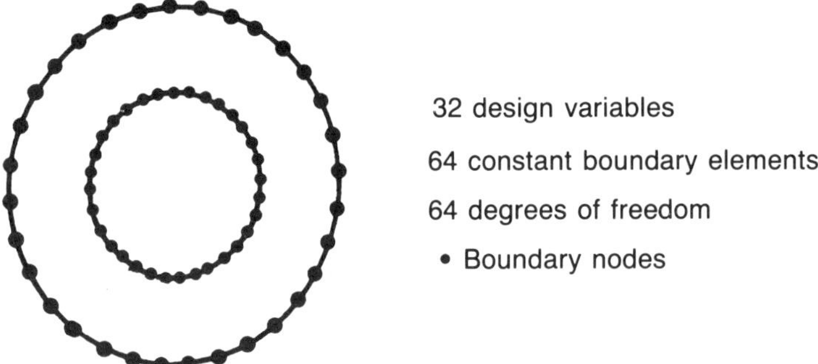

Figure 16. Boundary element model for hollow shafts.

applications, the final shape is found in a few iterations and without redefining the mesh. Some results are presented in Figure 17, and these values are in accordance with the exact solutions of Banichuk [31] and the finite element results of Hou, Haug and Benedict [36]. The iteration processes for two particular shafts are shown in Figures 18 and 19.

The linear boundary element model for hollow shafts is applied to the shape optimal design of shafts with an elliptical hole and subjected to constraints on the

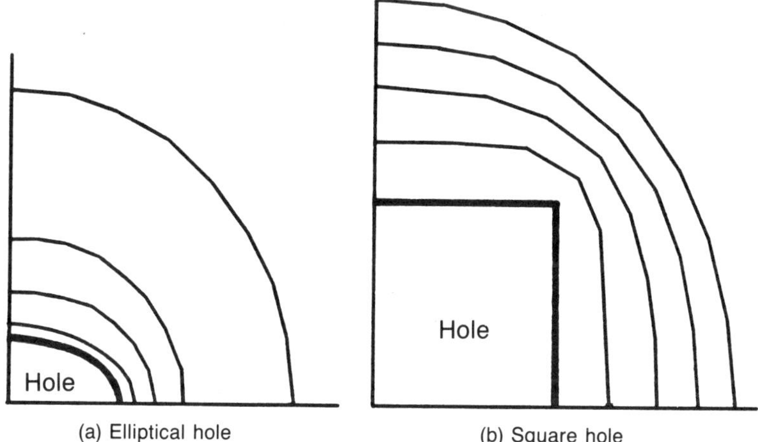

Figure 17. Hollow shafts results for several cross-section areas.

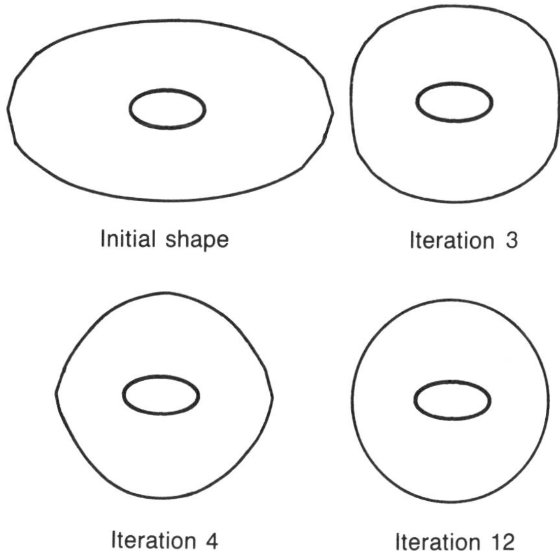

Figure 18. Modification of the shape of shaft with iteration process: elliptical hole.

cross-section area. The boundary element model has 48 degrees of freedom and 24 design variables. The results are in accordance with Figure 17(a).

The boundary element model for structures based on minimum compliance is applied to the shape optimal design of several elastic structural components. The square plate of Figure 20 is subject to uniform in-plane tensile loads along its four edges. The design objective is to find the shape of an initially square hole (area = 1.0 percent of the plate) which minimizes compliance.

References pp. 226–228

The boundary element model is represented in Figure 21. The model has 12 quadratic boundary elements, 12 linear design boundary elements, 13 design variables, and 72 degrees of freedom. The final design is achieved after seven iterations, 14 structural analyses, and five CPU minutes on a PRIME 750 super minicomputer. The evolution of the design of the hole is represented in Figure 22. It should be noted that after only two iterations, the design is almost optimal.

The final design of the hole is almost a circle whose radius is constant to within 1.8 percent. Also, the tangential stress at the boundary of the hole is constant to within 1.0 percent. The stress concentration factor of the hole is 2.02, with an error of only 1.0 percent. The final design is almost identical to the analytical solution given by Banichuk [31].

The same problem is also solved using a simpler model with eight quadratic elements, eight linear design elements, nine design variables, and 48 degrees of freedom. A practically identical design is achieved.

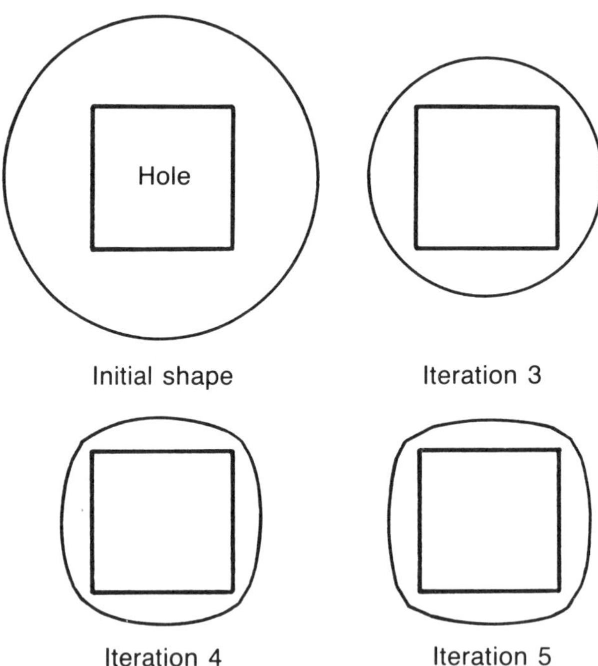

Figure 19. Modification of the shape of shaft with iteration process: square hole.

The formulation developed is applied to shape optimal design of the fillet shown in Figure 23. The boundary element model is shown in Figure 24. Starting from an initially straight line design and with a constraint on the area, the design process takes six iterations and 11 boundary element analyses to achieve the final design. The evolution of shape of the fillet is shown in Figure 25. The iterative process takes 1.5 CPU minutes on a PRIME 750 super minicomputer. The stress concentration factor of the final design is 1.37. Figure 26 presents optimal shapes of the fillet for prescribed areas.

The boundary element solutions are similar to finite element results presented in references [37–38]. The number of degrees of freedom of the boundary element model, however, is about five times smaller than the equivalent finite element model. The computer time required by the boundary element method is a factor of approximately eight times less than that required by the finite element method.

In each application with the compliance model, the final design is essentially achieved in a few iterations. However, due to the lack in sensitivity of compliance due to a small perturbation in a remote boundary, the final iteration process converges slowly. This fact puts a limitation on using compliance as an objective function, especially for very stiff structures for which the applied forces are very far from the design boundaries.

In most applications, stress constraints are more meaningful design criteria than compliance. The application of the stress model is under investigation and will be published shortly [39].

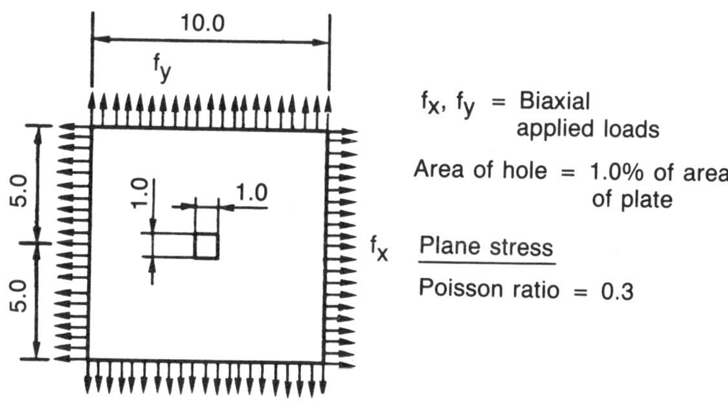

Figure 20. Square plate subjected to biaxial applied loads.

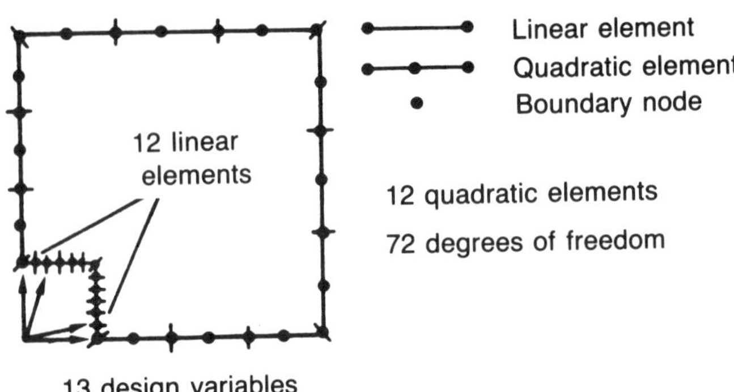

Figure 21. Boundary element model of one quarter of plate.

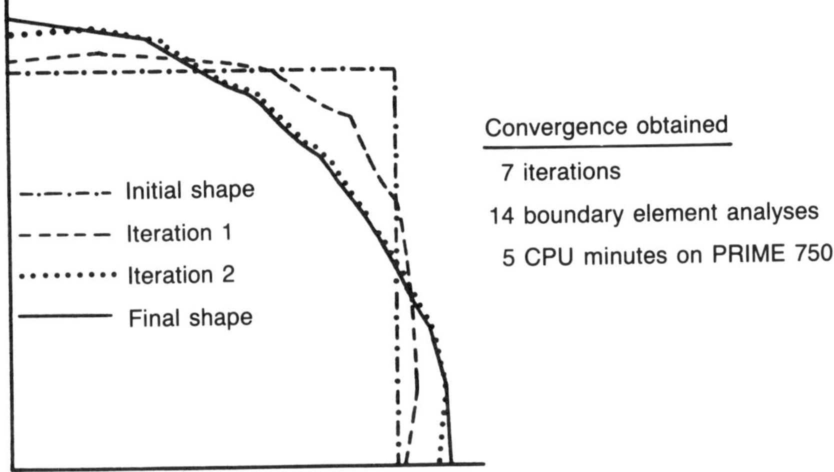

Figure 22. Evolution of design of hole with iteration process.

f_a, f_b = Applied loads

Γ_2 = Design boundary: unknown

Modulus of elasticity = 2.26×10^{11} N/m^2

Poisson ratio = 0.3

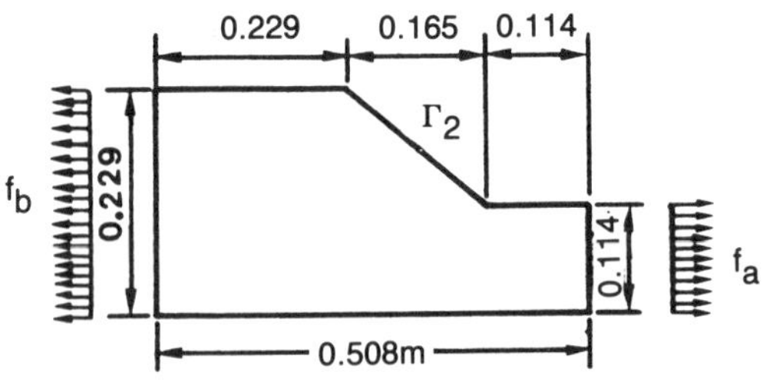

One-half of fillet

Figure 23. The fillet problem.

9 design variables
14 quadratic elements
10 linear elements
76 degrees of freedom

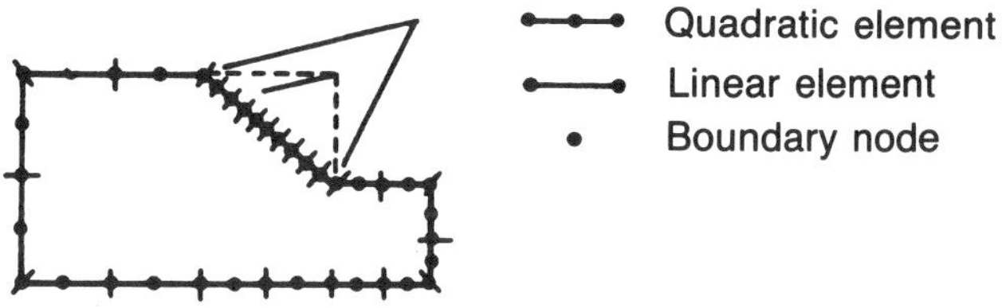

Quadratic element
Linear element
Boundary node

Figure 24. Boundary element model for fillet.

Number of iterations = 6
Number of analyses = 11
Time of computation = 1.5 CPU minutes
Stress concentrations factor = 1.37

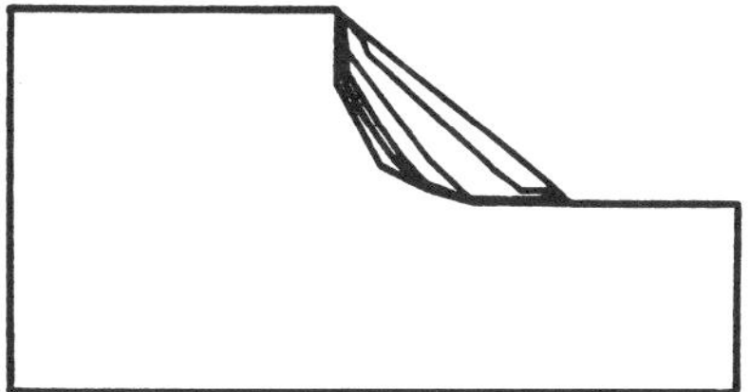

Figure 25. Evolution of design of fillet with iteration process.

$$\text{Area reduction} = \frac{\text{Allowable area}}{\text{Initial area (14.625)}}$$

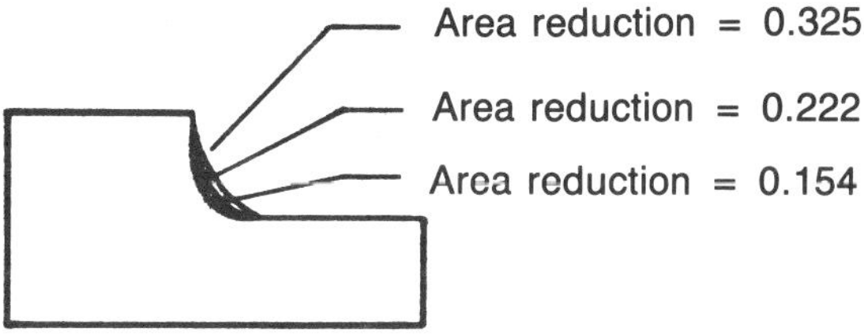

Area reduction = 0.325
Area reduction = 0.222
Area reduction = 0.154

Figure 26. Optimal shape of fillet for different areas.

References pp. 226–228

REFERENCES

1. O. Pironneau, *Optimal Shape Design for Elliptical Systems*. Springer-Verlag (1984).
2. E. J. Haug, K. K. Choi and V. Komkov, *Design Sensitivity Analysis of Structural Systems*. Academic Press (1985).
3. O. C. Zienkiewicz and J. S. Campbell, Shape optimization and sequential linear programming, pp. 109–126 in *Optimum Structural Design* (Edited by R. H. Gallagher and O. C. Zienkiewicz). Wiley, New York (1973).
4. C. A. Mota Soares, H. C. Rodrigues, L. M. Oliveira Faria and E. J. Haug, Optimization of the geometry of shafts using boundary elements. *J. Mech. Transm. Autom. Des.* **106**, 199–203 (1984).
5. C. A. Mota Soares, H. C. Rodrigues, L. M. Oliveira Faria and E. J. Haug, Optimization of the shape of solid and hollow shafts using boundary elements, pp. 883–889 in *Boundary Elements* (Edited by C. A. Brebbia). Springer-Verlag (1983).
6. C. A. Mota Soares, H. C. Rodrigues, L. M. Oliveira Faria and E. J. Haug, Boundary elements in shape optimal design of shafts, pp. 155–175 in *Optimization in Computer Aided Design* (Edited by J. S. Gero). North-Holland (1985).
7. H. C. Rodrigues and C. A. Mota Soares, Shape optimization of shafts. 3rd National Congress of Theoretical and Applied Mechanics. Lisbon (1983) (in Portuguese).
8. H. C. Rodrigues, Shape optimization of shafts using boundary elements. M.Sc. Thesis, Technical University of Lisbon (1984) (in Portuguese).
9. T. Burczynski and T. Adamczyk, Multiparameter shape optimization of a bar in torsion by the boundary element method. *Proc. 23rd Symposium on Modelling in Mechanics*, The Polish Society of Theoretical and Applied Mechanics. Gliwice (1984) (in Polish).
10. A. Zochowski and K. Mizukami, A comparison of BEM and FEM in minimum weight design, pp. 901–911 in *Boundary Elements* (Edited by C. A. Brebbia). Springer-Verlag (1983).
11. C. A. Mota Soares, H. C. Rodrigues and K. K. Choi, Shape optimal design of elastic structural components using boundary elements. *10th Int. Cong. on the Applications of Mathematics in Engineering Science*, pp. 80–82. Weimar (1984).
12. C. A. Mota Soares, H. C. Rodrigues and K. K. Choi, Shape optimal structural design using boundary elements and minimum compliance techniques. *J. Mech. Transm. Automat. Des.* **106**, 518–523 (1984).
13. R. P. Leal, Boundary elements in bidimensional elasticity. M.Sc. Thesis, Technical University of Lisbon (1985) (in Portuguese).
14. T. Burczynski and T. Adamczyk, The application of the boundary element method to optimal design of shape of structures. *Proc. 4th Conf. on Methods and Instrumentations of Computer Aided Design*. Warsaw (1983) (in Polish).
15. T. Burczynski and T. Adamczyk, The boundary element formulation for multiparameter structure shape optimization. *Appl. Math. Modelling* **9**, 195–200 (1985).
16. T. Burczynski and T. Adamczyk, The boundary element method for shape design synthesis of elastic structures. *7th Int. Conf. on Boundary Element Methods* (Edited by C. A. Brebbia). Springer-Verlag (1985).
17. D. Eizadian, Optimization of the shape of bidimensional structures by the boundary integral equation method. Ph.D. Thesis, National Institute of Applied Science of Lyon (1984) (in French).

18. D. Eizadian and Ph. Trompette, Shape optimization of bidimensional structures by the boundary element method. *Conf. on CAD/CAM, Robotics and Automation in Design.* Tucson, AZ (1985).

19. T. Futagami, Boundary element and linear programming method in optimization of partial differential systems, pp. 457–471 in *Boundary Element Methods* (Edited by C. A. Brebbia). Springer-Verlag (1981).

20. T. Futagami, Boundary element and dynamic programming method in optimization of transient partial differential systems, pp. 58–71 in *Boundary Element Methods in Engineering* (Edited by C. A. Brebbia). Springer-Verlag (1982).

21. T. Futagami, Boundary element method—Finite element method coupled with linear programming for optimal control of distributed parameter systems, pp. 891–900 in *Boundary Elements* (Edited by C. A. Brebbia). Springer-Verlag (1983).

22. M. R. Barone and D. A. Caulk, Optimal arrangement of holes in a two-dimensional heat conductor by a special boundary integral method. *Int. J. Numer. Meth. Eng.* **18**, 675–685 (1982).

23. R. A. Meric, Boundary integral equation and conjugate gradient methods for optimal boundary heating of solids. *Int. J. Heat Mass Transfer* **26**, 261–267 (1983).

24. R. A. Meric, Boundary element for static optimal heating of solids. *ASME J. Heat Transfer* **106**, 876–880 (1984).

25. R. A. Meric, Boundary element methods for optimization of distributed parameter systems. *Int. J. Numer. Meth. Eng.* **20**, 1291–1306 (1984).

26. P. K. Banerjee and R. Butterfield, *Boundary Element Methods in Engineering Science.* McGraw-Hill (1981).

27. G. Fairweather, F. J. Rizzo, D. J. Shippy and Y. S. Wu, On the numerical solution of two-dimensional potential problems by an improved boundary integral equation method. *J. Comput. Physics* **31**, 96–112 (1979).

28. C. A. Brebbia, J. Telles and L. Wrobel, *Boundary Element Techniques: Theory and Applications in Engineering*, Springer-Verlag (1984).

29. B. N. Pshenichny and J. M. Danilin, *Numerical Methods in Extremal Problems.* MIR, Moscow (1978).

30. I. M. Gelfand and S. W. Fomin, *Calculus of Variations.* Prentice-Hall (1961).

31. N. W. Banichuk, *Problems and Methods of Optimal Structural Design.* Plenum Press (1983).

32. F. Hartmann, Elastostatics, pp. 84–167 in *Progress in Boundary Element Methods* Vol. I (Edited by C. A. Brebbia). Wiley (1981).

33. E. Alarcon, L. Avia and A. Revester, On the possibility of adaptative boundary elements. *Int. Conf. on Accuracy Estimates and Adaptive Refinements in Finite Element Computations*, pp. 25–34. Lisbon (1984).

34. J. Rencis and R. L. Mullen, A self-adaptive mesh refinement technique for boundary element solution of the Laplace equation. *21st Annual Meeting of the Society of Engineering Science.* Blacksburg, VA (1984).

35. K. K. Choi, E. J. Haug, J. W. Hou and V. M. Sohoni, Pshenichny's linearization method for mechanical systems optimization. *J. Mech. Transm. Autom. Des.* **105**, 97–104 (1983).

36. J. W. Hou, E. J. Haug and R. L. Benedict, Shape optimization of elastic bars in torsion, pp. 31–55 in *Sensitivity of Functionals with Applications to Engineering Problems* (Edited by V. Komkov). Springer-Verlag (1984).

37. E. J. Haug, K. K. Choi, J. W. Hou and Y. M. Yoo, A variational method for shape optimal design of elastic structures, pp. 105–137 in *New Directions in Optimum Structural Design* (Edited by E. Atrek, R. H. Gallagher, K. M. Ragsdell and O. C. Zienkiewicz). Wiley (1984).

38. R. J. Yang, K. K. Choi and E. J. Haug, Numerical considerations in structural component shape optimization. *J. Mech. Transm. Autom. Des.*, to appear.

39. C. A. Mota Soares and K. K. Choi, Shape optimal design of structures based on stress constraints and boundary elements. *Advanced Study Institute on Computer Aided Optimal Design: Structural and Mechanical Systems.* Portugal (1986), to appear.

DISCUSSION

D. Vasilopoulos *(General Motors Research Laboratories)*

How does a boundary element analysis perform better with respect to the computational difficulties that arise from the singular adjoint loads?

Mota Soares

Although body forces are concentrated, singular adjoint loads in the mathematical equation are the same as the adjoint boundary tractions. We use exactly the same subroutines for both. However, we have had some difficulty with very small elements. We have to integrate them with nine Gauss points instead of four Gauss points, which is the normal integration. But the element is not singular; for example if it is a rectangle, we can integrate it analytically in some sides but we have to integrate with nine Gauss points in other sides.

Yet this means that there is some limitation of the size of the constraint. If it is very small, it becomes numerically unstable because we have all these very high self-equilibrating forces—at least 1,000 times the normal real force—concentrated in a small area. So there are some restrictions in the dimensions of our area constraint.

L. Schmit *(University of California–Los Angeles)*

Are stress constraints harder to handle than stiffness constraints?

Mota Soares

Much harder. We did the torsion problem in about one month. It took about three months to do the torsion problem with the hole, and probably four months were needed to do the compliance constraint. However, we have been working for more than one year and we have not been very successful with the stress constraints. Stress constraints are much more challenging–ten times more, I would say.

R. Haber *(University of Illinois at Urbana–Champaign)*

One issue that has come up repeatedly is that it is difficult to obtain accurate traction information from a finite element solution on the boundary of a structure—and that's clearly where the significant information is in the shape optimization problem. Actually, there are new methods (not the commonly used stress extrapolation techniques) based on Galerkin concepts that yield highly accurate traction information. All that is required is an inexpensive postprocessing computation following a conventional displacement solution. Surface tractions—not the stress tensor—are computed directly in this method. The finite element method does not pose insoluble problems for getting good traction information.

Mota Soares

We have done some investigation on that with reference to the hole in the plate problem. In that particular case the von Mises stresses vary rapidly with the radial distance. The stress concentration factor very quickly goes from one to two. So when using the finite element method to calculate the stresses at the boundary or near the boundary of the hole, the error must be big, even for very fine discretizations. In this particular problem, the boundary element method calculates a stress factor of two even with coarse models. It doesn't surprise me that in this problem the optimal solution has a very high rate of variance. The optimal solution and the stress at the optimum vary rapidly, and I don't think the finite element model can accurately calculate the boundary stresses in this case.

C. A. Mota Soares

Haber

We have been able to handle Griffith's problem with the new traction recovery method. If you know the form of the singularity, you can incorporate it in your trial functions and get very accurate results. There is one catch, which is interesting in light of the earlier discussion: the traction recovery method does not work well at a corner (unbounded curvature).

Mota Soares

I would say it does not work well because we cannot calculate the stresses accurately at corners. However, if the corner is placed where the stress is not very high, it can be handled. We have solved some problems with corners and got good solutions.

R. Bernhard *(Purdue University)*

You mentioned that you once had a little trouble with linear and parabolic elements. Can you describe that experience?

Mota Soares

The constant element problem is integrated analytically, so the program is no more than 20 to 40 FORTRAN statements. Clearly, it is accurate. However, when we attempted to integrate it numerically with parabolic elements, we had some trouble which we did not investigate further at the time. We have been pursuing it in the last three months.

B. Prasad *(Electronic Data Systems)*

Is the number of boundary points important to give you a good preliminary design? There has not been a lot of discussion on the adaptive node refinement. Should this be applied in boundary element techniques?

Mota Soares

All my experience indicates that we do not need very sophisticated models. For instance, on the plate with a hole, we started with 72 degrees of freedom but solved it with 40 or 50 degrees of freedom. I think that the problems lie more with the design variables and not with our model itself. The design variables give too much freedom to the design. Then it just tries to find a solution where the stresses are acceptable, when in practice they are not because of the kinks. But on the average it is correct. There is no doubt that you have to avoid using nodal coordinates as design variables and use distributed parameters instead.

E. Atrek *(Engineering Mechanics Research Corp.)*

You said there was some trouble with stress constraints, but I am not sure if it was due to formulation or to stress concentration.

Mota Soares

I think my problems came from having to evaluate stresses and adjoint strains on a small element, not because of the boundary elements themselves.

Atrek

If you were working with a smooth boundary, would your problem still be there?

Mota Soares

No. When we calculate the sensitivities, going from a smooth hole to another smooth hole, the sensitivities are good. Otherwise, the sensitivities are bad.

SESSION III
APPLICATIONS

Session Chairman
R. H. GALLAGHER

Worcester Polytechnic Institute
Worcester, Massachusetts

SHAPE OPTIMIZATION OF THREE-DIMENSIONAL STAMPED AND SOLID AUTOMOTIVE COMPONENTS

M. E. BOTKIN, R. J. YANG and J. A. BENNETT

Engineering Mechanics Department
General Motors Research Laboratories
Warren, Michigan

Abstract

The shape optimization of realistic, three-dimensional automotive components is discussed, stressing the integration of the major parts of the total process—modeling, mesh generation, finite element and sensitivity analysis, and optimization. Stamped and solid components are treated separately. For stamped parts, a highly automated capability has been developed; the problem description is based upon a parameterized boundary design element concept for the definition of the geometry. Automatic triangulation and adaptive mesh refinement are used to provide an automated analysis capability which requires only boundary data and takes into account sensitivity of the solution accuracy to boundary shape. For solid components, a general extension of the two-dimensional boundary design element concept has not been achieved. In this case the parameterized surface shape is provided by using a generic modeling concept based upon isoparametric mapping patches which also serve as the mesh generator. Emphasis is placed upon the coupling of optimization with a commercially available finite element program. To combine these, it is necessary to modularize the program architecture and to obtain shape design sensitivities using the material derivative approach so that only boundary solution data are needed. Several realistic component designs demonstrate the effectiveness of both capabilities.

INTRODUCTION

Although structural optimization for sizing variables has been treated extensively in the literature for many years [1, 2], the problem of designing the shape of a structure for minimum mass is a comparatively new research topic [3–5]. Although earlier work [6–8] stressed the need for automatically modifying the mesh as the structural shape changes, limitations in the boundary representation and mesh generation kept the capability from being truly automatic. Ultimately, one would like merely to describe the function of the structure to the computer in some convenient manner and then allow the program to produce the optimum design automatically [9]. The basic requirements to do this are as follows: 1) the design model that describes the initial shape of the structure, the loads and constraints, and the design requirements; 2) the analysis model, including the finite element mesh created using fully automatic mesh generation and improved by adaptive mesh refinement; and 3) the design modification, employing a numerical optimization process that iteratively improves the design until it converges to the optimum. Each of these topics and their implementation in the design program will be discussed.

Previous authors have not addressed the problem of handling the more general case of designing nonplanar parts. Here the major difficulty is in modeling, in a parametric sense, all of the three-dimensional geometry. To do this it is necessary to extend the existing capability for flat parts using an assembly process of the two-dimensional segments. Furthermore, the ability to add curvature to planar segments can be provided through the superposition of surface interpolation and transformation capabilities.

For solid components, very little research has been reported [7, 8]. In this chapter emphasis will be placed upon two major aspects of the problem. The first is the efficient calculation of the sensitivities of the displacement and stresses. Second, the idea of using one of the many commercially available finite element codes is attractive in order to alleviate the burden of software support of an analysis program sophisticated enough to handle solid models. Both of these issues are addressed.

The integrated design processes described in this paper stress the necessity for treating realistic, three-dimensional design problems typical of those found in automotive design. For this reason, the shape design element descriptions would be best suited for interfacing with the computer-aided drafting systems on which the geometry is initially created. In addition, it is absolutely necessary to have a capability which is as automatic as possible to free the engineer from the burden of finite element creation and design modification.

SHAPE OPTIMIZATION OF SHEET METAL PARTS

Design Model Description—A significant number of structural components, such as the typical part shown in Figure 1, are produced from a single sheet of uniform thickness material. Using conventional optimization techniques in which element thicknesses are the design variables, little reduction of mass can be achieved. To reduce the mass further, the shape of the part and the location of the cutouts must be represented by design variables. The resulting design model must provide the description of the boundary geometry as a function of the design variables and also must create the finite element structural model. To be most effective in impacting the design process, this information must be efficiently gen-

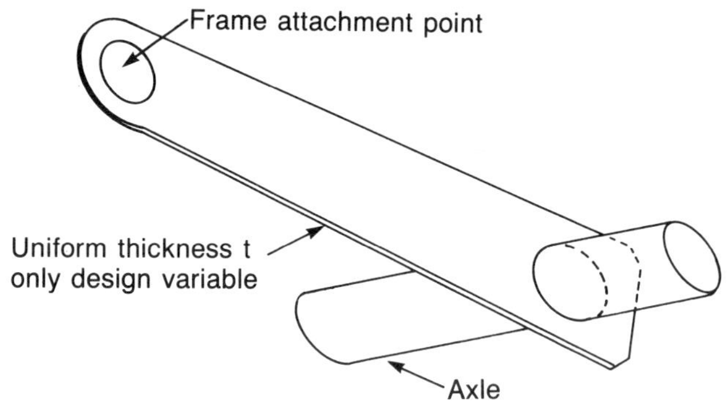

Figure 1. Typical part.

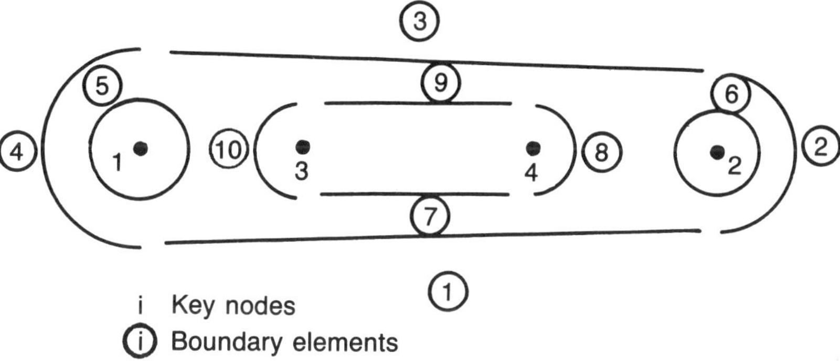

Figure 2. Boundary elements for typical part.

erated from conceptual sketches of the part or obtained through an interface to a computer-aided drafting (CAD) system. For that reason, the approach represented in Figures 2 and 3 has been chosen. The part shown in Figure 1 has been modeled in Figure 2, using what will be referred to as *boundary design elements*. As well as associating the boundary with design variables, the boundary design elements are also used to define the stress constraints. Each boundary design element will be associated with at least one stress constraint that will be computed from the maximum stress of all the finite elements touching that boundary design element. The loads and structural boundary conditions are related to a set of reference nodes which are shown as key nodes in Figure 3. This information is in turn automatically transferred to the finite element model once it has been generated.

Mesh Generation—Other work [6, 10] has stressed the need for automatically modifying the mesh as the structure changes shape, but it was observed that the commonly used mesh generation techniques based upon coarse isoparametric or transformal mapping patches imposed limitations on the ability to treat large variations in shape. While these techniques do redistribute interior nodes as boundaries

References pp. 255–257

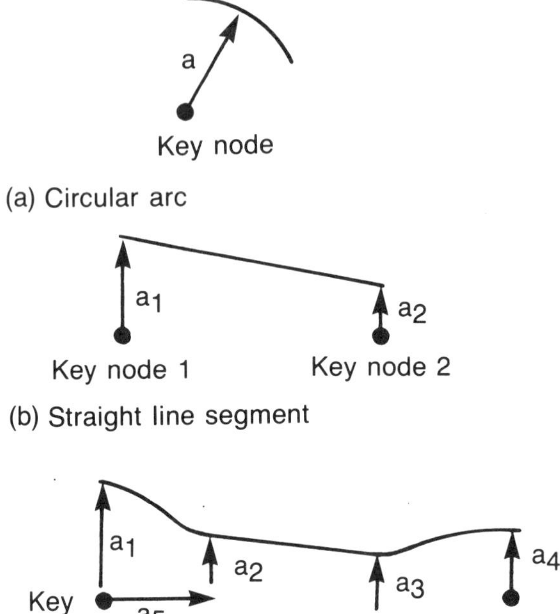

Figure 3. Boundary design elements.

move, aspect ratios may become objectionably large as the shape becomes significantly different than the initial shape. Mesh grading and solution accuracy are also difficult to control.

As an alternative to more traditional mesh generation methods, the use of fully automatic mesh generation based only upon boundary points coupled with adaptive refinement has been proposed [11]. This technique is capable of generating a nearly uniform initial mesh of triangular elements given a set of uniformly spaced boundary points. Thus, as the design changes, uniform triangular meshes can be recreated at any time.

After the design model has been created, the boundaries are automatically discretized into uniform segments called the characteristic length (CL) which is an input value. Automatic triangulation [12, 13] is used to create a nearly uniform mesh from the set of boundary points and from a set of points placed uniformly throughout the region's interior in approximately the same density as the boundary points. This process of creating the uniform mesh is repeated at each step in the design for which a new boundary description has been generated.

Adaptive Mesh Refinement—When the configuration of the boundary changes, it is not possible to assure the accuracy of the mesh as the shape changes, since the accuracy of various portions of the mesh will change. The concepts of adaptive mesh refinement can be incorporated to help resolve this difficulty [11].

The mesh refinement process is based upon the variation in strain energy density (SED) as a measure of the error in an element. Once SED variations have been determined for all elements, those elements that have undesirably high values must be selected for subdivision. Elements so selected define refinement regions which can be easily identified by graphical contouring. Since it is not practical from a computational standpoint to consider more than a two-step refinement process during the optimization (one initial and one refined analysis), a concept of multiple refinement regions has been implemented in an attempt to enhance convergence. As an example of the process Figure 4(a) represents a uniform finite element mesh, created using the triangulation technique described previously. Several refinement regions can be specified, as shown in Figure 4(b), so that the resulting mesh in Figure 4(c) will be more uniformly graded from coarse to fine. The elements in the region of highest SED variation, represented by the smallest dots in Figure 4(b), are approximately one-fourth the size of the initial grid. The region represented by the larger dots contains elements approximately one-half the initial grid size. As many as six regions can be specified, uniformly graded down to one-eighth of the original grid size. The size of the regions can be varied depending upon the selection of an input parameter.

Obviously the accuracy due to any refinement is unknown in advance. Although numerous papers have been written [14, 15] on error estimates of total strain energy, this work has not been extended to stresses and displacements. For the iterative design process described in this paper, it is desirable to have a conservative estimate of the converged finite element solution. This information may be obtained in an approximate manner using linear extrapolation, graphically represented in Figure 5. This figure shows a typical relationship between a solution quantity and mesh size in the absence of a singularity. Several steps of refinement are shown, with each step having reduced the element size in half. The solution will eventually converge to S_e, and the slope of the curve reflects the rate of convergence. A conservative estimate of the converged solution, represented by points S_i and S_o, may be obtained by extrapolating data points produced by one unrefined analysis and one refined analysis. The extrapolated values will be used as stress constraints.

In order that more realistic three-dimensional plate structures can be analyzed, accurate refinements are necessary for finite elements with bending deformation. In general, refinement works best for conforming elements such as the constant strain triangle already described. Meshes composed of these elements are always too stiff, and solution convergence is predictable as shown in Figure 5. On the other hand, meshes composed of nonconforming elements may switch from too stiff to too flexible as the refinement progresses. However, the triangular bending element used in this study [16] has been formulated in such a way as to reduce the degree of nonconformity, and convergence studies show that for uniformly refined meshes the element is always too stiff. Several examples presented in reference [11] indicate

References pp. 255–257

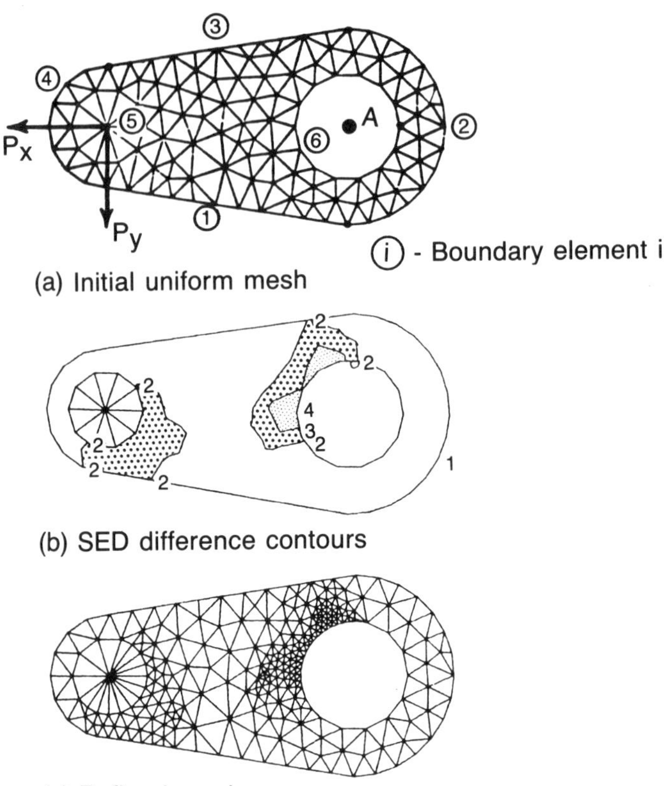

Figure 4. Mesh refinement.

that although the results are not as predictable as for the constant strain triangle, they are quite satisfactory.

Extension To Nonplanar Parts—The design process described has been extended to handle more realistic stamped sheet metal parts [17]. This was accomplished by treating the part as an assembly of the two-dimensional segments described above. Each segment has one completely closed exterior boundary which may contain one or more interior cutouts. Segments may be joined along straight sides to form more complex assemblies. Furthermore, segments may be rotated along the joined edges to form three-dimensional geometry, as shown in Figure 6(a). Because each segment is represented by two-dimensional boundary information only, the addition of surface curvature to a planar segment for added stiffness must be addressed separately. Large curvature, such as the cylinder in Figure 6(b), is handled through the definition of a cylindrical coordinate system for the segment alone. All nodes in that segment are transformed to the new surface. Small curvatures are treated by direct projection as shown in Figure 6(c). The final assembly process can be seen in Figure 7, in which all the three-dimensional

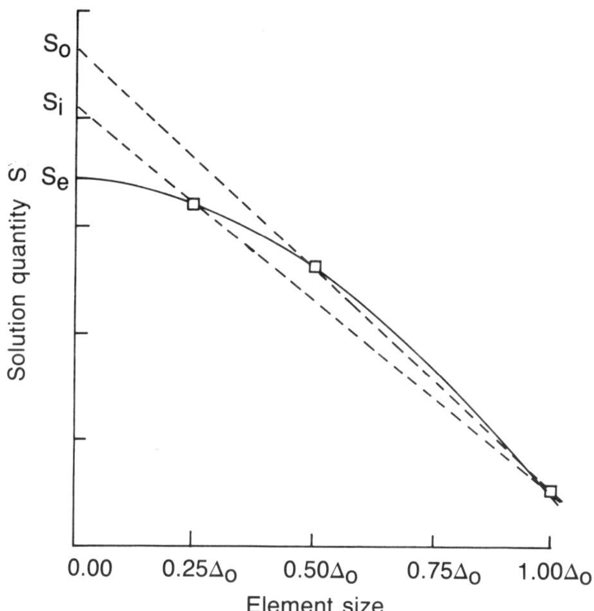

Figure 5. Typical solution convergence.

geometry has been expressed in terms of a small number of parameters that can be treated as design variables.

Interactive Graphics Geometrical Modeling—The need to model more complex geometries makes it obvious that some form of model preparation based upon graphics-oriented preprocessing is necessary. Unfortunately, existing finite element preprocessors cannot be used directly because they offer no means of parameterizing the shape of the model. Although some of the more recently developed modelers do include boundary functions such as splines, there are no design parameters available externally for use with other programs. Furthermore, since the finite element mesh must change to reflect shape changes, the loads and constraints must be associated with boundary functions instead of being directly applied to the finite element mesh, as in the typical modeling system.

As a result, a special graphics preprocessor for shape optimization was developed [18] which allows a user to create a paramaterized finite element model. A part is modeled as a collection of planar part segments, which are then assembled to form a three-dimensional plate structure. Design variables define the shape of each part segment. Loads and constraints are applied to finite element nodes through boundary functions instead of being applied directly to the nodes.

Next, commands and cross-hairs are used to create the key nodes and boundary design elements that define the geometry of the part to be optimized. Figure 8(a) shows the six key nodes needed to define the boundary of a planar triangular bracket. Three exterior key nodes locate the perimeter of the part, while

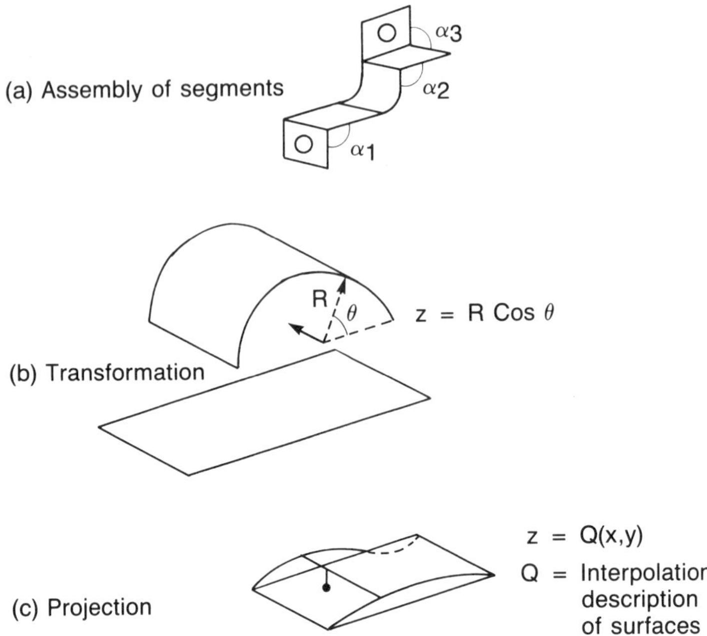

Figure 6. Three forms of nonplanar structures.

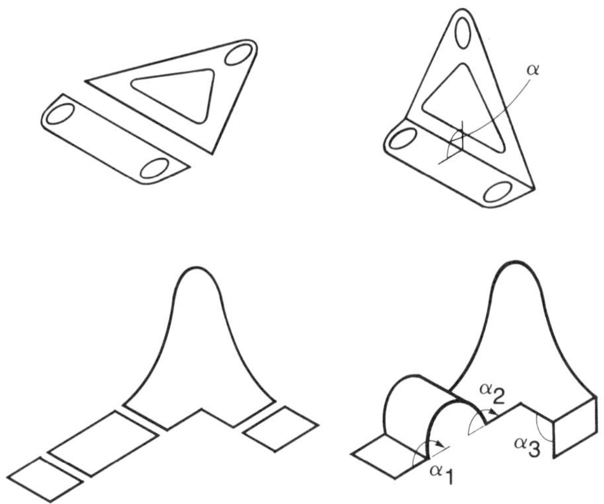

Figure 7. Assembly and rotation of segments.

three interior key nodes locate an interior cutout boundary. Associated with each key node is a radius, represented as a circle in Figure 8(a). The radius, as well as the x and y coordinates, are automatically designated as design variables.

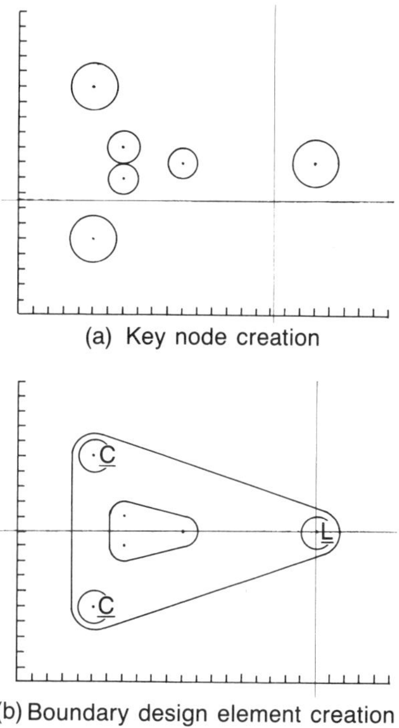

(a) Key node creation

(b) Boundary design element creation

Figure 8. Interactive creation of design model.

Once the necessary key nodes have been created, the cross-hairs are used to connect the key nodes and create the boundary design elements, as shown in Figure 8(b). If the same key node is selected twice, a circular arc boundary design element is created. A circular arc element can be used to represent a round boundary, a fillet or a circular hole. If two different key nodes are selected, the user can choose to connect the two key nodes with either a straight boundary design element or a double cubic boundary design element, as in Figure 3. All design variables specified for a particular element type are automatically assigned when the element is created. Commands are available to link design variables as required for the design model.

Other commands are available to be used for applying constraints or loads to a given boundary. The terminal cross-hairs are first used to select the boundary to be supported or loaded. The user is then prompted for a constraint type or a load magnitude and direction. The constrained boundaries are indicated by the letter C, while the loaded boundaries are indicated with an L, as shown in Figure 8(b). At the time when loads are applied, optimization constraints on displacements can also be specified.

Most real production parts, however, have more complex geometries than these examples. For instance, a common manufacturing operation used to add stiffness to a planar part involves adding a lip, or flange, along the edge of the part.

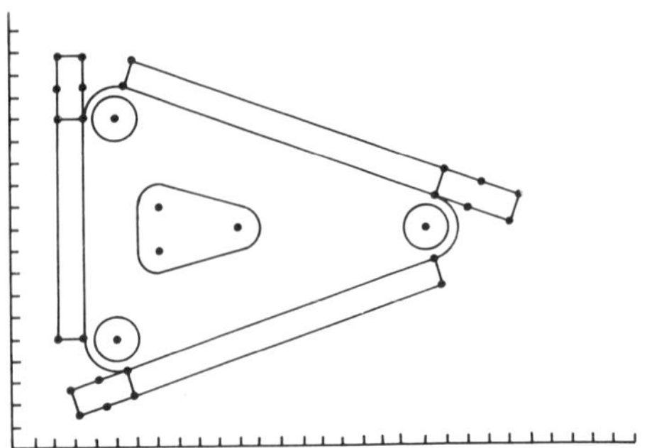

(a) Flanges added to triangular bracket model

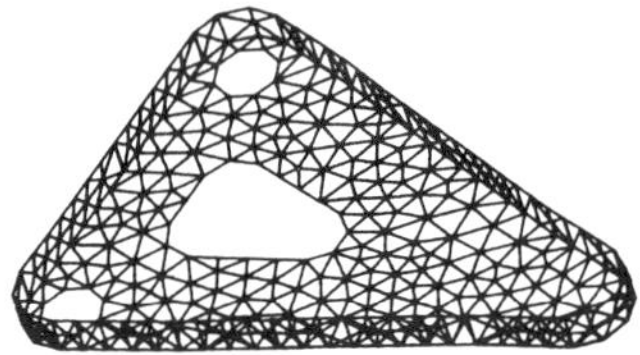

(b) Triangular bracket finite element mesh

Figure 9. Interactive creation of bracket with flanges.

Modeling such a part with a conventional finite element preprocessor is relatively simple; but if the design of the part is to be automated, the geometric model of the part must fulfill the requirements already mentioned. Commands are available to create the multiple part segments illustrated in Figure 7. An additional command can be used specifically for creating flanges, which automates some of the steps in creating multiple segments.

Figure 9(a) shows six flanges added around the perimeter of the triangular bracket. A flange is added by using the cross-hairs to locate the portion of the boundary for which a flange is desired. The user is then prompted to specify the flange height at each end. The model is completed by specifying the angle that each flange is rotated relative to the base part to form a three-dimensional model. This angle is normally 90 degrees. Each of the six flanges, as well as the base triangular bracket, is a separate part segment on which a finite element mesh is generated. Figure 9(b) shows the assembled finite element model of the triangular bracket, generated from the boundary shape information created with the preprocessor.

THREE-DIMENSIONAL SOLID COMPONENTS

Only a limited amount of work has been accomplished in three-dimensional shape optimization using solid finite element analysis [7, 8]. Because a fully automatic mesh generation scheme which relies only on surface data [19] has yet to be developed, the boundary description format for thin parts cannot be implemented for solid three-dimensional parts. Instead, it will be assumed that surface representation and mesh generation will be handled by a generic modeling

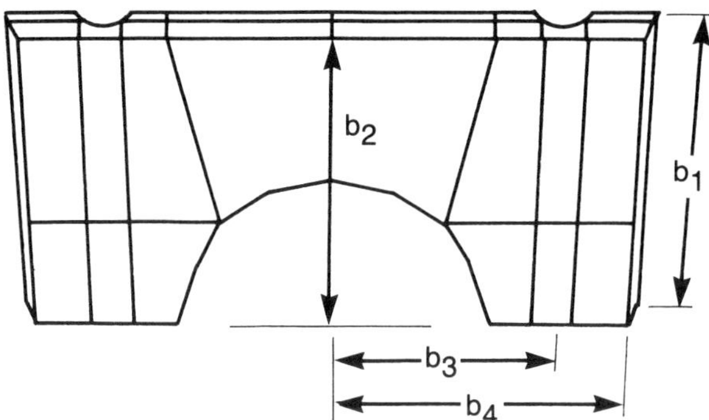

Figure 10. Generic model of engine bearing cap.

scheme based upon isoparametric mapping patches, such as the one described in [8] and shown for a typical part in Figure 10.

In this section, design sensitivities and program architecture are addressed. Work in both these areas was largely driven by the desire to use a variety of structural analysis programs (NASTRAN, ANSYS, ADINA, etc.) with a relatively small amount of additional program development. In this study, NASTRAN was used for analysis.

Design Sensitivity Analysis—The variational design sensitivity theory uses the material derivative concept of continuum mechanics and an adjoint variable method to obtain computable expressions for the effect of shape variation on the functionals that arise in the shape design problem. The resulting expressions provide analytical sensitivities of structural response.

The variation of the displacement functional ψ with respect to shape change is derived by differentiating the variational equilibrium equation and employing the adjoint variable method, to obtain [20–22]

$$\partial\psi/\partial\mathbf{b} = -\int_{\Gamma} \sigma^{ij}(\mathbf{z})\varepsilon^{ij}(\boldsymbol{\lambda})n^T \partial\mathbf{r}/\partial\mathbf{b}\, d\Gamma. \qquad (1)$$

This equation is an integral along the perturbed boundary in which the required data for evaluation are the stresses from the actual load σ^{ij}, the strains from the adjoint load ε^{ij}, the position vector $\mathbf{r}$, and the design variable vector $\mathbf{b}$. It should be pointed out that in equation (1), assumptions have been made in the derivation so that the kinematically constrained boundaries and loaded boundaries are assumed to be fixed, and the variation of the displacement functional is only affected by the normal movement of the boundary of the physical domain. Physically, the adjoint solution required in equation (1) is calculated by applying a unit load at the point where the displacement is of interest.

To see the advantage of equation (1) a comparison can be made with the well-known expression for design sensitivities resulting from the implicit differentiation of the finite element equations (see Chapter 3 in this volume):

$$\partial \mathbf{z}/\partial \mathbf{b} = -\mathbf{K}^{-1}\partial \mathbf{K}/\partial \mathbf{b}\, \mathbf{z}. \tag{2}$$

This equation evaluates the displacement derivative by computing derivatives of the terms of the stiffness matrix. There are two shortcomings to this approach. First, obtaining analytical expressions for the stiffness matrix derivatives is very difficult for boundary movements. These expressions are, in general, different for each element type, thereby requiring special computer code for each different element type. For this reason, a finite difference method is generally used to obtain stiffness derivatives, and this usually requires a judicious choice of the step size to maintain accuracy. Finally, if it is desired to use a program for finite element analysis—and if the source code is not available, it is very difficult to manipulate the stiffness matrices to compute the needed derivatives. For these reasons, equation (1) is a more desirable expression for computing displacement sensitivities. Most programs store the needed stresses and strains on files to be used by a postprocessing routine to obtain the derivatives.

The stress variation also can be derived to obtain an expression similar to equation (1), except that the discontinuity of the stresses along the interelemental boundaries has to be properly handled. A characteristic function that averages stress over a small region is introduced to treat stress constraints in references [23] and [24]. This approach is similar to using the finite element center as the stress constraint point if the element is chosen as the small region and may lead to a misleading constraint value. It may also result in an undesirable or inaccurate optimum shape if the finite element model is inadequate [24].

An alternative that avoids this problem is to obtain the stress sensitivity at a point, using the definition of stress computation in finite element analysis. The elemental stresses are computed by using

$$\boldsymbol{\sigma} = \mathbf{D}\mathbf{B}\mathbf{z}^e \tag{3}$$

where $\mathbf{D}$ is the elasticity matrix, $\mathbf{B}$ is the strain recovery matrix that contains the derivatives of shape functions, and $\mathbf{z}^e$ is an elemental displacement vector. Differentiating equation (3) with respect to the design variables b, one obtains

$$\boldsymbol{\sigma}_i' = \mathbf{D}(\mathbf{B}(\mathbf{z}_i^e)' + \mathbf{B}_i'\mathbf{z}^e) \tag{4}$$

where the subscript i with a prime superscript indicates the derivative with respect to the ith design variable. Notice that the first term on the right side of equation (4) is only a combination of displacement gradients; it can be obtained by applying a combined adjoint load to the system and using the same formula of equation (1).

The primed matrix of the second term of equation (4) can be evaluated from the derivative of the nodal coordinates with respect to shape design parameters [25]. The matrix can be computed analytically or by using a finite difference method. For a linear shape function element, such as the constant stress triangular element, the matrix $\mathbf{B}'$ vanishes; for a quadratic element, the $\mathbf{B}'$ matrix is constant. Therefore, the finite difference method is sufficient to evaluate the $\mathbf{B}'$ matrix except when a higher-order element is used. In this study, analytical derivatives are used for $\mathbf{B}'$ and the eight corner points of the solid element are chosen as the stress constraint points.

Modularized Program Architecture—It was desired to have a system which uses a commercial finite element code as the analysis capability because these codes have gained widespread acceptance in the structural analysis community. A major drawback to achieving this goal is that most commercial finite element codes cannot be used as a subroutine. This problem was addressed by building a system of independently executable program modules in which the overall execution is controlled by Job Control Language.

The modularized system consists of a mesh generator, the finite element code (NASTRAN), the adjoint load and constraints definition program, a design sensitivity analysis module and an optimization module. Each of the components is an independent program and is treated as a module. The flow chart of the system is shown in Figure 11. Initially, one has to generate a generic model for the structural component and then create a NASTRAN data deck for the NASTRAN run. The whole cycle of the system proceeds as follows: run the NASTRAN code for the actual load; calculate the cost function, constraints, and the adjoint loads using the NASTRAN output; rerun the NASTRAN code for the adjoint loads; and perform the design sensitivity analysis and optimization to obtain a new design. Finally, a new finite element mesh and NASTRAN data deck for the new design are generated.

The MSC/NASTRAN version 63 finite element code is employed for analysis. The new feature of the NASTRAN data base is used to save computing time for reanalysis of the adjoint loads. This data base, created by the first NASTRAN run, preserves the stiffness and boundary condition information, and it results in easier input data preparation and less computing time for the reanalysis. The displacements, stresses and geometric information that are needed for design sensitivity calculation are obtained by using an ALTER feature in NASTRAN to write that information on a file for postprocessing.

The ADJLOD module (Figure 11) is used to define the cost function and constraints for the design problem, and to calculate the adjoint loads for the constraints which are active or violated. The displacements, stresses and geometric information from the NASTRAN output are first read to define the constraints for the structural component. A NASTRAN deck containing the adjoint loads is then created for reanalysis.

The SENSTY module (Figure 11) performs the design sensitivity analysis for the cost and the active constraints, and then performs the optimization process by calling the optimizer (CONMIN[26]) as a subroutine. Before executing the module, the NASTRAN output files for the actual load and the adjoint loads should be available. The module then changes the design and creates new input data for the

References pp. 255–257

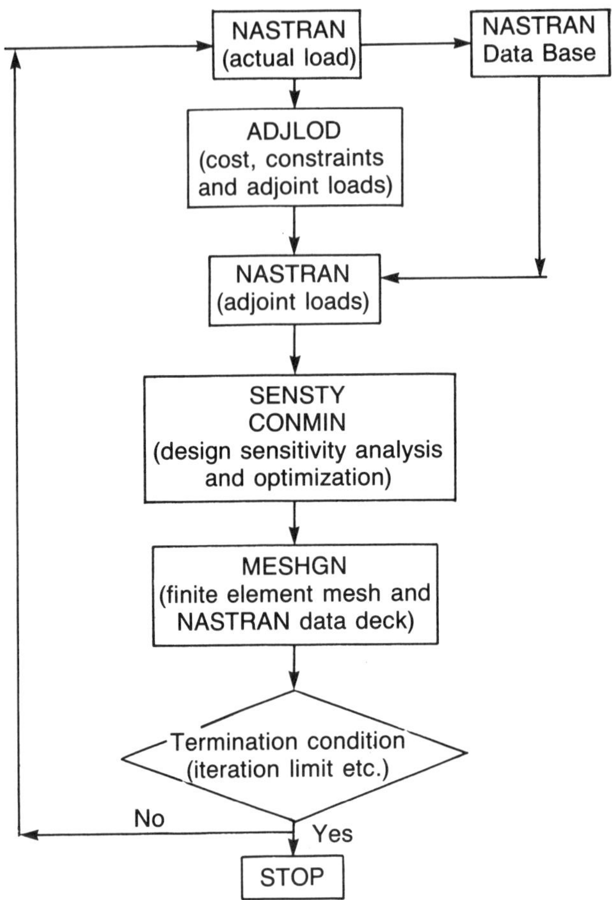

Figure 11. Flow chart of modularized system.

MESHGN module which will generate a new mesh and a new NASTRAN data file for the next design iteration, if necessary.

DESIGN EXAMPLES

Three-Dimensional Sheet Metal Part—Figure 12 shows the initial shape and dimensions of a realistic design example of a sheet metal part [17]. The model was initially created in two dimensions and then segments 2 and 4 were transformed into the third dimension. Structural boundary conditions were imposed around holes C and D. Loads P_1 and P_2 were applied at hole A in the y and z directions, respectively. Load P_3 was applied at hole B in the y direction. The design criteria were a stress limit on all boundaries and a displacement limit at hole A. CL was chosen to be 0.80 cm for the initial mesh.

The current model is similar to an earlier sheet metal part [17], except that

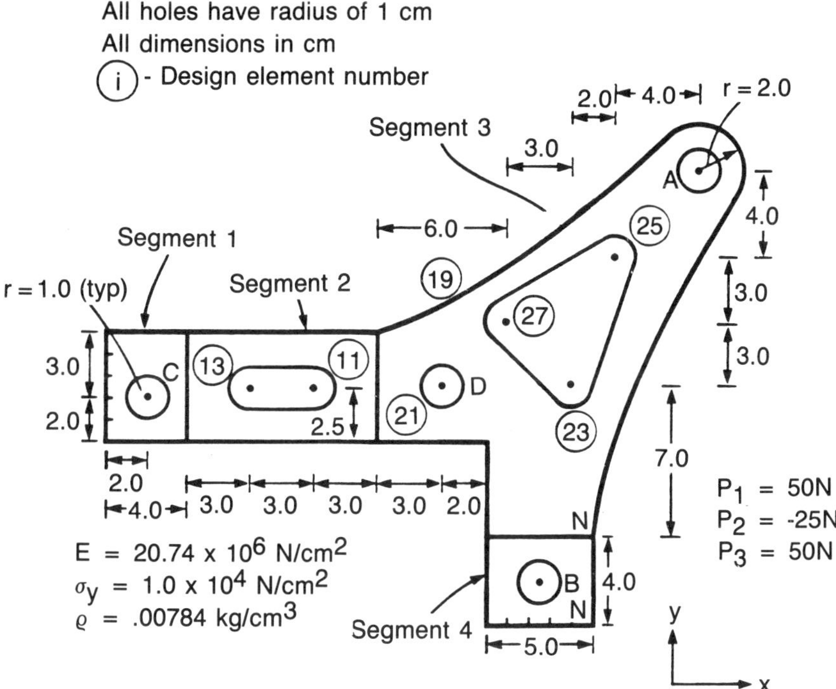

Figure 12. Dimensions of example.

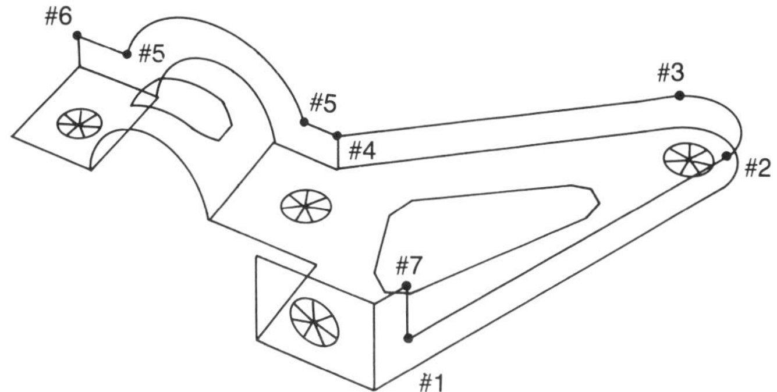

Figure 13. Flange design variable locations for example.

flanges on the new model add seven flange design variables to the problem. The locations of these design variables are shown in Figure 13. A total of 19 design variables were used to parameterize the part's shape. Figure 14 shows the initial, unrefined finite element mesh.

This part was modeled to determine how the program would reduce the mass and tailor the flanges, subject to a displacement constraint. A displacement constraint was applied to hole A, such that the displacement of the point was limited to one mm in the $-z$ direction. Figure 15 shows the initial and final part

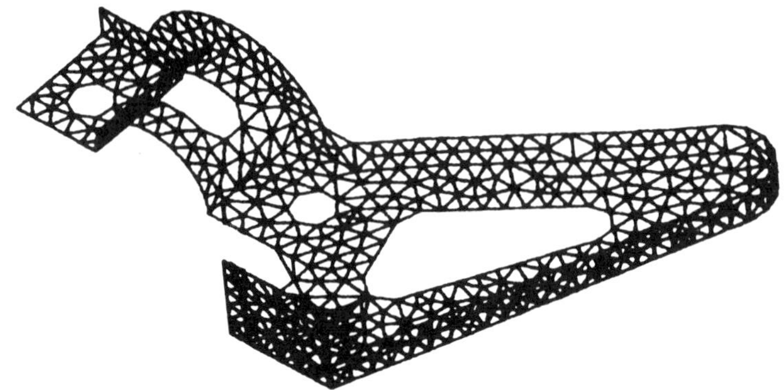

Figure 14. Transmission bracket finite element mesh.

designs. The program removed material from the interior cutouts on the base triangular part segment and from the cylindrical part segment. A small amount of in-plane curvature was added along the edges of the triangular part segment to which flanges are attached. Flange heights were reduced to less than half the initial values everywhere except along the upper edge of the triangular part segment. The flange heights along this edge are controlled by flange design variables 3 and 4 as shown in Figure 13. This edge serves as the primary load path for the structure because it transfers the load from the tip of the triangular part segment to the support points. As a result, one would expect the flange along this edge to be the most important in maintaining the stiffness of the part. The flange design variable values for the initial and final designs are given in Table 1.

Figure 16 shows the design history for this part. A design variable move limit of 5 percent was used for the first ten steps, followed by a move limit of 2.5 percent for the last fourteen steps. The characteristic length was reduced from .8 to .6 in the last four steps to obtain more accurate displacement values in the unrefined analyses. Reduction of the characteristic length eliminated the design oscillations that emerged once the displacement constraint became active. The initial unrefined finite element mesh included 3,000 degrees of freedom, while the initial refined mesh contained 4,000 degrees of freedom.

Finally, some comments are in order concerning the results. First, the design history in Figure 16 does not show traditional convergence behavior. The optimizer was turned off when it was judged that further mass reduction would require an excessive amount of computer time. Second, the finite element accuracy in the flange areas might be questioned. Constant strain elements were used, and only one or two elements were used to span the depth of each flange in the unrefined mesh. Bending of the flanges could result in stress variations that would not be picked up by so few constant strain elements. For this reason, the automatic mesh refinement technique described above was also used in the flanges to minimize this error.

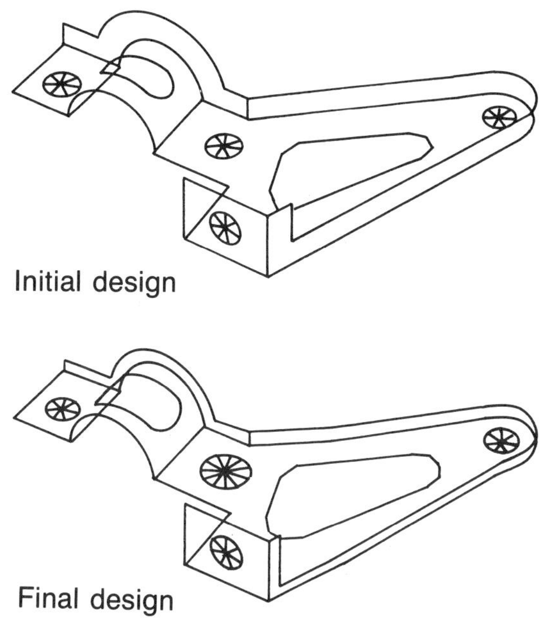

Figure 15. Initial and final designs for example.

Table 1
Flange design variables for transmission bracket

No.	Initial	Final	Lower bound	Upper bound
1	2.12	0.91	0.5	3.0
2	1.50	0.68	0.5	3.0
3	1.50	0.89	0.5	3.0
4	1.50	0.89	0.5	3.0
5	2.12	0.96	0.5	3.0
6	1.50	0.68	0.5	3.0
7	1.50	0.67	0.5	3.0

Three-Dimensional Solid Part—An idealized engine connecting rod, which connects the crank shaft and piston pin of an engine and transmits an axial compressive load during firing and a tensile load during the intake cycle of the exhaust stroke, is employed as the example. Shape optimization of similar components have been studied by Yoo et al. [27] and Yang et al. [28] assuming a plane stress state. We will discuss a fully three-dimensional shape optimization for the connecting rod.

Figure 17 shows the generic model for the connecting rod. For simplicity, the right hole of the connecting rod which connects the piston pin is fixed to eliminate rigid body motion; and the arbitrarily selected pressure of 3,000 MPa is applied to

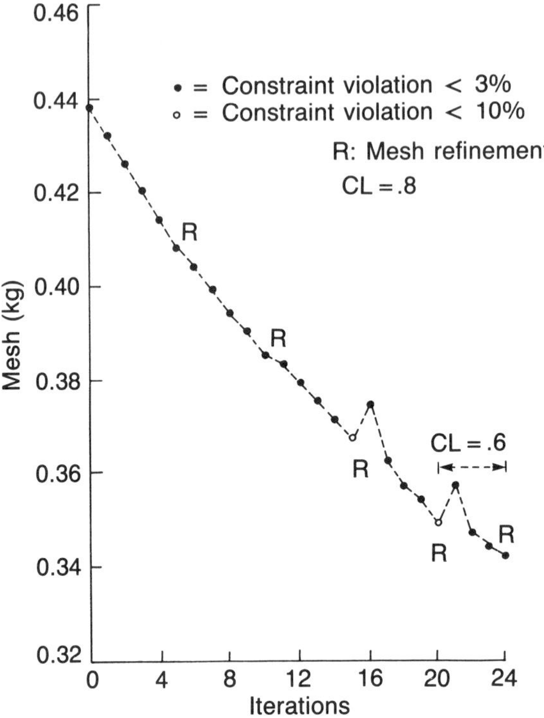

Figure 16. Design history for example.

the left hole (from 0 to 90 degrees) to simulate the firing forces. The von Mises stress constraint is imposed at each node in the finite element model of the connecting rod. The critical yield stress used for analysis is chosen as 3,000 MPa. Young's modulus and Poisson's ratio are 10.0 x 10E6 MPa and 0.3, respectively. The numerical data were selected to demonstrate the use of the system and are not representative of a specific production part.

Using the symmetrical conditions, only a quarter of the structure needs to be analyzed. The design variables are shown in Figure 17. In this model, eight design variables are chosen; five parameters define the shape of the shank and neck regions, two are the outer radii of the right and left holes and one parameter defines the height of the web. The finite element model, displayed in Figure 18, contains 105 solid (20 node) elements, 928 nodal points and 2,126 degrees of freedom.

The initial values of the design variables appear in Table 2. Initially, the volume is 15,686.7 cu mm with no stress violation. After 20 design iterations, the volume is reduced to 7,217.8 cu mm with no stress violation. The final design variables and the final shape are shown in Table 2 and Figure 17, respectively. Figures 19 and 20 show the design histories for the cost and the maximum constraint values, respectively, of the idealized connecting rod. In Figure 19 one observes that the convergence rate is reasonably good. From design iterations 10 to 17, the

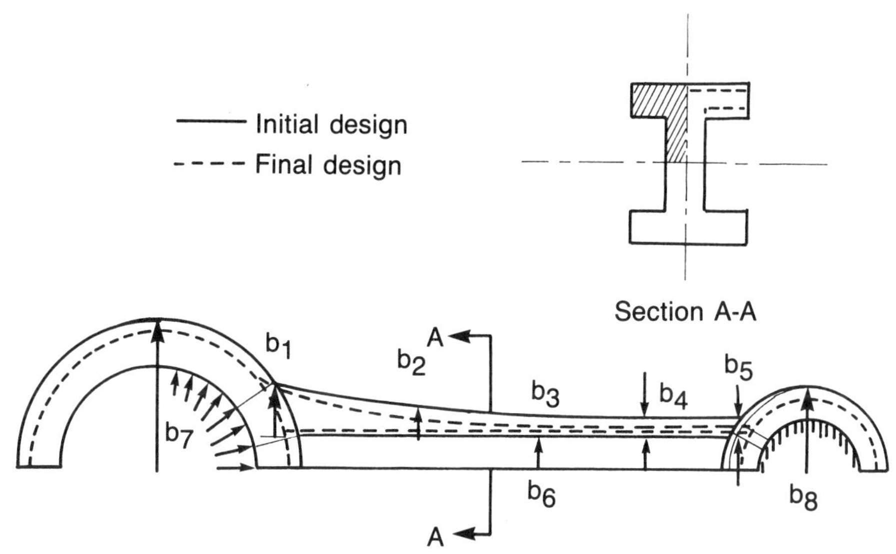

Figure 17. Generic model of engine connecting rod.

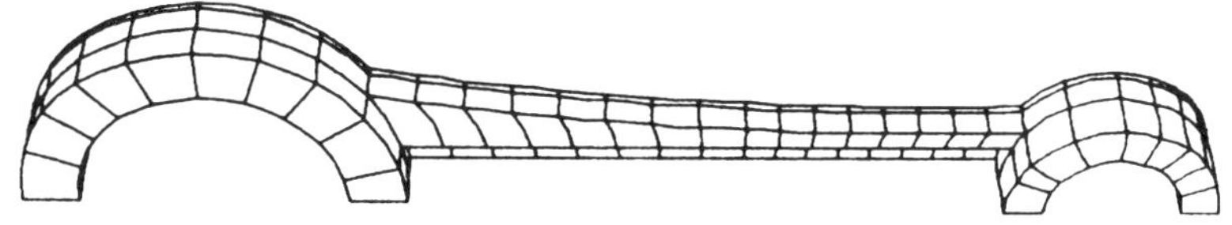

Figure 18. Finite element mesh of connecting rod.

optimizer tries to force the design into the feasible region. The slow correction for stress violation shown in Figure 20 may result from Taylor's series expansion approximation for the constraint functions.

Table 2
Design variables for engine connecting rod

No.	Initial	Final	Lower bound	Upper bound
1	10.956	12.512	0.1	100.0
2	6.37	2.8478	0.1	100.0
3	3.9667	1.4220	0.1	100.0
4	3.0024	1.0964	0.1	100.0
5	3.2711	1.2733	0.1	100.0
6	6.8156	7.2219	0.1	100.0
7	31.271	26.461	24.0	100.0
8	17.553	13.300	13.3	100.0

References pp. 255–257

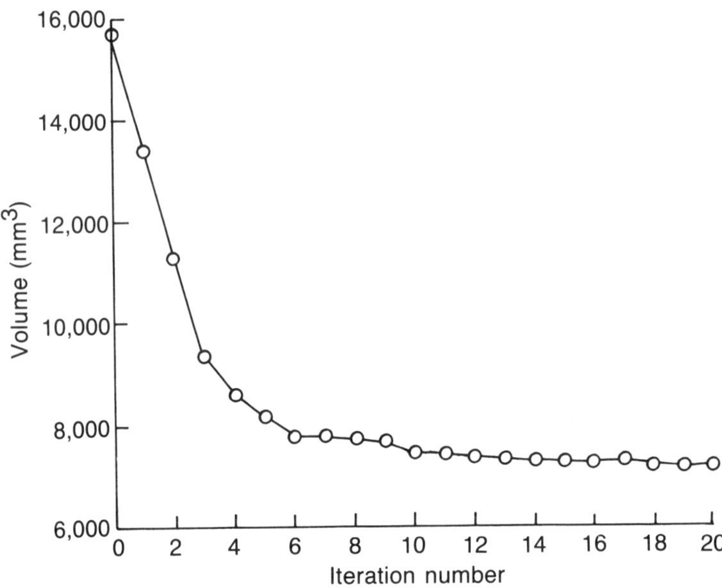

Figure 19. Mass design history of engine connecting rod.

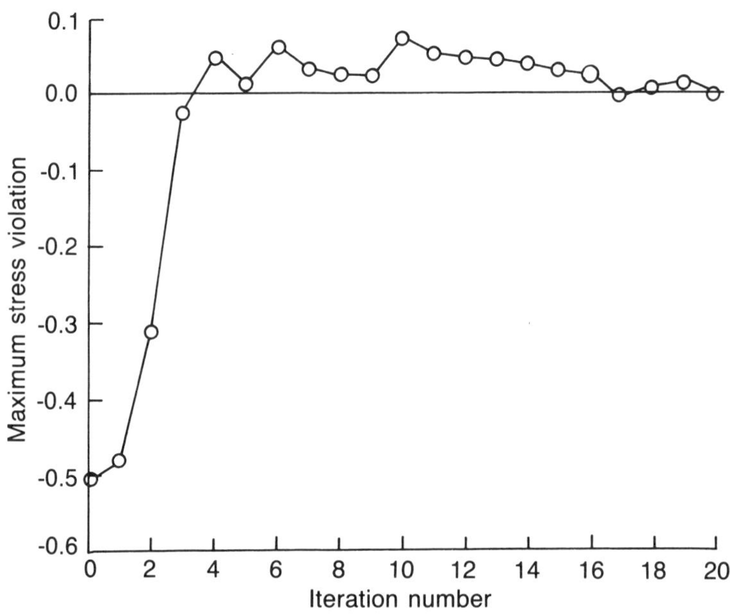

Figure 20. Constraint design history of engine connecting rod.

SUMMARY

An integrated approach to the shape design problem has been described for sheet metal parts in which the problem description is stated in a simple format, the finite element mesh is generated automatically and its accuracy is improved by adaptive mesh refinement. Nonplanar structures can be treated using an assembly process of two-dimensional segments in such a way that all three-dimensional geometry is expressed in terms of a relatively small number of parameters. Surface curvature variations can be added to the planar subassemblies through the superposition of a variety of surface transformation and mapping options. All geometric problem description has been formulated so that it is particularly suitable for interfacing with modern CAD systems.

For the design problem in which the boundaries of the part are moving, it was found that the accuracy of the finite element mesh must be continuously assessed and updated. Strain energy density variations within an element were used as a measure of error. Elements with errors greater than a specified value in an unrefined analysis were refined by adding nodes, and a new mesh was created using automatic triangulation. Results of the refined analysis were combined with the unrefined results to compute stress intensification factors, which were then used to approximate a refined solution for intermediate designs in which refinement did not take place.

The development of a modular computer program for the shape optimization of three-dimensional solid components was also discussed. Our program uses NASTRAN for analysis and CONMIN for optimization. Since design sensitivities with respect to shape variables are not available in NASTRAN, to obtain these sensitivities a module had to be written, based on the material derivative concept applied to the variational state equation. Parameterized surface definitions and the finite element mesh were obtained from a module based upon generic modeling concepts. Each program module constituted a separately executable program but all modules can be executed sequentially using Job Control Language. A realistic design example has been provided to demonstrate the capabilities of the program.

In general, it has been shown that it is possible to automate the structural design process for determining the shape of quite complicated three-dimensional components through the integration of a parameterized geometric description, automatic mesh generation, finite element analysis, design sensitivity analysis and optimization. The resulting capabilities eliminate the need for tedious data transfer inherent in existing trial-and-error design approaches as well as eliminating many of the repetitive steps involved.

REFERENCES

1. L. A. Schmit, Structural synthesis by systematic synthesis. *Proc. 2nd Conf. on Electronic Computation ASCE*, pp. 105–122. New York (1960).

2. G. N. Vanderplaats, Structural optimization—Past, present, and future. *AIAA J.* **20** (7) 992–1000 (1982).

3. O. C. Zinkiewicz and J. S. Campbell, Shape optimization and sequential linear programming, Chap. 7 in *Optimum Structural Design* (Edited by R. H. Gallagher and O. C. Zienkiewicz). John Wiley & Sons, New York (1973).

4. E. J. Haug, K. K. Choi, Y. M. Hou and Y. M. Yoo, A variational method for shape optimal design of elastic structures, in *Optimal Structural Design II* (Edited by R. H. Gallagher). John Wiley & Sons, New York (1983).
5. R. T. Haftka and R. V. Grandhi, Structural shape optimization—A survey. *The 26th AIAA SDM Conf.* CP No. 85–0772, pp. 617–628 (1985).
6. M. E. Botkin, Shape optimization of plate and shell structures. *AIAA J.* **20** (2), 268–273 (1982).
7. M. H. Iman, Three-dimensional shape optimization. *Int. J. Numer. Meth. Eng.* **18**, 661–673 (1982).
8. M. H. Imam, Minimum weight design of 3-D solid components. *Proc. 2nd ASME Computers in Engineering Conf.* Vol. 3, pp. 119–126 (1982).
9. J. A. Bennett and M. E. Botkin, Structural shape optimization with geometric problem description and adaptive mesh refinement. *AIAA J.* **23** (3), 458–464 (1985).
10. V. Braibant and C. Fleury, Shape optimal design using b-splines. *Comput. Meth. Appl. Mech. Eng.* **44**, 247–267 (1984).
11. M. E. Botkin, An adaptive finite element technique for plate structures. Tech. Note, *AIAA J.* **23** (5), 812–814 (1985).
12. J. C. Cavendish, Automatic triangulation of arbitrary planar domains for the finite element method. *Int. J. Numer. Meth. Eng.* **8**, 679–696 (1974).
13. W. H. Frey and J. C. Cavendish, Fast planar mesh generation using the Delaunay triangulation. *Society for Industrial and Applied Mathematics Meeting.* Seattle, WA (July 16–20, 1984).
14. I. Babuška and W. D. Rheinbolt, Adaptive approaches and reliability estimates in finite element analysis. *Comput. Meth. Appl. Mech. Eng.* **17/18**, 519–549 (1979).
15. M. S. Shephard, R. H. Gallagher and J. F. Abel, The synthesis of near-optimum finite element meshes with interactive computer graphics. *Int. J. Numer. Meth. Eng.* **15**, 1021–1039 (1980).
16. J. J. Conner and G. Will, A triangular flat plate bending element. M.I.T., Department of Civil Engineering, Rep. TR–68–3. Cambridge, MA (1968).
17. M. E. Botkin and J. A. Bennett, Shape optimization of three-dimensional folded-plate structures. *AIAA J.* **23** (11), 1804–1810 (1985).
18. M. E. Botkin and G. S. Gressel, Shape optimization of sheet metal components with flanges. *Proceedings of 6th SAE International Vehicle Structural Mechanics Conf.* Detroit, MI, SAE Publication p–178, Paper No. 860803, 29–38 (Apr. 1986).
19. M. A. Yerry and M. S. Shephard, Automatic three-dimensional mesh generation by the modified-octree technique. *Int. J. Numer. Meth. Eng.* **20**(11), 1965–1990 (1984).
20. E. J. Haug, K. K. Choi, J. W. Hou, and Y. M. Yoo, A variational method for shape optimal design of elastic structures, in *New Directions in Optimum Structural Design* (Edited by E. Atrek et al.). John Wiley & Sons, New York (1984).
21. K. K. Choi and E. J. Haug, Shape design sensitivity analysis of elastic structures. *J. Struct. Mech.* **11** (2), 231–269 (1983).
22. E. J. Haug, K. K. Choi and V. Komkov, *Design Sensitivity Analysis of Structural Systems.* Academic Press, New York (1985).
23. R. J. Yang and M. E. Botkin, The relationship between the variational approach and the implicit differentiation approach to shape design sensitivities. *AIAA SDM Conf.* CP No. 85–0774. Orlando, FL (Apr. 1985).

24. R. J. Yang, K. K. Choi and E. J. Haug, Numerical considerations in structural component shape optimization. *J. Mech. Transm., Autom. Des.* Paper No. 84–DET–219 (1984).
25. C. V. Ramakrishnan and A. Francavilla, Structural shape optimization using penalty functions. *J. Struct. Mech.* **3** (4), 403–422 (1974–1975).
26. G. Vanderplaats, CONMIN—A FORTRAN program for constrained function minimization: User's manual. NASA, TMX-62,282 (1973).
27. Y. M. Yoo, E. J. Haug and K. K. Choi, Shape optimal design of an engine connecting rod. *J. Mech. Transm. Autom. Des.* **106**, 415–489 (1984).
28. R. J. Yang, K. K. Choi and E. J. Haug, Finite element computation of structural design sensitivity analysis. Rep. CCAD No. 84–3, The University of Iowa (1984).

DISCUSSION

R. Haftka *(Virginia Polytechnic Institute and State University)*

The integral of the shape sensitivity shown seems to have been a surface integral rather than a domain integral. Have you gone into any of the sensitivity accuracy problems that Professor Haug described?

Botkin

In the three-dimensional problems, we haven't observed any inaccuracies that would adversely affect convergence. They may very well be there, but so far we have not observed any problems along those lines. It is my feeling that if we were having problems with accuracy, they would be showing up in the convergence. We have done a few parts so far and we seem to be obtaining satisfactory accuracy for this class of parts.

M. E. Botkin

C. Fleury *(University of California–Los Angeles)*

In your previous work you used the so-called semianalytical approach of sensitivity analysis, which means finite difference is employed to get the derivative of the stiffness matrix at the element level. Now, when using NASTRAN, why did you choose to abandon this approach? It seems to me that it could be feasible to maintain the same type of approaches using NASTRAN—just making a loop over the generation of the stiffness matrices. Of course it will be much easier when they put shape sensitivity in NASTRAN.

Botkin

If you want to use NASTRAN in that context, it is extremely difficult to do element-level operations. Now for our two-dimensional parts, which were described in some of our papers, all the pseudo load vectors are assembled using element-level operations. We go through each element and find its variational stiffness, and multiply it by the displacements for that particular element. This is much more efficient than actually calculating a global stiffness matrix and differencing.

Fleury

I was referring just to the element matrices, not the global stiffness matrix.

Botkin

Even though we have a lot of experience with NASTRAN, we don't know how to do that using NASTRAN. Perhaps you can do it, but we were not able to figure it out.

Fleury

I suppose it is difficult.

O. C. Zienkiewicz *(University of Wales)*

In the iterative process for two-dimensional problems where you have adaptive mesh refinement, do you refine at every stage of the design or do you do this only once?

Botkin

We have an indicator in the program that lets us refine every n steps. We refine in that step and save those extrapolation parameters, which I showed in my presentation, for each stress constraint. That gives us a factor. If we want to modify the intermediate designs (such as this stress constraint that is 2.5 times higher than in the unrefined mesh) we save them, and we use that build-up factor for all intermediate meshes. This causes the bumping along during convergence. But because of the computational requirements of trying to do all that refinement, we tried to come up with a methodology for handling this.

K. Kline *(Wayne State University)*

Are frequency constraints ever going to be important in the parts that you are attempting to optimize?

Botkin

Definitely. Engineers tell us they have a class of parts that are frequency constrained, so we plan to have frequency constraints in the program. There is a question whether you need mesh refinement, and if so, how you handle mesh refinement when you have frequency constraints and stress constraints at the same time. That's a real problem.

Moderator—R. H. Gallagher

You are dealing with stamped parts, and it seems that instability constraints might appear in that situation. Would looking at frequency constraints and instability constraints alter your choice of the sensitivity requirement?

Botkin

At this point it would alter our choice for sensitivity calculation because I am not sure if there's a variational approach for both frequencies and buckling. Professor Haug may know more about this.

E. J. Haug *(University of Iowa)*

For the natural frequencies and buckling loads, the variational approach is ideally suited.

Botkin

In that case it wouldn't alter my choice of the sensitivity requirement.

J. A. Swanson *(Swanson Analysis Systems, Inc.)*

You mentioned tying into commercial CAD systems. Do you find that they have the required parametric representation available to feed into the optimization process, or do you have to modify them?

Botkin

They have enough parametric representation because we have very simple representations of the boundary. We have a circular segment and a cubic segment. In fact, some models have been created for me on CADAM. CADAM has everything we need—and the spline as well.

Swanson

Do these systems have the parametric representations in the data file, which they pass to you? (IGES format, for example?)

Botkin

It's in there. In this case, it is in the CADAM database. All we want is a geometrical description.

D. Grierson *(University of Waterloo, Canada)*

I see the extension to frequency constraints and then to buckling constraints. What about residual stresses that you might get for these types of components? Is there a concern in your work for inelastic behavior and inelastic stress?

Botkin

So far we haven't really considered these things.

H. A. Kamel *(University of Arizona)*

In the first two slides where you discussed the shape design element, you pointed at some weaknesses. Now I observed that you still require a parametric representation of your mesh. What you have done is to substitute a more efficient and generally more easily performed mesh generation scheme instead of a regionally subdivided mesh. You also remarked that the original idea would not take care of different configurations.

Botkin

I said it was difficult to go from one configuration to another, which we eventually would like to be able to do automatically. When going from one configuration to another, you want to have an approach for redefining only the boundaries if you are trying to change topologies. We can't automatically change the boundary description and also change the complicated connectivity of mapping elements.

Kamel

You might think of the boundary selection as just a global or a rougher mesh generation operation.

Botkin

Yes, but you would have to think in terms of triangles—a coarse mesh of triangles. The only way I can think of doing it is by triangulation of those patches.

C. Wilson *(MacNeal-Schwendler Corporation)*

How do you select your points for monitoring stress constraints? Do you do it over the entire part?

Botkin

For two-dimensional parts, for each shape and boundary design element, we define a single stress constraint per design element. For long boundaries we actually have ways of defining additional stress constraints. Now, for three-dimensional parts, I think we defined a stress constraint for every node of the shape design element. Ren Jye Yang would have to tell you if that was the case.

R. J. Yang *(General Motors Research Laboratories)*

For each finite element.

Z. Mróz *(Polish Academy of Science)*

In handling the design problems for engines, you should face at some time a problem of temperature gradients, such as in piston design. You have the action of pressure and temperature gradient causing thermal stresses. This is a conflicting situation because for strength you could make thicker cross-sections whereas for temperature gradient you have to make them thinner. You have conflicting demands for initial strain loading and for external or traction loading. There is some trade-off between compliance and stiffness. You cannot design too strong a section because it is not good enough for temperature loading. Have you faced this problem in your application?

Botkin

We have not considered those things at this point. Right now we are just trying to get static analysis to be able to calculate stresses so that all of this will work. Piston design is an extremely complicated process, and we realize that there are things we cannot handle right now.

J. Kane *(University of Bridgeport)*

Can you translate the two slides that involve number of iterations into the number of analyses?

Botkin

Every iteration required an analysis to set up the approximate problem. In between iterations, we used the first-order Taylor series expansion.

MULTIDISCIPLINARY SHAPE OPTIMIZATION

G. N. VANDERPLAATS
Department of Mechanical & Environmental Engineering
University of California
Santa Barbara, California

Abstract

Shape optimization is expanded here beyond the specific discipline of structural synthesis to consider the spectrum of design tasks which fall into the general multidisciplinary category. This logical extension of optimization is a fruitful area of research and applications.

A principal application of shape optimization in a multidisciplinary environment is that of combining structural and aerodynamic design. While the most clear-cut example is that of an aircraft wing, the same design task exists in automotive and marine vehicles. In each case the optimum structural design and the optimum aerodynamic design are different; thus the optimum system is not the sum of optimum parts. Therefore, techniques must be devised to formally treat the interaction among the parts. Recent work in multilevel and multidisciplinary optimization provides the groundwork for a dramatic expansion of design capabilities.

A variety of applications of shape optimization in addition to structural design are identified, and the present state of the art is assessed. The mathematical and numerical aspects of multidisciplinary shape optimization are discussed and critical research needs are identified. Finally, it is noted that distributed computing offers a unique challenge as well as an opportunity to use optimization in multidisciplinary design. By fitting the optimization process into the traditional design environment, user acceptance is improved while formally automating the system synthesis task. The immense computational power now available, together with the maturing of optimization technology, can make formal industrial use of optimization a reality.

INTRODUCTION

Shape optimization is perhaps the most fundamental of optimization problems. This is often seen in nature as both an aesthetic and a utilitarian solution to nature's "design task." One example is the shape of a trout, identified by Sir George Cayley [1] as a logical shape to minimize resistance, and later compared to an "optimum" airfoil shape by von Karman [2]. Another example is that of a tree trunk, which satisfies a multiobjective function criterion, as shown by Stadler [3]. Turning to man-made products, a classical example is the shape of a cardboard box to contain a given volume while using a minimum amount of material.

In structural shape optimization, most early work used variational methods to find the optimum shape of frames, plates, engine disks and numerous other components. With the advent of the finite element method and mathematical programming techniques for structural synthesis, there has been an ever-increasing breadth of applications in structural shape optimization. Simultaneously, there have been increasing applications of mathematical programming in other engineering disciplines, motivated and directed to a large degree by research in structural synthesis.

Schmit's classical work introduced structural synthesis using nonlinear programming in 1960 [4]. Schmit also introduced us to a new design tool. Mathematical programming, or numerical optimization, has since been recognized as a uniquely computer-oriented design methodology with wide-ranging applications. The majority of these nonstructural applications include shape variables in some form, and so the multidisciplinary shape optimization task is a common and natural corolary to structural shape optimization.

The purpose here is to identify a variety of design tasks where numerical optimization has been used with success. While the examples cited here are by no means exhaustive, they are sufficient to indicate the breadth and complexity of design problems that can be solved by numerical optimization, and to indicate the vast amount of research remaining before optimization becomes a standard tool for multidisciplinary engineering design. In addressing this topic, it should be emphasized that the term *multidisciplinary* is used here to include both the design process wherein several disciplines are considered simultaneously, *and* the design of single components related to only one discipline other than structure.

We will begin by identifying the common feature of the multidisciplinary design problem. Following this, a variety of examples are cited from disciplines ranging from aerodynamic design to thermal system design to complete aircraft and ship synthesis. A logical conclusion from this discussion is that the breadth of design tasks that are amenable to solution using numerical optimization is almost inexhaustible.

FEATURES OF THE MULTIDISCIPLINARY PROBLEM

While all optimization tasks share a commonality in the basic formulation of the problem, multidisciplinary optimization presents some unique differences, both in the nature of the problems and in the level of maturity of the user community. To begin with, the majority of optimization tasks are defined by relatively few design variables, typically less than ten. These may include integer variables, such as the number of tubes in a heat exchanger. Also, the design variables may be fundamentally different in nature. For example, in aircraft design, the thickness to chord ratio of a wing may be treated simultaneously with maximum engine thrust as design variables.

The analysis task, used as a subproblem to design, is often complex and may be very expensive. A typical example is the solution of Euler's equation or even the Navier Stokes equations in aerodynamic analysis. In a quite different application, calculation of the condensates in a steam condenser is a complex and often expensive iterative process. Furthermore, gradients of the response (analysis) quantities such as lift coefficient with respect to the shape design variables are virtually never available for use in optimization. This is not to say that gradient information cannot be calculated. It is more a recognition of the state of the art and the fact that optimization users in the multidisciplinary environment have not yet undertaken the major effort needed to provide this information.

In a general sense, the multidisciplinary optimization task has no clear mathematical structure. Whereas in linearly elastic finite element structural analysis the mathematical form of the problem is relatively well understood, in multidisciplinary optimization such as aircraft or ship synthesis, where numerous disciplines are combined to create the design objective and constraint functions, the interactions among these disciplines are neither clear nor well understood except perhaps by a very few highly experienced designers. This lack of mathematical structure, together with the fact that optimization concepts are themselves new to the generalist, leads to the common approach of treating optimization as a "black box" which is coupled with a more familiar analysis program for the purpose of obtaining an answer. While this may not be the most desirable approach to multidisciplinary optmization, it is perhaps the most efficient way to disseminate this powerful tool into the engineering community. At least it recognizes the fact that it is a tool, just as any other, to be used together with trade-off studies, experimentation and insight.

Despite this lack of formal study and problem formulation, applications of numerical optimization techniques in the multidisciplinary environment have been quite broad and were often surprisingly successful. In the following sections, various applications are described which indicate the depth and breadth of shape optimization in the multidisciplinary environment. The details of the optimization tasks as well as the underlying analysis may be found in the references.

AERODYNAMIC OPTIMIZATION

One of the most attractive areas where optimization can be applied outside of structural synthesis is aerodynamic shape optimization. The first modern application of optimization here was by Schmit and Thornton in 1965 [5]. The application was to a double wedge airfoil where the objective was to minimize a combination of weight and drag, and the design variables included both structural member sizing and aerodynamic shape (thickness). The analysis was based on supersonic piston theory. This work was followed in 1968 by that of Hague, Rozendaal and Woodward [6], in which a biconvex section was designed and only aerodynamic drag was minimized. Again the analysis was based on supersonic piston theory. In 1974, Hicks, Murman and Vanderplaats [7] designed the first nonsymmetric airfoils using numerical optimization. Both subsonic and supersonic aerodynamics were considered and the analysis was based on inviscid, transonic, small disturbance fluid flow. Later, this was extended to include boundary layer effects, and in 1979 Lores, Smith and Hicks [8] extended this to three-dimensional surfaces.

Figure 1 shows the progress in aerodynamic shape optimization in recent years. In recent applications, only aerodynamic functions were treated as objectives and constraints. Figure 2 shows a measure of progress in design efficiency over the last ten years. The numbers of analyses given in the figure are for an example of low-speed airfoil optimization. In the initial work [7], polynomial shapes were used to define the airfoil. While this provided design improvements, the optimization suffered both from the numerical coupling between the shape variables and from the fact that good airfoils are not well defined by polynomials.

Borrowing the reduced basis concept from structural synthesis [9], in 1976 the design efficiency was improved by a factor of more than two. At the same time, the optimum objective was improved in comparison with the previous shapes [10]. In 1978, again borrowing concepts first developed in structural synthesis [11], approximation concepts were applied to airfoil optimization [12]. Again, efficiency improved by a factor of two, and there was an improvement in the optimum. Here, while first-order Taylor series expansions in reciprocal space were effectively used for structural synthesis, airfoil optimization required development of second-order approximations based on function values alone. While this approach appears to be computationally costly, efficiency was gained by the fact that only a few design variables were needed. Also, using this approach, it was possible to save the results from one optimization to be used as a database in the next. Hence if, in a design study, it is first desired to maximize lift with a constraint on drag and then later to minimize drag with a constraint on lift, the airfoils which were analyzed in the first optimization were useful in directing the second.

Building on the concept of using basis airfoils, in 1983 Aidala, Davis and Mason [13] created what they termed a "smart aerodynamic optimizer." Here, airfoil shape modifications were defined to produce the desired effect as shown in Figure 3. This now allows the aerodynamic designer to use his knowledge of aerodynamics to guide the optimization process, leading to more realistic and usable design as well as enhancing design efficiency.

In summary, aerodynamic optimization related to airfoils has seen rapid progress in the last ten years. Also, during this time two important lessons have been

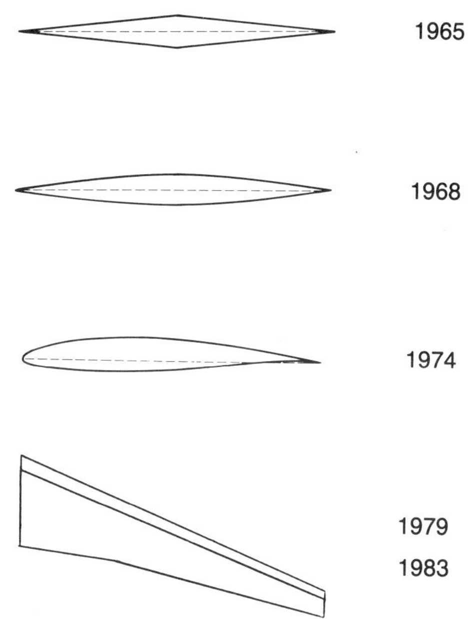

Figure 1. Progress in aerodynamic optimization.

learned. First, the use of inviscid aerodynamics as a design tool is often misleading because the optimization takes advantage of the inherent weakness of the analysis to produce unrealistic airfoil sections. Second, airfoils (and wings) are commonly designed for only one flight condition. This can lead to an airfoil with very high cruise performance but poor takeoff, climb or maneuver performance. An essential step in future applications is to consider multiple flight conditions, just as multiple loading conditions are always considered in structural synthesis.

Aerodynamic optimization is not limited to airfoils and wings; two additional examples may be found in references [14] and [15]. In reference [14], again using inviscid aerodynamics, the optimum shape of an aircraft turret is sought to minimize the phase distortion of a laser beam at both subsonic and supersonic speeds. In reference [15], the shape of a hill is found to concentrate wind energy available to a windmill. Finally, reference [16] gives an indication of the power of optimization to design experiments. In this application, the purpose was to find the pitch settings on the vanes of an aircraft engine compressor to maximize adiabatic efficiency. In this work, Garberoglio, Song and Boudreaux show that optimization can be effectively used to predict the vane settings in the next test to improve efficiency. The unique feature of this work is that the analysis was not performed on the computer. Instead the objective function was evaluated by running the engine on a test stand. In their work they show that not only can the final design be improved, but the number of data points needed to reach the optimum was typically reduced by 40 percent. This represents a unique and innovative use of optimization on a design problem of extreme complexity for which reliable computer analysis is not available.

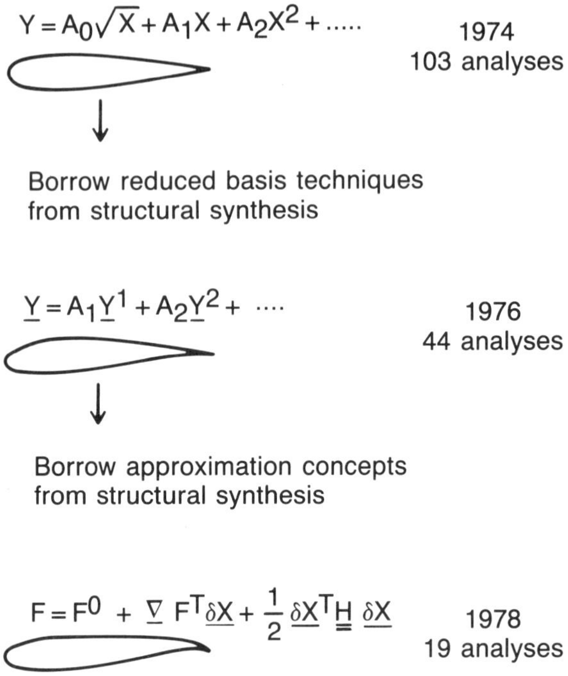

$Y = A_0\sqrt{X} + A_1X + A_2X^2 + \ldots..$

1974
103 analyses

↓

Borrow reduced basis techniques
from structural synthesis

$\underline{Y} = A_1\underline{Y}^1 + A_2\underline{Y}^2 + \cdots$

1976
44 analyses

↓

Borrow approximation concepts
from structural synthesis

$F = F^0 + \underline{\nabla} F^T \underline{\delta X} + \frac{1}{2} \underline{\delta X}^T \underline{\underline{H}}\ \underline{\delta X}$

1978
19 analyses

Figure 2. Efficiency improvements in aerodynamic optimization.

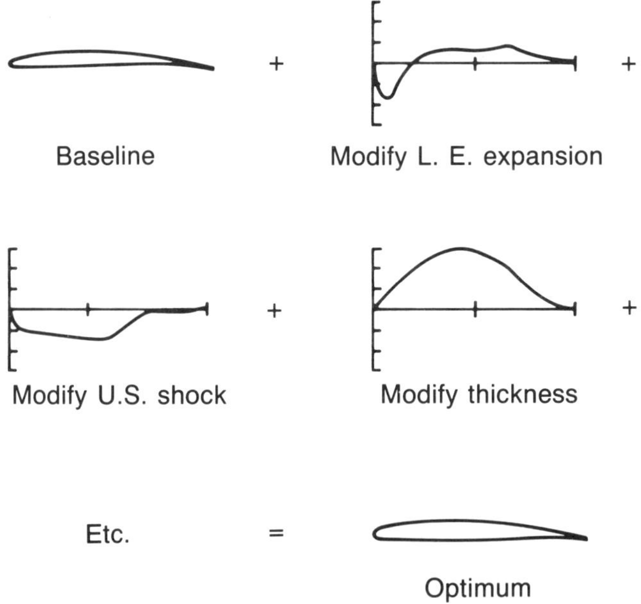

Figure 3. Smart aerodynamic optimization.

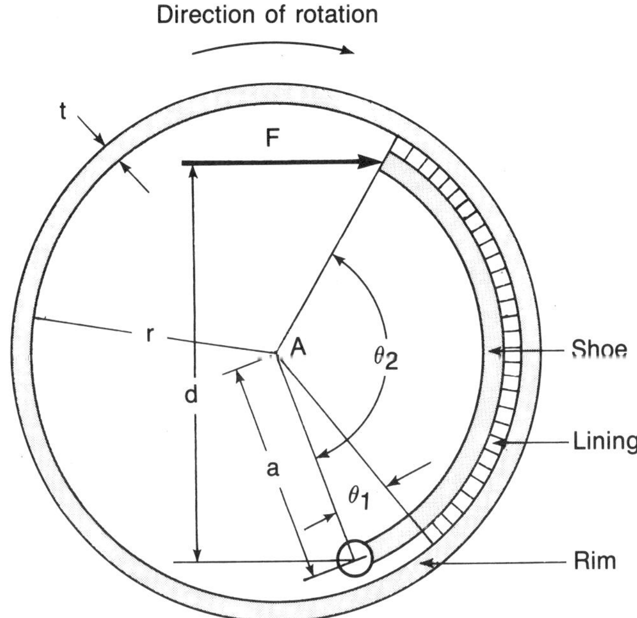

Figure 4. Internal expanding brake optimization.

MECHANICAL COMPONENTS

Numerous mechanical components fall into the general category of shape optimization. These include such applications as robotic manipulators, various linkages, clutches and brakes, record player tone arms and others. Figure 4 is a typical example of optimization of a brake drum [17]. While this may include sizing of the drum for strength, a fundamental part of this task is related to nonstructural considerations. Here it is desired to find the width and thickness of the drum, subject to stopping ability and thermal constraints between the shoe and drum surface. The design variables include shoe location, brake material position and stopping force. This example indicates the ease with which optimization can be applied to an everyday mechanical design problem and shows that optimization has a real use for design of components which may not otherwise be thought of as exotic.

THERMAL SYSTEM DESIGN

Design of thermal systems represents a major opportunity for optimization which has been addressed only to a minor degree up to now. Examples here include the design of combustion chambers, heat exchangers, steam condensers, thermal doublers, electronic cooling, solar energy conversion and a multitude of others. Design of such systems should properly include structural and sometimes fluid mechanics considerations, and so it is truly a multidisciplinary design task. To date, these multidisciplinary aspects have not been considered, but several applications including thermal and fluid aspects have been pursued.

References pp. 277–279

Figure 5 shows the design of a common heat exchanger [18]. A hot liquid, commonly water, is passed through a series of tubes and cooled by the cold air passing by the outside of the tubes. The design objective is to minimize the size of the heat exchanger. The principal constraint is to meet heat dissipation requirements. Geometric constraints are imposed to maintain structural integrity. The design variables include the number of tubes and the number of passes, as well as the tube diameters, wall thickness, cooling fin dimensions and tube layout. The number of tubes and the number of passes are integer design variables. However, the number of tubes is usually large and can be treated as a continuous variable. On the other hand, the number of passes the hot liquid makes through the system is a small integer number and so must be dealt with directly. One option here is to treat this in a formal mathematical programming way, but in this case it was found more advantageous simply to optimize several different heat exchangers, each with a different number of passes. This avoids the complexity of nonlinear integer programming, which is not presently well established. Also it provides a series of optimum designs, allowing the final selections to be made on considerations other than minimum size alone. References [19]–[21] contain a variety of examples similar to this.

Figure 6 shows the general layout of a steam condenser for marine applications [22]. Again, the objective is to minimize system volume, and the design variables include the number of tubes, tube geometry and tube layout. Constraints include geometrical limits as well as limits on the amount of condensation. The number of tubes is large enough so that this can be treated as a continuous variable in the optimization. As with many nonstructural optimization tasks, the analysis portion of the optimization is critical. Good computational analysis of steam condensers is not generally available, so that much of the design effort is experimental. Therefore, considerable care must be exercised to create an analysis capability that is reliable and well defined for optimization. Here one example of the complexity of the problem is how to deal with designs where all of the steam is condensed out before the last row of tubes is reached. Once this happens, the heat transfer coefficients change, and this must be accounted for. Also, as steam condenses out and the condensate drops onto tubes below, the heat transfer characteristics are changed, again complicating the analysis not only from a heat transfer point of view, but simply as a result of geometrically following the condensate.

Other examples of optimization of thermal systems include thermal doublers, cooling fins, solar energy conversion systems and so on. It is clear that this discipline is an ideal candidate for extensive use of optimization. However, it is important that these applications include more than just thermal and fluid flow considerations. As these applications increase, it is clear that structural considerations must quickly become part of the design, leading again to a multidisciplinary design task.

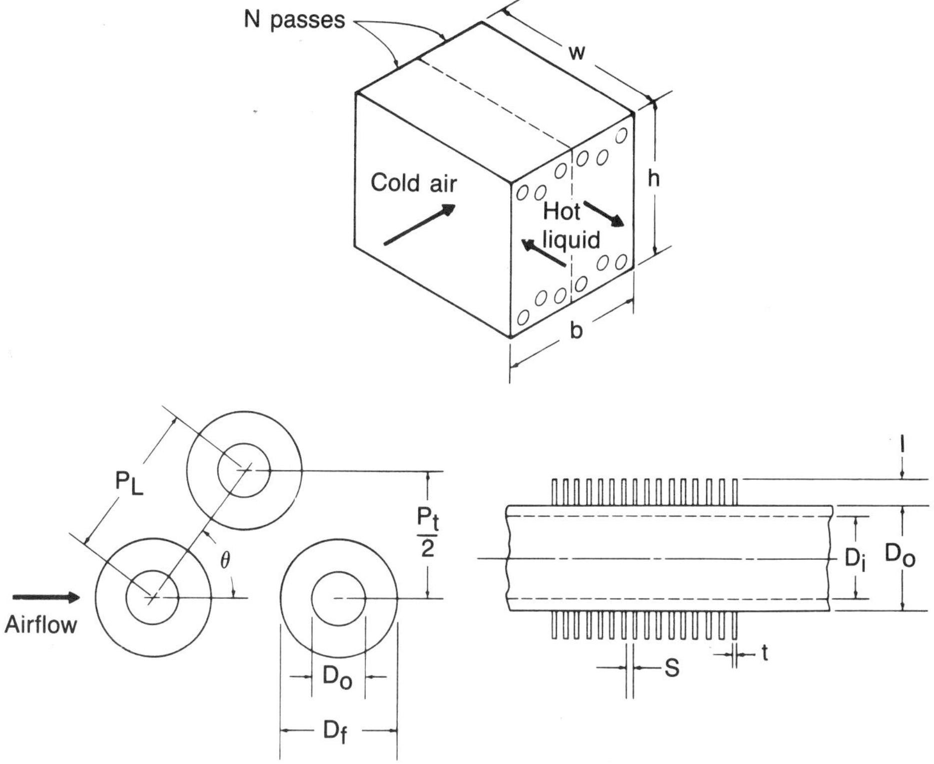

Figure 5. Heat exchanger optimization.

AIRCRAFT AND SHIP SYNTHESIS

Aircraft and ship synthesis represents a truly multidisciplinary design task. Here, we will begin by describing the aircraft synthesis process and we will identify some unique features. This will be followed by a brief discussion of the ship synthesis process, showing the similarities to that of aircraft design.

Aircraft synthesis has been the subject of considerable effort over the past 20 years. While the future aircraft synthesis task may utilize very sophisticated design tools including multilevel and parallel optimization (see for example references [23] and [24]), to date applications have been primarily at the conceptual design level. Here, optimization provides a unique capability to compare a variety of design concepts where each aircraft has been optimized to the specified mission. In order to achieve this, the analysis used by optimization is necessarily simple by today's standards. Geometric properties of the aircraft may be a set of cylinders and cones while aerodynamics are usually semiempirical but include the complete aircraft and often high angles of attack as well. Normally, structural design is not performed at this design level, but empirical weight-estimating relationships are

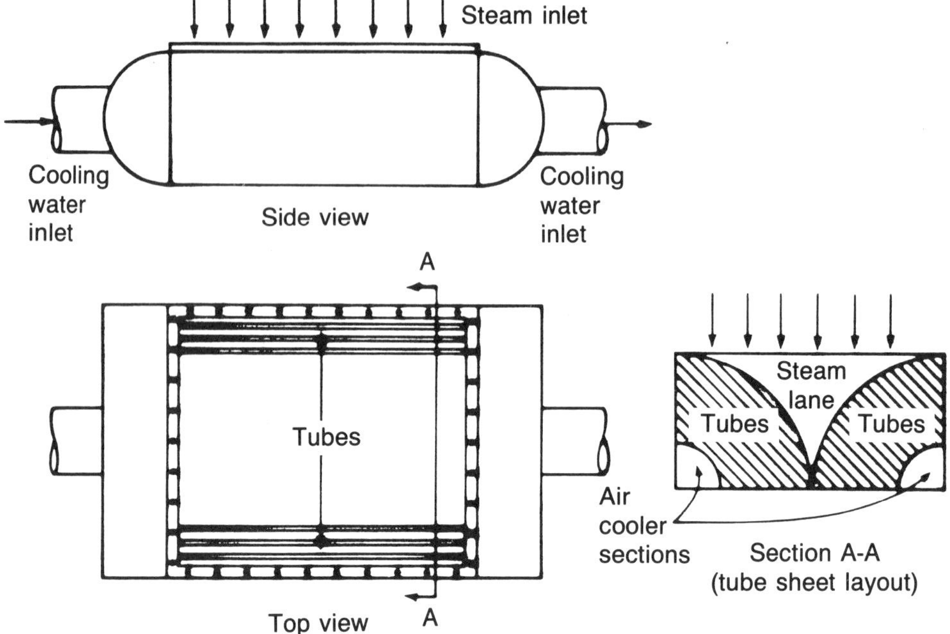

Figure 6. Steam condenser optimization.

developed instead. Similar levels of detail are considered in each discipline. Overall, the aircraft synthesis process includes consideration of geometry, aerodynamics, propulsion, weights and volumes, trajectory analysis, stability and control, and perhaps economic analysis as components of the design task.

The design objective may be minimum gross weight, minimum fuel consumption, maximum combat performance, minimum life cycle cost or any of a variety of other user-defined objectives. At the conceptual design level, typical design variables include wing dimensions such as aspect ratio, platform area, thickness to chord ratio, taper ratio, fuselage length and diameter, and maximum engine thrust. Constraints include volume requirements, performance limits such as maximum sustained load factor and specific excess power, and perhaps life cycle costs (e.g., design to cost). The total number of design variables is typically ten and usually fewer than ten constraints are imposed.

The conceptual aircraft synthesis task presents some unique difficulties. The analysis task normally uses semiempirical relations for aerodynamics, propulsion and weight estimation. If the analysis is not carefully correlated with existing aircraft of similar design, the optimization results can be quite misleading. Furthermore, the sensitivity of the analysis to the design variables must be carefully scrutinized. For example, historically, as the aspect ratio of aircraft wings has increased, the weight has been reduced. This weight reduction has come about by other technological advancements, most notably the use of high-strength materials. However, if an equation to estimate a wing weight is created based on historical data alone, the sensitivity of the weight to the aspect ratio will be incorrect, even though the weight estimate for a specific existing aircraft is accurate. Then, because the weight is predicted to be less for higher aspect ratios and because higher aspect ratios are aerodynamically more efficient, the optimization process will project wings of unbelievably high aspect ratios. This underscores the observation that optimization will take advantage of any weakness in the analysis, and that the analysis portion of the aircraft synthesis program must be based on proper physics.

Another major consideration in the aircraft synthesis problem is the choice of design variables. There is little mathematical structure available to identify which variables will yield a well-conditioned optmization problem. Furthermore, the traditional variables with which the aerospace designer is comfortable may be a poor choice from an optimization point of view. Traditionally, wing loading (aircraft weight divided by wing area) is considered to be a parameter which has major impact on the design, and so it is chosen as a design variable. As a subproblem, the wing area is calculated as the aircraft weight divided by the design variable, W/S. However, using this as a design variable for optimization can lead to inconsistent results and even prevent proper convergence of the optimization. Figures 7 and 8 demonstrate this graphically. Figure 7 shows the platform of an oblique winged, remotely piloted surveillance vehicle that was the subject of a design study. The objective was to minimize the weight, and the only critical constraint was that the fuel and equipment fit within the available wing volume. The design variables were initially chosen as the thickness to chord ratio of the wing, TC, and the wing loading, WS. The top half of Figure 8 shows the two-variable design space for this example. Here, two observations are noteworthy. First, the optimum design is not well defined, since various combinations of WS and TC give nearly the same objective. Secondly, the design space is clearly nonconvex. When the design variable is changed from wing loading, WS, to the actual wing platform area, S, the design space of the lower half of Figure 8 results. Now the design space is convex, and the optimum is well defined and easily reached. If the wing loading is required to be within specified bounds for takeoff and landing or other performance considerations, this is easily accomplished by treating it as a constraint. The net effect is that, with a simple reformulation of the design task, the optimization efficiency and reliability is greatly enhanced.

Figure 9 shows another example of aircraft synthesis, this time for a high performance combat aircraft [25]. The objective is again to minimize weight, and the critical constraint is performance at the combat condition. The design variables include body length and diameter, wing area, aspect ratio, sweep and thickness to chord ratio, and engine thrust. The initial geometry was chosen so as to be similar to an existing supersonic aircraft, and the optimum geometry was a counterintuitive result as shown in the figure. The wing was relatively unswept and quite thin. When the reason for this was investigated, it was found that the large loiter time requirement was the cause, and that the result was indeed reasonable. The aircraft must carry the loiter fuel throughout the entire mission and so, in effect, the design minimized that. When the loiter requirement was relaxed, the more traditional

References pp. 277–279

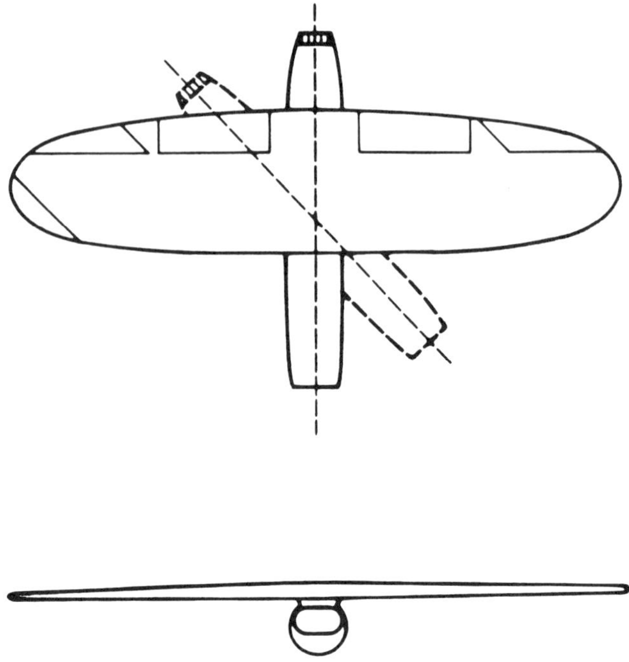

Figure 7. Oblique wing remotely piloted vehicle.

supersonic shape resulted. This underscores the power of optimization to provide trade-off information rapidly, based on optimized aircraft rather than the more traditional approach of comparing point designs.

The use of optimization at the conceptual design level has been quite widespread over the years (see references [26]–[32]), with varying degrees of success. In identifying the reasons why optimization has sometimes been unsuccessful, it has invariably been the result of inaccurate analysis. This leads again to the observation that a major part of the task of developing aircraft synthesis tools lies in correlating good designs and insuring a proper sensitivity to design parameters.

The use of optimization at the preliminary or detailed design phase is almost nonexistent at the complete system level. However, as structural, aerodynamic and engine component optimization techniques become more firmly established, this can be expected to be a topic of immense interest and effort.

Another example of system synthesis is in ship design. Naval architects often refer to the design spiral shown in Figure 10(a). It will be noted that the design considerations are not unlike those of aircraft design. The objective is still minimum weight in most cases; constraints include fuel limits, performance and volume limitations. The design variables are now length, beam, height, prismatic coefficient and section coefficient.

In automating the traditional design spiral, the computer code usually con-

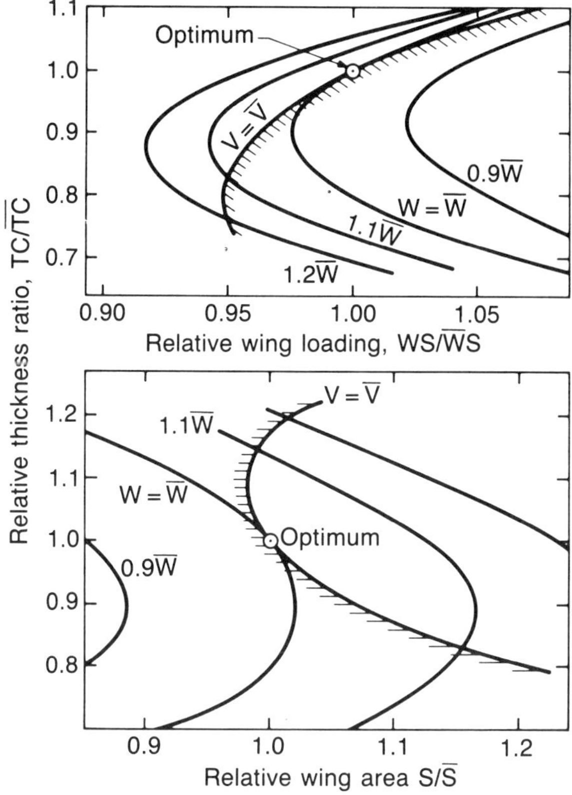

Figure 8. Choosing the proper design variables.

tains a multitude of nested DO loops to deal with the conflicting requirements (constraints). By using optimization, most of this effort can be removed from the code and treated as constraints in the optimization [33]. The net effect is that the analysis portion of the design code is dramatically simplified and the computational cost at that level is reduced, often by an order of magnitude. Now the optimizer deals with the conflicting constraints to efficiently produce the optimum system.

In complete ship design, optimization is still not being used to any significant extent. However, as these tools become more familiar to the designer, we can expect this to be an area of immense development.

SUMMARY

In applying optimization to design problems outside the structures field, it is almost routine to consider shape variables in one form or another. Varied examples of this have been offered here to demonstrate that optimization is a powerful design tool in a multidisciplinary environment and that there is condsiderable ongoing activity in disciplines other than structural optimization.

The multidisciplinary problem is in many ways different from structural

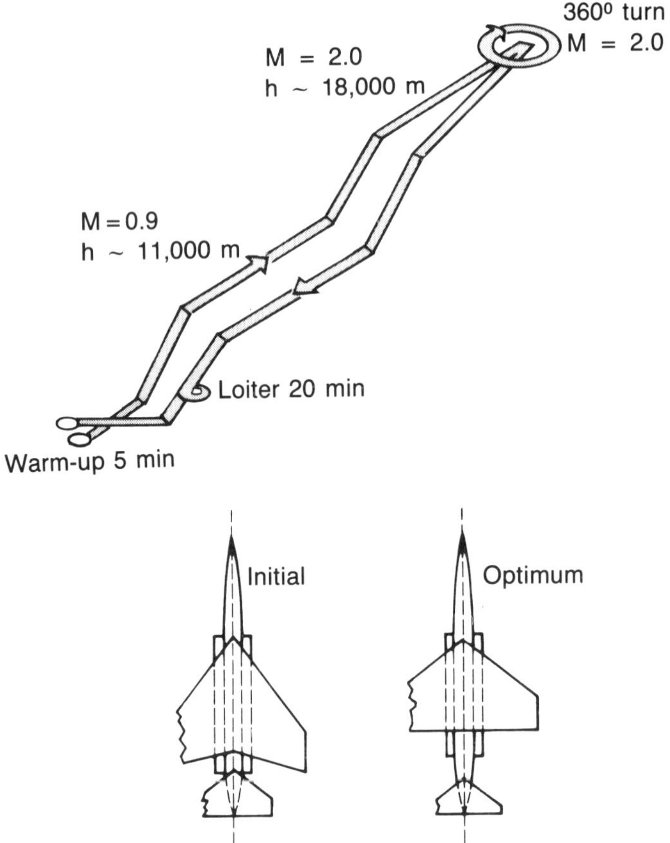

Figure 9. Supersonic cruise aircraft optimization.

optimization. Yet it has benefited immensely from research in structures, as noted specifically in the discussion of aerodynamic optimization. In general, however, the multidisciplinary problem offers unique complexities. First, there are usually fewer than ten independent design variables. The analysis subproblem is often expensive and complex, and may be highly nonlinear. Also, the analytic gradients are seldom available, demanding that the optimization be particularly efficient, and there is seldom a clear mathematical structure that can be exploited to enhance design efficiency. Finally, the practitioner is seldom expert in optimization techniques or theory, and thus optimization is viewed as a "black box" design tool. It can be anticipated that as this black box approach gives way to a more knowledgeable use of optimization, nonstructural applications will see the same interest that has marked the structural optimization field in recent years.

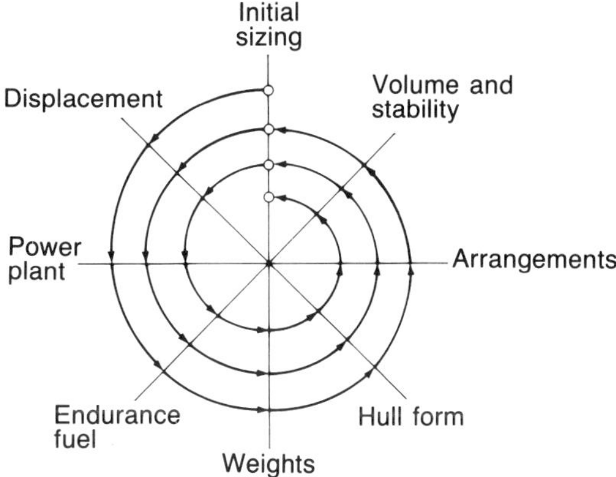

(a) Conventional design spiral

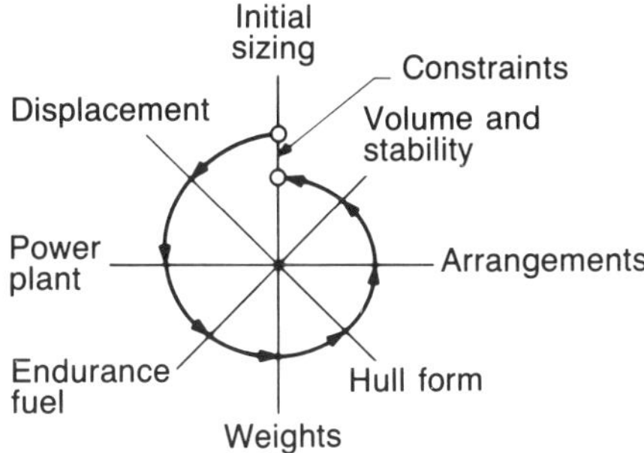

(b) Design using optimization

Figure 10. Ship synthesis.

REFERENCES

1. Cayley, Sir George, *Aeronautical and Miscellaneous Note-Book (ca. 1799–1826).* W. Heiffer and Sons, Cambridge, England (1933).

2. T. von Karman, *Aerodynamics.* McGraw-Hill (1963).

3. W. Stadler, Natural structural shapes (The static case). *Q. J. Mech. Appl. Math.* **31** (2), 169–217 (May 1978).

4. L. A. Schmit, Structural design by systematic synthesis. *Proc. 2nd Conf. on Electronic Computation* pp. 105–122. ASCE, New York (1960).

5. L. A. Schmit, and W. A. Thornton, Synthesis of an airfoil at supersonic mach number. NASA CR-144 (Jan. 1965).

6. D. S. Hague, H. L. Rozendaal and F. A. Woodward, Application of multivariable search techniques to optimal aerodynamic shaping problems. *J. Astronaut. Sci.* **15** (6), 283–296 (Nov.–Dec. 1968).

7. R. M. Hicks, E. M. Murman and G. N. Vanderplaats, An assessment of airfoil design by numerical optimization. NASA TM-X 3092 (July 1974).

8. M. E. Lores, P. R. Smith and R. M. Hicks, Supercritical wing design using numerical optimization and comparisons with experiment. *Proc. AIAA 17th Aerospace Sciences Meeting.* AIAA Paper 79-0065, New Orleans, LA (Jan. 1979).

9. R. M. Pickett, Jr., M. G. Ruginstein and R. B. Nelson, Automated structural synthesis using a reduced number of design coordinates. *AIAA J.* **11** (4), 489–494 (1973).

10. G. N. Vanderplaats and R. M. Hicks, Numerical optimization using a reduced number of design coordinates. NASA TM-X 73, 151 (1976).

11. L. A. Schmit and H. Miura, Approximation concepts for efficient structural synthesis. NASA CR-2552 (1976).

12. G. N. Vanderplaats, Efficient algorithm for numerical airfoil optimization. *AIAA J. Aircraft* **16** (12), 842–847 (1979).

13. P. V. Aidala, W. H. Davis, Jr. and W. H. Mason, Smart aerodynamic optimization. *Proc. AIAA Applied Aerodynamics Conf.* AIAA Paper No. 83-1863. Danvers, MA (July 1983).

14. G. N. Vanderplaats, A. E. Fuhs and G. A. Blaisdell, Optimized laser turrets for minimum phase distortion. *Proc. Aero-Optics Symposium on Electromagnetic Wave Propagation from Aircraft.* Ames Research Center (Aug. 1979). NASA Conf. Publication 2121 (Apr. 1980).

15. G. N. Vanderplaats and A. E. Fuhs, Aerodynamic design of a conventional windmill using numerical optimization. *AIAA J. Energy* **1** (2) (Mar.–Apr. 1977).

16. J. E. Garberoglio, J. O. Song and W. L. Boudreaux, Optimization of compressor vane and bleed settings. *Proc. 27th Int. Gas Turbine Conf. and Exhibit.* ASME Paper No. 82-GT-81. London (Apr. 1982).

17. M. Peer, Numerical optimization of internal expanding brake. M. S. Thesis, Naval Postgraduate School, Monterey, CA (Mar. 1981).

18. C. P. Hedderich, M. D. Kelleher and G. N. Vanderplaats, Design and optimization of air-cooled heat exchangers. *J. Heat Trans. ASME* **104**, 683–690 (Nov. 1982).

19. J. W. Palen, T. P. Cham and J. Taborek, Optimization of shell-and-tube heat exchangers by case study method. Chemical Engineering Progress Symposium Series, **70** (138), pp. 205–214 (1974).

20. K. A. Afimiwala, Interactive computer methods for design optimization. Ph.D. Thesis, Mechanical Engineering Department, State University of New York–Buffalo (1976).

21. H. J. Fontein and J. G. Wassink, The economically optimal design of heat exchangers. *Engineering and Process Economics* **3**, 141–149 (1978).

22. C. M. Johnson, G. N. Vanderplaats and P. J. Marto, Marine condenser design using numerical optimization. *ASME J. Mech. Des.* **102**, 469–475 (1980).

23. R. E. Fulton, J. Sobieszczanski and E. J. Landrum, An integrated system for preliminary design of advanced aircraft. *Proc. AIAA Aircraft Design, Flight Test, and Operations Meeting.* Los Angeles, CA (Aug. 1972).

24. J. Sobieszczanski-Sobieski, A linear decomposition method for large optimization problems. NASA TM 83248 (1982).

25. G. N. Vanderplaats and T. Gregory, A preliminary assessment of the effects of advanced technology on supersonic cruise tactical aircraft. *Proc. Super Cruise Military Aircraft Design Conference.* Colorado Springs, CO (Feb. 1976).

26. G. N. Vanderplaats, Automated optimization techniques for aircraft design. *Proc. AIAA Aircraft Systems and Technology Meeting.* AIAA Paper No. 76-909. Dallas, TX (Sept. 1976).

27. V. A. Lee, H. G. Ball, E. A. Wadsworth, W. J. Moran and J. D. McLeod, Computerized aircraft synthesis. *J. Aircr.* **4**, 402–408 (Sept.–Oct. 1967).

28. R. R. Heldenfels, Integrated computer-aided design of aircraft. AGARD CP 147 (1973).

29. R. E. Wallace, Parametric and optimization techniques for airplane design synthesis. AGARD LS 56 (1972).

30. E. R. Schuperth and L. Celniker, Synthesizing aircraft design. *Space/Aeronautics* **51** (4), 60–66. (Apr. 1969).

31. A. W. Bishop, Optimization in aircraft design - The whole aircraft. *Proc. Symposium on Optimization in Aircraft Design.* London (Nov. 1972).

32. W. L. Straub, Managerial implications of computerized aircraft design synthesis. *J. Aircr.* **11**, 129–135 (Mar. 1974).

33. J. L. Jenkins, Application of optimization techniques to naval surface combatant ship synthesis. Master's Thesis, Naval Postgraduate School, Monterey, CA (Oct. 1982).

DISCUSSION

R. Haber *(University of Illinois – Champaign/Urbana)*

You touched on the use of smart basis vectors for optimization in a subspace. I think these basis vectors are very significant in terms of the issues a designer should be considering when performing an optimization. They provide insight about the design problem—not just a number coming out of a black box. Can you comment on how you obtained the smart vectors? Was there some generalizable concept for obtaining the vectors, or were they the result of someone's personal expertise in aerodynamics?

Vanderplaats

It came almost exclusively out of expertise, but keep in mind there is a large expertise base. A first-rate aerodynamicist such as Aidala would ask himself what he would need to do to get the desired results. Most of these basis vectors were created by an inverse method: if this is the pressure distribution I want to get, what should the airfoil look like, and how do I modify an airfoil to get that? Inverse methods are popular, but they have the weakness that occasionally you get an airfoil that looks like a fish with a tail and a few minor problems like that. Aidala used the inverse method to get the basis vectors.

D. Grierson *(University of Waterloo, Canada)*

At the end of your steam condenser optimization problem, you noted the difficulty with the integer variables. I'm not clear what you mean by an integer variable. Are you referring to discrete sizes being available or a number of items?

Vanderplaats

In this case it was the number of tubes. In fact it could be discrete sizes coming from tables. One obvious example in structural optimization is picking wide flanges out of the AISC code.

G. N. Vanderplaats

Grierson

We have been rather successful in applying Fleury's DUAL-1 optimization codes to the particular problem you just raised. Were other applications more complicated?

Vanderplaats

It is more complicated if you do not have the separability that Claude Fleury had in DUAL-1. For the integer problem the branch and bound codes are available, but they are extremely inefficient. On the other hand, recognizing that when you get near the optimum you may be able to approximate the problem as separable may lead to a good approximation. There may be some keys there to lead us to a more efficient approach.

Grierson

In the hopes that he may buy me a beer later, I would like to say that Fleury's codes are extremely efficient. We put the entire set of Canadian commercial sections into a database, and the codes handle it beautifully for both continuous and discrete optimization.

L. A. Schmit *(University of California–Los Angeles)*

I would like to comment on Grierson's point about how well DUAL-1 worked for him; namely, it depends how discrete the problem is, in a sense. For example, if you have a discrete variable for fiber composites where the number of plies might be somewhere under ten, and you have to pick a number between one and ten to determine the number of individual plies, you will have to use a true, strictly discrete variable algorithm.

On the other hand, if you have as many intermediate points as you have with a typical set of commercially available sections for steel design, or if you are dealing with laminates that go up to 100 or more plies, then you can use methods such as Fleury's DUAL-1 algorithm with more confidence. It is a question of how discrete the problem is.

OPTIMAL SHAPE DESIGN OF AXISYMMETRIC STRUCTURES

PH. TROMPETTE, J. L. MARCELIN and C. LALLEMAND

Structural Mechanics Laboratory
French National Institute of Science
Villeurbanne, France

Abstract

This paper is concerned with the optimal shape design of axisymmetric structures loaded symmetrically or nonsymmetrically. The objective is to make the tangential stress uniform along a part of the boundary to minimize the stress concentrations. The boundary of the structure is made of straight or circular segments defined by input data of master point coordinates and radius values. The design variables are easily deduced from the data. The analysis of the structure is performed by the finite element method using triangular (six nodes) or quadrilateral (eight nodes) isoparametric elements.

The concept of mobile or fixed substructures is used and associated to an automatic mesh generator. Several improvements in connection with the calculations of stress and stiffness matrix derivatives are proposed. An application of the program to the shape optimization of a part of a helicopter rotor demonstrates the efficiency of the process.

INTRODUCTION

An increasing number of papers have been recently devoted to the shape optimization of structures. Almost of them are concerned with two-dimensional structures. This restriction is due to the large number of design variables needed to describe the structure and its evolution, and this is the main obstacle to further development of three-dimensional shape optimization.

Except for a very few papers which use the boundary element method [1–3], almost all of the authors have selected the finite element method [4–7] to analyze the structure and to calculate the sensitivities. Even though in shape optimization applications the finite element method is not as attractive as in other applications because mesh rezoning may be compulsory, it is obviously the only numerical method which allows for the study of a large variety of industrial structures. This is the reason why this method has been retained in the present paper, which is concerned with shape optimization of axisymmetric structures. The objective is to make the tangential stress uniform along one, several or all parts of the boundary. The loads and displacements are expresssed as Fourier's series so that unsymmetric loads can be taken into account. Special attention has been given to stress derivatives and sensitivity calculations in order to save computer time and to obtain accurate results. A semiautomatic mesh generator has been included in the computer program to simplify the finite element description and to control its evolution and change it if necessary. In addition the tangential stress along the modified parts of the boundary can be displayed graphically. The optimal shapes obtained from each Fourier term can be calculated. By weighting each result with the associated strain energies, it is possible to determine the final optimal shape. A real example—i.e., a part of a helicopter gear box—is presented which demonstrates the capabilities of this shape optimization computer program, named AXIOPT.

DEFINITION OF THE SHAPE

A shape optimization process begins by the definition of the geometry, its parameters and limitations and their dependencies on the design variables. This first step can be called the *shape algorithm.* Here the shape algorithm is identical to those described in several previous papers [3–5]. It has been selected because of its ability to describe a large variety of two-dimensional (or axisymmetric) structures using the variables identified by the designers. Note that the initial data are a set of nodal points connected by straight segments. Each nodal point is identified by its two cylindrical coordinates (r, z), a real $R > 0$ which represents a radius when it is used to define a circular arc, and a binary mobility code digit. The design variables are easily deduced from the data, as shown in Figure 1 and Table 1. The computer calculations give the coordinates of any boundary point and especially the tangent points (Figure 1: $T_i, i = 1, 2, ...$) necessary to define the circular arc lengths.

Table 1
Initial input data for a fillet boundary definition

Point	Coordinates			Mobility	
A	X_A	Y_A	0	0	
B	X_B	Y_B	0	0	
C	X_C	Y_C	$R_C(-)$	1	design variables
D	X_D	Y_D	0	0	
E	X_E	Y_E	$R_E(+)$	0	
F	X_F	Y_F	0	0	

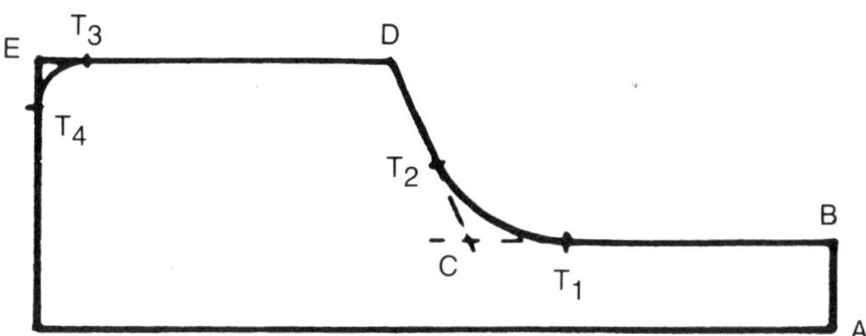

Figure 1. Description of a fillet.

CONSTRAINTS

All constraints originate from geometric considerations since the objective function is formulated to smooth the tangential stress, which means that the first and the following geometries are admissible from a stress point of view. These constraints allow a mobile nodal point to move on a straight line or in a given area. The length of either a straight segment, a circular segment or a radius value cannot become null without starting a new optimization problem.

ANALYSIS OF THE STRUCTURE

Analysis is performed by the finite element method in which the special character of an optimization process has been considered, to ease the calculations, to improve the accuracy of the results and to save computer time.

First, because just a few parts of the structure must often be modified, the substructure concept is used to separate the "fixed" and the "mobile" parts. The fixed parts are calculated twice: once at the beginning and also at the end of the optimization process. Only the reduced stiffness matrices of these substructures are added to the matrices of the mobile parts. Related to this division, an automatic mesh generator creates the finite element mesh of each substructure considered as a macro finite element. These macro elements are either triangular (six nodes) or quadrilateral (eight nodes). Following a well-known technique, the same subdivision is used in the parent space to obtain the mesh itself, which is obviously made out of the same types of elements. During the optimization process this mesh is controlled and a new discretization can be chosen if necessary.

OBJECTIVE FUNCTION

The optimization process attempts to smooth the tangential stress distribution along the variable parts of the boundary. This objective is obtained by minimizing the integral:

$$J = \int_{\Gamma_m} (\sigma_t - \sigma_0)^2 d\Gamma_m \qquad (1)$$

in which Γ_m is the variable part (possibly made of different parts), and σ_0 is a calculated or a given value of σ_t which can be different for compression or traction.

References p. 292

NUMERICAL ASPECTS

In any optimization procedure both an accurate determination of J and of its derivatives are needed to obtain a good convergence. This accuracy, and also the computer time, are strongly dependent on the type of finite element used, the mesh density and the numerical methods. Some details are given here on all these aspects.

Stress Values—As equation (1) is calculated by the Gaussian integration method, the stress values are needed at the Gauss points on the boundary of every concerned finite element. The method used is described in [8] and will be briefly summarized here. First the stress tensor is determined at each Gauss point of the finite element by the usual representation of the displacement field, i.e.,

$$\boldsymbol{\sigma}^e = \mathbf{DBU^e} \tag{2}$$

where $\mathbf{D}$ is the elasticity matrix, $\mathbf{B}$ is the differential matrix operator applied to the shape functions, and $\mathbf{U^e}$ is the nodal displacement vector. Then the stresses are calculated at the nodal points by a local smoothing and averaged at a node common to several finite elements. Lastly they are calculated at each Gauss point on the boundary by using one degree lower shape functions.

From these values equation (1) is written as

$$J = 2\pi \sum_{e=1}^{N^e} \sum_{q=1}^{N^g} \{(\sigma_t^e(\xi_{Ng}) - \sigma_o)^2 r^e(\xi_{Ng}) L^e(\xi_{Ng}) W_{Ng}\} \tag{3}$$

in which e is the index of the element, ξ_{Ng} represents the parametric value of the Gauss point position, and W_{Ng} is the weight factor.

The derivatives of the objective function $J = J(u(\alpha), \alpha)$ with respect to the design variable α_k is obtained by the self adjoint state method [9] which requires the derivative calculations of J considered as only function α_k and of $\partial|\mathbf{K}|/\partial\alpha_k$. Indeed it has been demonstrated that

$$\frac{\partial J(u(\alpha_k), \alpha_k)}{\partial \alpha_k} = \frac{\partial J}{\partial \alpha_k} - < \boldsymbol{\lambda} > \left| \frac{\partial \mathbf{K}}{\partial \alpha_k} \mathbf{u} \right. \tag{4}$$

with $\boldsymbol{\lambda}$ vector solution of

$$\mathbf{K}\boldsymbol{\lambda} = \begin{vmatrix} \partial J/\partial u_1 \\ \cdot \\ \cdot \\ \cdot \\ \partial J/\partial u_n \end{vmatrix} \tag{5}$$

The first term of (4) may be written

$$\frac{\partial J}{\partial \alpha_k} = 2\pi \sum_{e=1}^{N^e} \sum_{q=1}^{Ng} W_{Ng} \{[2 \frac{\partial \sigma_t^e}{\partial \alpha_k} r^e L^e + (\sigma_t^e - \sigma_0)(\frac{\partial r^e}{\partial \alpha_k} L^e + r^e \frac{\partial L^e}{\partial \alpha_k})](\sigma_t^e - \sigma_0)\} \qquad \xi = \xi_{Ng} \tag{6}$$

with

$$\frac{\partial \sigma_t^e}{\partial \alpha_k} = \frac{\partial \sigma_t^e}{\partial r^e}\frac{\partial r^e}{\partial \alpha_k} + \frac{\partial \sigma_t^e}{\partial z^e}\frac{\partial z^e}{\partial \alpha_k} \tag{7}$$

In equations (4), (6) and (7) the derivatives of the shape functions with respect to the nodal coordinates are needed to calculate $\partial r^e/\partial \alpha_k, \partial z^e/\partial \alpha_k$ and $\partial \mathbf{K}/\partial \alpha_k$. This is done in the program from the analytical formulations of all these derivatives which are calculated simultaneously with the shape functions.

In (6) and (7) it must be pointed out that the derivatives of the stress tensor must also be calculated at the Gauss points of the boundary. Since high accuracy is required, it is not advisable to differentiate twice the displacement shape functions of degree 2. A new numerical method to avoid this double differentiation has been proposed and published in [10]. It is used here.

OPTIMIZATION PROCEDURE

The mathematical optimization method selected to find the minimum of J is the augmented Lagrange multiplier method [11]. The associated unconstrained minimization problem is solved by the Davidon-Fletcher-Powell variable metric method.

RESULTS

This program has been widely tested on several examples. Some results obtained in investigating the optimal shape of a part of a helicopter gear box (Figure 2) are presented here. The part is located just below the rotor blades. It is estimated that the main contribution of the excitation load is contained in the first term of the Fourier series and to a lesser degree in the fifth term. The structure is long enough to be considered as clamped at its base. The design variables and the constraints given by the manufacturer (Aérospatiale) are reported in Table 2. Figures 3–6 show the initial and final boundary and the tangential stress distributions. Table 3 gives some numerical values of the optimization procedure. The influence of σ_0 has also been investigated on both the final σ_t distribution and the convergence. Table 3 shows that the choice of σ_0 is not insignificant, and it is thus advisable to try at least one other value of σ_0 when the first convergence is obtained.

References p. 292

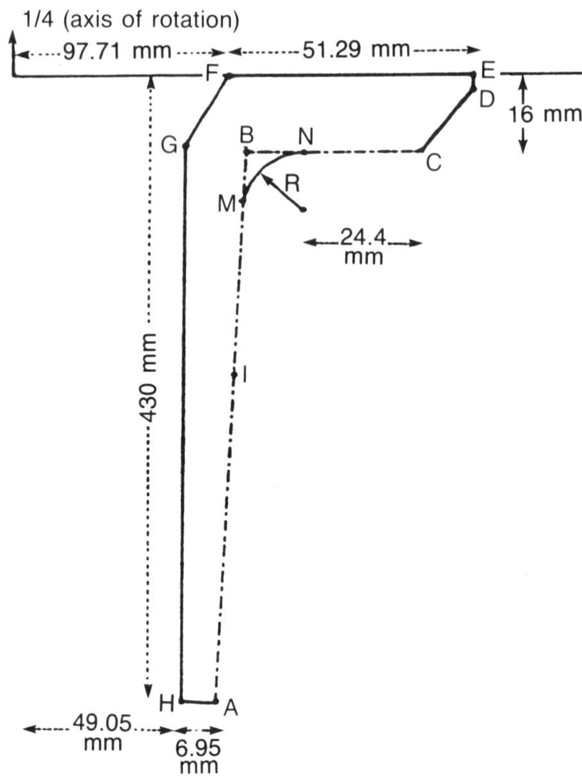

Figure 2. Description of the rotor blade support.

Table 2
Initial design variable and constraint values of the rotor blade support

Design variables	Initial values	Constraints	Constraints
X_B	62.0	$60 < X_B < 70$	
			$\|\|CN\|\| > 15$
Y_B	-16.0	$-20 < YB < 10$	
			$Z_C = Z_B$
R_B	-12.0	$-20 < R < 0$	

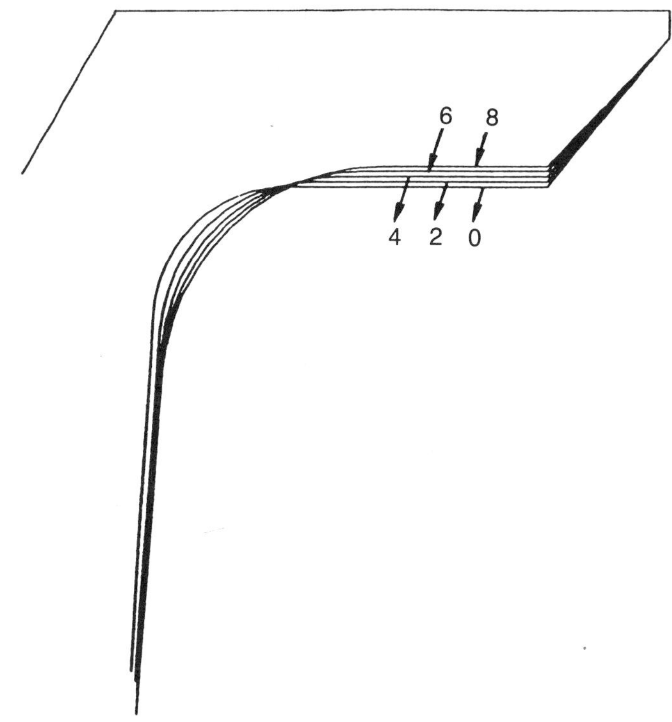

Figure 3. Successive optimal shapes: $\sigma_0 = 20N/mm^2$.

Table 3
Optimal design variable values with different σ_0

	Design variables		σ_0		
σ_0	Initial	Final	Initial	Final	Weight
	62.128	65.242			
15	-16.000	-15.920	70.318	42.519	
	-12.000	-19.130		-40%	+20%
		63.557		50.605	
20	id.	-14.130	id.		
		-20.000		-28%	+2%
		62.432		57.080	
25	id.	-12.224	id.		
		-20.000		-14%	-15%

References p. 292

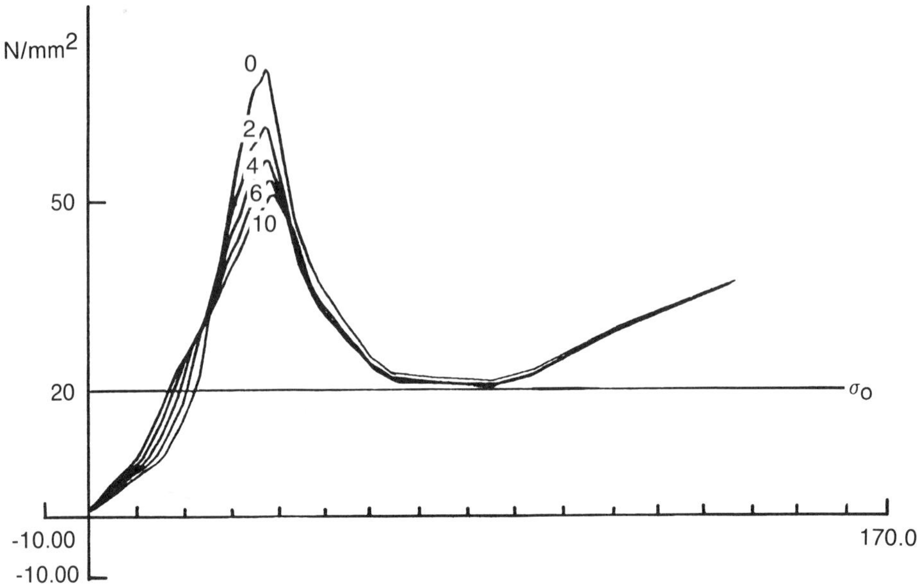

Figure 4. σ_t **variations along the moveable boundary.**

Lastly it has been observed that the new optimal shape is also efficient for the fifth term of the Fourier series. This is the reason why the process has been stopped here.

CONCLUSION

A general computer program to optimize the shapes of axisymmetric structures has been presented which is widely used in several research departments to improve the design process. The study of optimum sensitivity to the constraints and especially to the given value σ_0 will be undertaken in the near future.

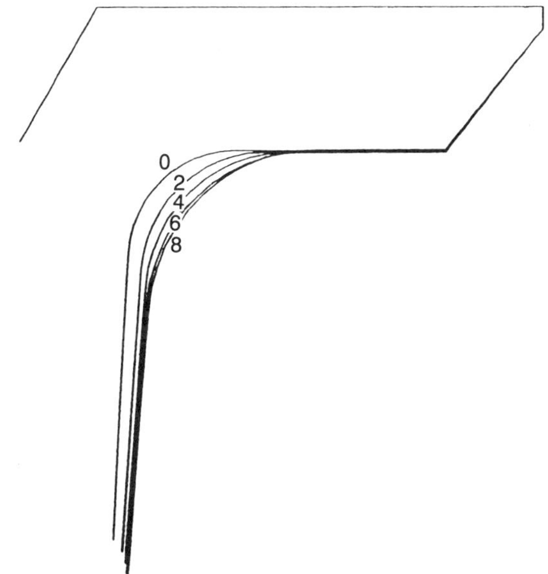

Figure 5. Successive optimal shapes: $\boldsymbol{\sigma}_0 = \mathbf{15}N/mm^2$.

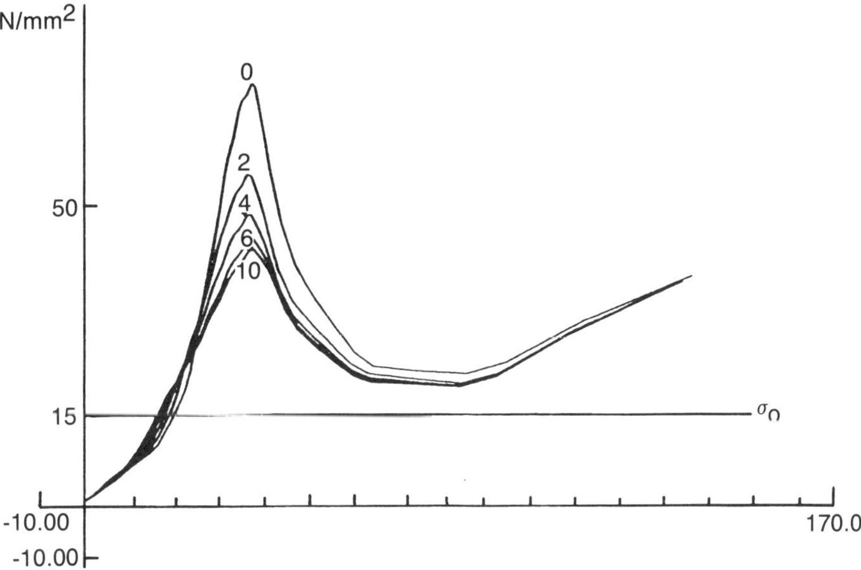

Figure 6. $\boldsymbol{\sigma}_t$ variations along the moveable boundary.

References p. 292

ACKNOWLEDGMENTS

The authors are grateful to Aérospatiale (French National Aerospace Company) for authorizing the publication of this paper.

REFERENCES

1. C. A. Mota Soares, H. C. Rodrigues, L. M. Oliviera Faria and E. J. Haug, Optimization of the geometry of shafts using boundary elements. ASME Paper 83-DET-39.
2. C. A. Mota Soares, H. C. Rodrigues and K. K. Choi, Shape optimal structural design using boundary elements and minimum compliance techniques. ASME Paper 84-DET-57.
3. Ph. Trompette and D. Eizadian, Shape optimization of bidimensional structures by the boundary elements method. CAD/CAM, Robotics, and Automation Int. Conf., Tucson, AZ (Feb. 1985).
4. S. S. Bhavikatti and C. V. Ramakrishnan, Shape optimization of structural systems using finite elements and sequential linear programming (pp. 224–235) in *Large Engineering Systems*. Pergamon Press (1977).
5. J. P. Quéau and Ph. Trompette, Two dimensional shape optimal design by the finite element method. *Int. J. Numer. Meth. Eng.* **15**, 1603–1612 (1980).
6. A. Francavilla, C. V. Ramakrishnan and O. C. Zienkiewicz, Optimization of shape to minimize stress concentration. *J. Strain Analysis* **10** (2) 63–70 (1975).
7. E. Schnack, An optimization procedure for stress concentration by the finite element method. *Int. J. Numer. Meth. Eng.* **14** 115–124 (1979).
8. E. Hinton and J. S. Campbell, Local and global smoothing of discontinuous finite element functions using a least squares method. *Int. J. Numer. Meth. Eng.* **8** 461 480 (1974).
9. E. J. Haug and J. S. Arora, Design sensitivity analysis of elastic mechanical systems. *Computer Meth. Appl. Mech. Eng.* **15**, 35–62 (1978).
10. A. M. Baudron, J. L. Marcelin and Ph. Trompette, On accurate stress derivatives determination. *Commun. Appl. Numer. Meth.* **1**, 249–253 (1985).
11. L. E. Madsen, Engineering design optimization by the augmented Lagrange multiplier method. M. S. Thesis, Naval Postgraduate School, Monterey, CA. (1981).

DISCUSSION

E. Atrek *(Engineering Mechanics Research Corporation)*

In your rotor example, you showed that for different values of the limiting stress, we get different final designs; however, I noticed that only some part of the shape was being changed. Let us suppose that the entire shape was open to change and that we have uniform stress constraints—in other words, the same limiting stress—over the shape. Would you expect the final design to be sensitive to change in the limiting stress?

Trompette

With respect to change in which parts of the final design?

Ph. Trompette

Atrek

I am supposing that I can change the entire shape, that the entire boundary is open to change. Now suppose that I have 25 N/mm^2 limiting stress and I obtain a certain optimum shape. Would you expect a different shape if the limiting stress were 15 N/mm^2?

Trompette

I don't know what happens if all the boundaries are free to move, as I have not been able to explain the changes in the shapes for different values of σ_0. Perhaps in the future we will try to predict the sensitivity of the optimum to these parameters, but right now we do not know.

Atrek

In my experience, the optimum shape is independent of the value of the limiting stress if the entire system is allowed to change under a uniform stress constraint. But when you fix some parts of the system and allow the rest to change, then one would obtain different shapes.

J. Taylor *(The University of Michigan)*

Does Dr. Atrek's last question allude to the possibility of a fully stressed design as the optimal solution? If so, in practical application one can seldom expect the problem to be so free of constraints that it would allow a fully stressed design.

Atrek

What I implied by *uniform stress constraints* was not a fully stressed design, but rather to prescribe at the outset what the limiting stresses are for each region of the system. In that case, I would have the same limiting stress. Whether the final design has uniform stress or not would depend on the system itself.

D. Grierson *(University of Waterloo, Canada)*

What would happen if your structure was symmetrical in loading and structure? Would you expect the same phenomena when you changed σ_0?

Trompette

We only calculate the structure for Fourier numbers 1, 3 and 5. If the structure was completely symmetric, it means that we have to calculate it for the Fourier number 0 and we do not do that, as the main values for the loading were Fourier numbers 1, 3 and 5. Changes in the loading may lead to changes in the objective function. For example, you can choose another stress function to take into account the torsion problem. You have to take the stress σ_{00} in place of the tangential stress σ. In the future, we will replace the tangential stress by the von Mises stress and show that we get another design.

B. Prasad *(Electronic Data Systems)*

Have you considered all those earlier stress functions 1, 3 and 5 simultaneously, or did you consider them one at a time?

Trompette

No, we do that in sequence. The shape obtained after one analysis for the Fourier number 1 was not optimal for the Fourier number 3, and vice versa. But the importance of the loading of 3 is so small compared to the loading for the Fourier number 1 that we kept the shape resulting from number 1. The shape for the Fourier number 3 was not bad, not good. In fact, you must weight both, but which number should be used for weighting? I think the strain energy of the structure for each number can be used.

R. Haftka *(Virginia Polytechnic Institute and State University*

It is not clear to me why you have used different values of σ_0. Since you only have one maximum of the stress, couldn't you simply have minimized the minimum value of the stress in that problem?

Trompette

In the past I have done that, and I encountered many problems. Some of them may interest a mathematician, but not me. For example, if you try to minimize the stress only at one point, you get a shape in which the stress is minimized but the computer tried to generate another design in which the minimization leads to the preceding design, and so on. So we tried to minimize at two points, and it was slightly better. But if you minimize at two points, and afterward at three and so on, why not minimize on the frontier?

G. N. Vanderplaats *(University of California - Santa Barbara)*

The solution to that is simply to minimize an artificial variable subject to all the constraints less than the artificial variable; then it can even be fully stressed. Rather than choose the maximum among all the stresses that you are going to change and have conflicting gradients, just minimize some extra variable and treat all the stresses as constraints that have to be less than that extra variable.

C. Fleury *(University of California - Los Angeles)*

You have a formula for stress sensitivity that looks very interesting, but it needs some accurate stresses to calculate the minimum. How do you get these accurate stresses?

Trompette

We used a method which has been published by A. M. Baudron, J. L. Marcelin and myself in *Communications in Applied Numerical Methods.* It involves a local smoothing of the stress obtained from values at the nodes and, in my opinion, it works well.

SHAPE OPTIMAL DESIGN BY THE CONVEX LINEARIZATION METHOD

C. FLEURY
University of California-Los Angeles
Los Angeles, California

Abstract

The main goal of the present paper is to discuss the choice of suitable numerical methods for shape optimal design of elastic structures discretized in finite elements. After a short description of the approach we followed to create an appropriate geometric model involving a relatively small number of design variables, attention is mainly directed toward the selection of an adequate optimization algorithm. To this aim the paper will briefly present the various attempts that we have successively undertaken before adopting the convex linearization method as the basic optimizer, not only for shape optimal design problems but also for all our other structural synthesis capabilities. Various examples of applications to optimum shape problems are offered to demonstrate the efficiency of this new algorithm. Finally some comments are made about future developments needed to effectively implement shape optimization concepts into the real design cycle.

INTRODUCTION

The main goal of this paper is to discuss various optimization algorithms that have been tried to solve shape optimal design problems. The approach followed to describe the structural geometry can be found in detail in a previous paper [1]. The class of shape optimal design problems that can be dealt with by this approach is restricted to two-dimensional structures in plane stress or plane strain, as well as axisymmetric structures. The method is highly efficient, mainly because it is based upon an internal parametric representation typical of modern techniques employed in computer aided geometric design. The two-dimensional structure is decomposed into a few subregions of simple geometry which are compactly described by using a limited number of control nodes (also named master nodes).

During the optimization process the geometry of conveniently selected subregions is allowed to change: these regions are called design elements. The movements of the corresponding control nodes are the design variables.

The design element boundaries are described by blending functions commonly used in computer graphics technology for interactive generation of curves (Bézier, B-splines) and surfaces (Coons patches) [2, 3]. Such an internal parametric representation permits determining the coordinates of any point inside the design element or on its boundaries. Indeed the design element can be mathematically defined, either as the Cartesian product of two families of curves, or by using Coons patches. In both cases an analytical interpolation scheme is provided. This formulation lends itself well to shape optimal design problems. Only a few design elements are generally sufficient to fully describe the regions that are modified during the optimization process. In addition it is relatively straightforward to generate a suitable finite element mesh and to maintain its integrity throughout the optimization process. A regular mesh is initially constructed in the curvilinear coordinate system of each design element. After each iteration the mesh is updated by applying coordinate transformations to the grid points within each design element. Finally the sensitivity analysis may be formulated in a fully analytical way. In other words it is possible to get the derivatives of the element matrices (i.e., stiffness, stress and consistent load) with respect to the positions of the control nodes, without having to resort to a finite difference scheme. Note however that analytical sensitivity analysis does not seem to be a critical capability for shape optimal design, as demonstrated by the excellent results reported in reference [4] on the basis of finite difference techniques.

To help fix ideas, let us consider a simple example. The plate shown in Figure 1 is stressed by uniform in-plane tensile loads applied along the four edges. For the initial design, stress concentration occurs along the boundary of the hole. The objective is therefore to determine the best possible shape of the hole that would minimize the total weight of the plate while avoiding stress concentration. The analytical optimal solution of this problem yields an elliptical hole with principal axis lengths proportional to the amplitude of the loads acting in the X and Y directions. In order to solve the problem numerically, the design model consists of two subregions: the outer region has a fixed geometry, while the region connected to the unknown hole is a design element of variable geometry. The design element borders are described by two periodic B-splines, each governed by 16 control nodes. Of course only the inner control nodes are allowed to move during the optimization process. Therefore the design variables represent the move distances of each variable control node (when nodes are attached to the internal boundary of the design element) with respect to the corresponding fixed control node attached to the external boundary of the design elements. The move distances are measured along prescribed directions, which are defined as the lines connecting each corresponding control node in the initial design. The finite element model is derived from the design model by linear interpolation of its inner and outer boundaries.

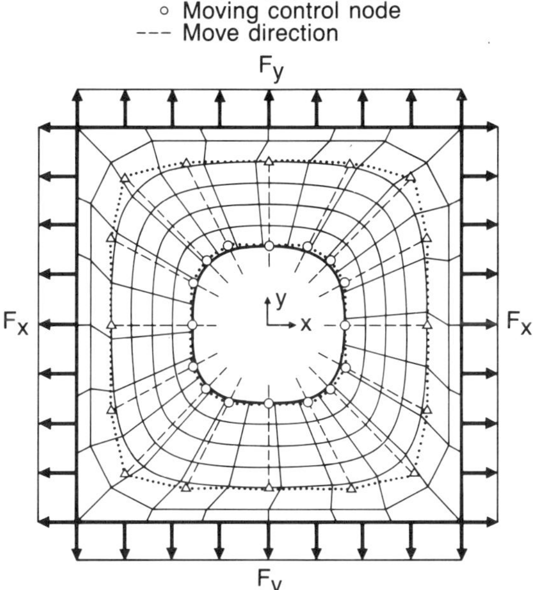

Figure 1. Example of design model.

THE NUMERICAL OPTIMIZATION PROBLEM

Since we can define a design model with only a few design variables, we can now envision efficient techniques to improve the characteristics of the structure by modifying its shape in an optimal way. Most often such a shape optimal design problem consists of minimizing some objective function subject to constraints, insuring the feasibility of the structural design. Mathematically the numerical optmization problem considered in this paper can be written in the following general form:

$$\text{minimize } f(\mathbf{x}) \tag{1}$$

subject to the constraints

$$c_j(\mathbf{x}) \leq \hat{\bar{c}j} \qquad j = 1, m \tag{2}$$

$$\underline{x}_i \leq x_i \leq \bar{x}_i \qquad i = 1, n \tag{3}$$

$$\left(\sum_i d_{ij} x_i \leq \bar{d}_j \right) \qquad j > m. \tag{4}$$

The objective function (1) is a nonlinear function of the design variables x_i. It usually represents a structural characteristic to be minimized (e.g., the weight). The nonlinear inequalities (2) are the behavior constraints that impose limitations on structural response quantities (e.g., stresses and displacements under static loading cases). The design variables must also be bounded by the side constraints (3), where $\underline{x}_i$ and $\bar{x}_i$ are lower and upper limits that reflect fabrication or analysis validity considerations. It should be noted that the side constraints (3) constitute a particular case of the linear constraints (4). However, they are written separately in our optimization problem statement, because the dual method approach described later can handle them more efficiently when considered apart from the general constraints. For various reasons the design variables may also be subjected to linear constraints such as (4). These requirements often reflect fabricational or rule of thumb considerations, such as a linear taper rule for the thicknesses of contiguous structural members in a sizing problem (e.g., increase in the number of layers in a laminated composite plate). In shape optimal design problems they may represent geometric considerations such as the imposition of tangential continuity along a moving boundary. Because the linear constraints in (4) are not always present in the problem statement, they have been put into parentheses.

The nonlinear programming problem (1–4) can be solved iteratively by using numerical optimization techniques. Each iteration begins with a complete analysis of the system behavior in order to evaluate the objective function and constraint values along with their sensitivities to changes in the design variables (i.e., first derivatives). Most often the analysis capability is based on a finite element discretization. A design iteration is completed by employing the results of these behavioral and sensitivity analyses in a minimization algorithm which searches the n-dimensional design space for a new primal point that decreases the objective function value while remaining feasible in terms of satisfying the constraints. Many such iterations are usually required before achieving the optimum design. This observation brings us to the essential difficulty in solving the nonlinear programming problem (1–4): the implicit character of the constraint functions $c_j(\mathbf{x})$. In other words these functions are not explicitly known in terms of the design variables. For each new design, they can only be evaluated numerically through a finite element analysis. The iterative nature of the optimization process implies that many structural reanalyses must usually be accomplished before finding an acceptable solution. Those repeated finite element analyses can lead to a prohibitive computational cost when dealing with large-scale problems.

One widely used approach to shape optimal design problems is to join together by brute force a general purpose optimizer and a finite element package having the required sensitivity analysis capabilities. During the initial development phase of our research efforts, this straightforward though not highly efficient approach was in fact the only possible choice. So, in a first step, direct nonlinear programming methods were compared to conventional linearization techniques using the SIMPLEX algorithm. The numerical experiments conducted for that purpose have demonstrated, as could be expected, that direct approaches such as gradient projection or feasible direction methods are inadequate for shape optimization problems in view of the large number of iterations required for convergence.

On the other hand recursive linear programming techniques, even though they necessitate difficult adjustments of move limits, prove to be computationally efficient. It was therefore decided to pursue the idea of sequential linearization, and to improve it by implementing more suitable approximation concepts similar to those used for sizing problems. The basic approach consists of replacing the primary optimization problem with a sequence of explicit approximate subproblems having a simple algebraic structure. Each subproblem is generated through Taylor series expansion of the objective function and constraints in terms of intermediate linearization variables. It was found that the efficiency of this approach is very sensitive to the mode of expansion employed for building the approximate subproblems: (1) first- or second-order expansion of the objective function in terms of the design variables, and (2) first-order expansion (i.e., linearization) of the constraints in terms of the design variables or their reciprocals.

Linearization of the constraints with respect to reciprocal variables is a well-recognized technique to solve optimal sizing problems. Although this linearization scheme cannot be physically justified for optimum shape problems, surprisingly it leads to remarkably good results. At this stage of our research efforts the question was why are reciprocal variables so miraculous in structural synthesis problems? The answer was later found in a new and rather general optimization method, called the convex linearization method (CONLIN). The key idea in the CONLIN algorithm is to perform the linearization process with respect to mixed variables (either direct or reciprocal) independently for each function involved in the optimization problem. A convex, separable subproblem is therefore generated that can be efficiently solved by a dual method formulation. Because it uses conservative approximations CONLIN has an inherent tendency to generate a sequence of steadily improving feasible designs.

Various examples of applications to optimum shape problems will be offered later in this paper to demonstrate the efficiency of this new algorithm. However, in order to give some sense of its power, let us now solve the simple plate design example introduced in Figure 1. The iteration history data provided by a general purpose optimizer (the well known CONMIN [5]) are compared in Figure 2 to the results generated by the new CONLIN optimizer. It can be seen that, when using CONMIN, convergence is not yet achieved after 20 iterations. Moreover it should be noted that each iteration involves several structural analyses in order to determine a suitable step length along the selected feasible direction (line search process). Although this line search is implemented in a very efficient way within CONMIN, 53 structural reanalyses are needed to accomplish the 20 iterations shown in Figure 2. The corresponding computational time is more than seven hours CPU on a VAX 11/780. On the other hand, when the convex linearization method is employed, only four structural analyses are needed to get an optimal design. Note that the theoretical ellipse is recovered by the numerical optimization algorithm.

APPROXIMATION CONCEPTS IN SHAPE OPTIMAL DESIGN

The approximation concepts approach to structural synthesis is now widely employed to solve optimal sizing problems [6]. Such problems consist of minimizing the weight of thin-walled structures modeled by bar and membrane elements.

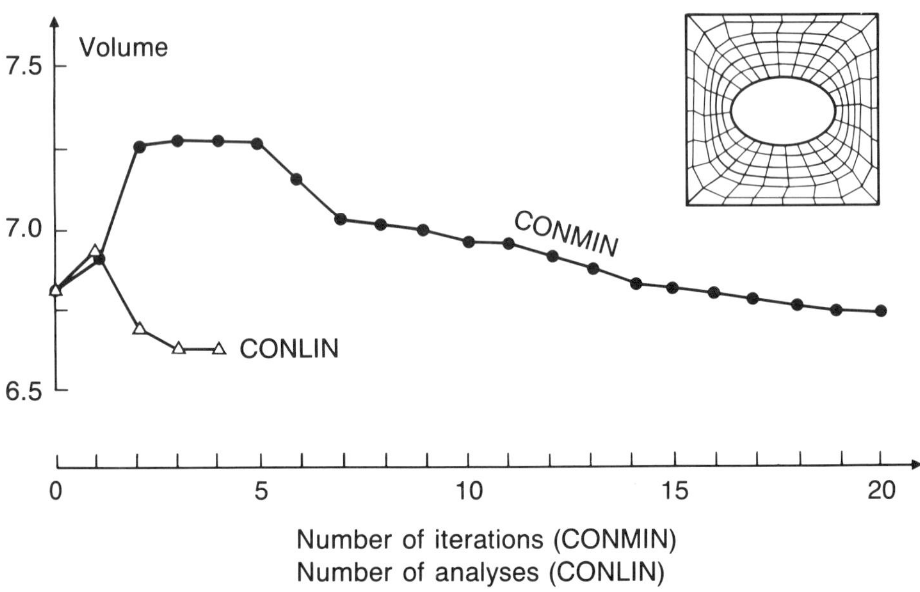

Figure 2. Iteration history for plate example.

Because the geometry is fixed, the design variables reduce to the transverse sizes of the structural members, i.e., bar cross-sections and membrane thicknesses. This approach consists basically of the following steps.

- A finite element analysis is performed for the initial trial design.
- From the results of the current structural analysis, an approximate optimization problem is generated. This step implies that sensitivity analysis capabilities be available in the finite element code. The approximate subproblem is created by linearizing the behavior constraints in terms of the reciprocals of the design variables while the objective function keeps its exact linear form in terms of the direct sizing variables.
- Because the subproblem is fully explicit, convex and separable, it can be efficiently solved by resorting to its dual formulation. When expressed in the reciprocal design variables, the objective function is strictly convex and the linearized constraints are linear.
- The solution of the approximate subproblem is adopted as a new starting point in the design space, and the optimization process is continued until convergence is achieved.

From extensive numerical experiments it can be argued that the approximation concepts approach converges to an optimum design in usually less than ten iterations or finite element analyses. These remarkable convergence properties are generally attributed to the fact that the behavior constraints have a tendency to be much less nonlinear in the space of the reciprocal variables. The convex linearization method described later in this paper provides a more rigorous explanation.

When dealing with shape optimal design problems, the foregoing considerations on the intuitive choice of reciprocal variables or other intermediate linearization variables are no longer valid. There is indeed no reason why the constraints should be more shallow with respect to the reciprocals of nodal coordinates. Furthermore, although the structural weight keeps a simple explicit form, it is no longer a linear function of the shape design variables. For those reasons various approximation schemes have been successively tried. In the sequel they will be classified depending upon the mode of Taylor series expansion employed to build the approximate subproblems. Four distinct problems will be discussed.

- ***Problem E1***. First-order expansion (i.e., linearization) of the objective function and constraint functions with respect to the direct variables.
- ***Problem E2***. Second-order expansion of the objective function and linearization of the constraint functions with respect to the direct variables.
- ***Problem E3***. Second-order expansion of the objective function with respect to the direct variables, and linearization of the constraint functions with respect to the reciprocal variables.
- ***Problem E4***. Linearization of the objective function with respect to the direct variables, and linearization of the constraint functions with respect to the reciprocal variables.

The numerical experiment described hereafter shows that the convergence properties of the optimization process mainly depend upon the quality of the constraint approximation. When the constraint approximation is too rough, divergence or oscillation frequently occurs. It can also happen that the solution of an approximate problem becomes infeasible. In such a case many subsequent iterations are usually required in order to recover a feasible design.

With regard to solving the explicit subproblem, three methods are available:

- The well known SIMPLEX algorithm can only be applied to the linear programming problem E1.
- The gradient projection method may be used in conjunction with problems E2, E3 or E4.
- In the special case where the objective function of problem E4 is strictly convex, the dual method approach can be applied.

Sequential Linear Programming—The sequential linear programming proceeds by replacing the primary nonlinear problem (1–4) with a sequence of linear subproblems. Each subproblem is generated by linearizing the objective function and the constraint functions with respect to the shape design variables x_i.

PROBLEM E1: Minimize

$$f(\mathbf{x}^0) + (\mathbf{x} - \mathbf{x}^0)^T \nabla f(\mathbf{x}^0)$$

subject to

$$c_j(\mathbf{x}^0) + (\mathbf{x} - \mathbf{x}^0)^T \nabla c_j(\mathbf{x}^0) \leq \bar{c}_j$$
$$\underline{x}_i \leq x_i \leq \bar{x}_i$$
$$\sum_i d_{ij} x_i \leq \bar{d}_j.$$

Note that $\mathbf{x}^0$ denotes the current design point, that is, the point in the design space where the objective function and the constraints are linearized. Because the explicit problem E1 is a linear programming problem, it can be efficiently solved by using the SIMPLEX algorithm. It is well known that the optimal solution point corresponding to each linear subproblem necessarily lies at a vertex of the design space. As a result the overall optimization process may either converge to a nonoptimal solution of the primary problem, or it may oscillate indefinitely between two or more vertices. In order to avoid this undesirable behavior, the so-called move limits strategy can be implemented. It consists of temporarily adding some artificial side constraints to the linearized problem so that the design points will not move too far away from the current linearization point $\mathbf{x}^0$. However, it is very difficult even for an experienced user to define adequate rules for setting the move limits, and to determine the way they should be updated after each iteration. Therefore, even though some meaningful results have been obtained in the case of simple problems through the use of sequential linear programming [7], this approach has not been retained in the subsequent developments of our research efforts.

Second-Order Expansion of the Objective Function—In an attempt to avoid the drawbacks of the sequential linear programming approach, a second-order expansion of the objective function has been envisioned. Because some curvature is introduced in the objective function contours, the solution of each quadratic subproblem does not necessarily lie at a vertex of the design space. It can even be expected that such an approximation scheme should not need the imposition of artificial move limits. In fact it can be proven that in the special case of two-dimensional structures considered in this paper, the structural volume, and so the structural weight, is an exact quadratic function of the design variables. (In the axisymmetric case, it is a cubic function.)

By resorting to this improved technique, an explicit approximate subproblem of the following form is created at each iteration:

PROBLEM E2: Minimize

$$f(\mathbf{x}^0) + (\mathbf{x} - \mathbf{x}^0)^T \nabla f(\mathbf{x}^0) + \frac{1}{2}(\mathbf{x} - \mathbf{x}^0)^T \nabla^2 f(\mathbf{x}^0)(\mathbf{x} - \mathbf{x}^0)$$

subject to

$$c_j(\mathbf{x}^0) + (\mathbf{x} - \mathbf{x}^0)^T \nabla c_j(\mathbf{x}^0) \leq \bar{c}_j$$
$$\underline{x}_i \leq x_i \leq \bar{x}_i$$
$$\sum_i d_{ij} x_i \leq \bar{d}_j.$$

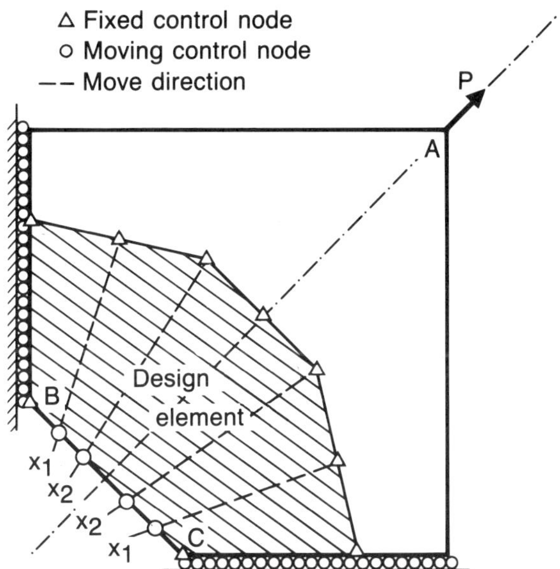

Figure 3. Two-variable example.

Such a quadratic programming problem can be efficiently solved by the gradient projection method for linear constraints. Numerical experiments have demonstrated that, unfortunately, this supposedly enhanced strategy behaves just like the sequential linear programming approach. Most often the optimization process converges to a vertex in the design space, or it oscillates between two vertices. Sometimes it completely diverges. Therefore move limits have again to be employed in order to maintain the design point within a prescribed box around the linearization point. This undersirable feature is due to the fact that the weight objective function is frequently quasi-linear. Indeed it can be shown that the Hessian matrix $\nabla^2 f(\mathbf{x})$ appearing in the Problem E2 statement has zero diagonal elements. Moreover most of the off-diagonal elements are also zero: they vanish whenever the two corresponding control nodes move along parallel directions and, of course, if the control nodes govern the shape of two different curves. Because the diagonal of the Hessian matrix is always zero, the weight objective function has a saddle point. However, by restricting the variations of the design variables to acceptable values, i.e., values yielding feasible geometries, the weight can be considered as a convex function on the restricted domain. Note that in this sequential quadratic approach, adopting an objective function other than the structural weight (e.g., a response quantity like a stress or a displacement) would lead to prohibitive computational costs when evaluating the required second derivatives.

To help fix ideas let us consider the example problem depicted in Figure 3. The goal is to minimize the structural weight (i.e., the area of the cross-hatched part in the drawing) by determining the shape of the moving boundary BC. One single behavior constraint is imposed: an upper bound is assigned to the displacement

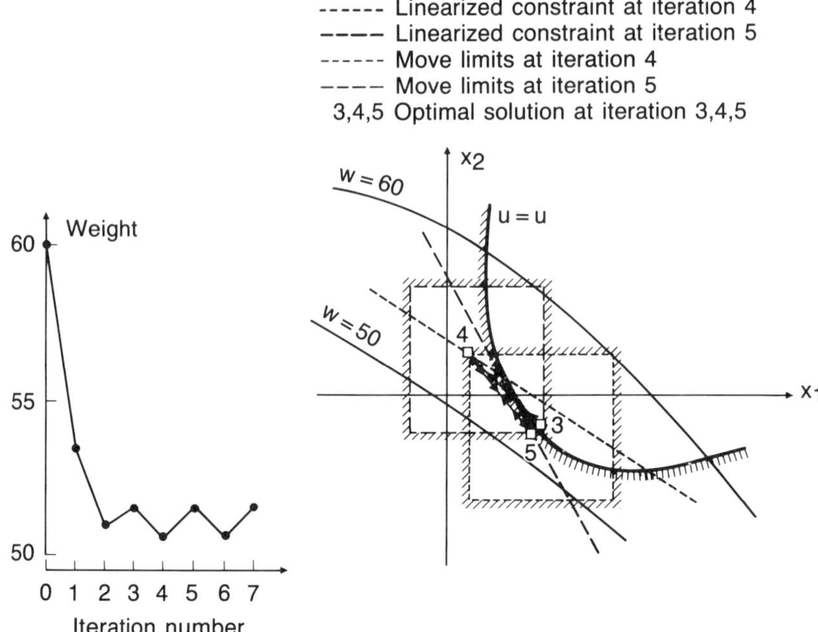

Figure 4. Constraint linearization (direct variables).

at node A in the direction of the applied load P. The unknown boundary BC is represented by a B-spline curve governed by six control nodes. The end control nodes B and C are fixed, and so the design variables correspond to the positions of the four remaining control nodes along the move directions shown on Figure 3. Due to symmetry considerations, only two design variables need to be taken into account, so the design space can be visualized as indicated in Figure 4. In this figure W denotes the weight objective function, while u refers to the displacement constraint. Note that the feasible domain is located on the right side of the restraint surface $u = \bar{u}$. The same figure also shows the approximate design spaces corresponding to the explicit subproblems constructed at iterations 4 and 5. It is clear that only the introduction of move limits can prohibit the design point from moving too far into the infeasible domain. These considerations explain the oscillatory behavior of the optimization process, whose iteration history is depicted in Figure 4. Note that the oscillations seem to stay approximately constant, because the move limits are not updated from one iteration to the next.

Constraint Linearization in the Reciprocal Space—It has now become almost certain that the best approaches to optimal sizing problems are those which make use of constraint linearization with respect to the reciprocal design variables [6]. There is an intuitive explanation for the success of these methods, in that stresses and displacements are exact linear functions of the reciprocal sizing variables in the case of a statically determinate structure. Unfortunately, for shape optimal design problems there is no such physical guideline for the selection of intermediate linearization variables. Nevertheless, as explained in this section, this change of variables continues to have a highly beneficial effect on the convergence properties of the shape optimization process.

The approach considered at this point consists of keeping the quadratic approximation of the weight objective function in terms of the original design variables, while linearizing the behavior constraints with respect to the reciprocals of the design variables. Momentarily the linear constraints of equation (4) will no longer be accepted in the optimization problem statement. This technique generates at each iteration an approximate subproblem of the following form:

Problem E3: Minimize

$$\frac{1}{2}\sum_i \sum_k x_i H_{ik} x_k + \sum_i f_i x_i$$

subject to

$$\sum_i c_{ij}/x_i \leq \bar{c}_j$$

$$\underline{x}_i \leq x_i \leq \bar{x}_i$$

where H_{ik} are the elements of the Hessian matrix of the objective function and f_i are related to its gradient components. The c_{ij} coefficients denote the first derivatives of the constraint functions c_j with respect to the reciprocal variables:

$$z_i = 1/x_i.$$

When restated in terms of the reciprocal variables z_i. Problem E3 still involves only linear constraints, and therefore it can again be solved by using a gradient projection algorithm. Note that this solution scheme can be applied because the linear constraints in (4) have been dropped from the problem statement for the time being. All the numerical experiments conducted with this strategy tend to exhibit convergence properties similar to those previously obtained for optimal sizing problems: less than ten finite element analyses are usually sufficient to achieve stable convergence of the optimization process without having to resort to artificial move limits.

Considering again the shape optimal design problem of Figure 3 and using now Problem E3 formulation, convergence becomes fast and stable, as shown in Figure 5. The corresponding reciprocal design space is also illustrated. Note that the weight decreases when the new design variables z_i increase. Because of the important curvature introduced in the objective function contours, the progression of the design point is inherently restricted to a region where the linearization process yields sufficiently accurate values of the approximate constraint.

Dual Solution Scheme—An additional step can be taken when it is realized that the second-order term in the approximate form of the weight objective function can be neglected without loss of accuracy or convergence speed. This feature has

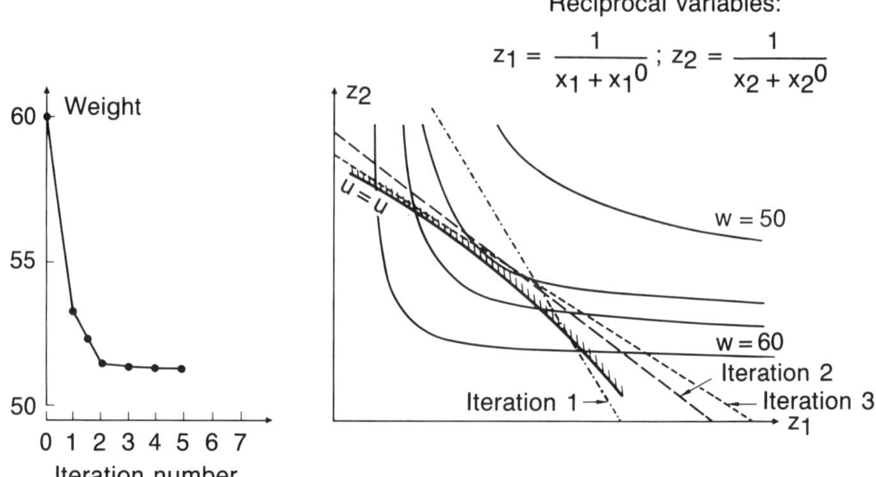

Figure 5. Constraint linearization (reciprocal variables).

been confirmed by many numerical experiments. A fourth explicit subproblem can therefore be stated:

PROBLEM E4: Minimize

$$\sum_i f_i x_i$$

subject to

$$\sum_i c_{ij}/x_i \leq \bar{c}_j$$

$$\underline{x}_i \leq x_i \leq \bar{x}_i$$

which is generated from the primary problem (1–4) by linearizing the objective function with respect to the direct design variables x_i, and by linearizing the behavior constraints with respect to the reciprocal variables z_i. When restating Problem E4 in terms of these reciprocal variables, a nonlinear problem with linear constraints is again generated (assuming of course that no explicit constraints of the form (4) must be taken into account).

This explicit problem can still be solved by using the same gradient projection algorithm as before. However, it can be recognized that Problem E4 is convex and separable, and therefore it lends itself well to solution by a dual method approach [8]. In fact Problem E4 has exactly the same explicit form as the subproblem used in the approximation concepts approach to optimal sizing problems. It thus becomes possible to resort to the previously developed DUAL-1 or DUAL-2 optimizers that have proven to be remarkably efficient [6].

When reintroducing the linear constraints (4), the explicit subproblem can no longer be directly solved by using a primal projection method. Formulated either in terms of the direct variables x_i or in terms of the reciprocal variables z_i, the subproblem always involves nonlinear constraints. A possible strategy would be to restate the explicit constraints (4) in terms of the reciprocal variables, and to linearize them in the same way as the behavior constraints (2). The resulting approximate subproblem then keeps the same form as Problem E4, amenable to solution by gradient projection or by dual algorithms. However it would be unfortunate, if not ridiculous, to approximate the only constraints that have a simple explicit form in the problem statement.

THE CONVEX LINEARIZATION METHOD

Simultaneously with our research efforts in shape optimal design, we were faced with another project aimed at improving the dual method approach to solve difficult optimal sizing problems that involved many linear constraints in an exact manner. The initial research effort has been focused on exploiting again the separable nature of the explicit subproblem to be solved, i.e., Problem E4 with addition of the linear constraints (4), and to resort to a specially devised dual method approach. Unfortunately, as shown in reference [9], this research was not successful. Because the required convexity properties are not satisfied, the dual problem happens to be nondifferentiable, and is therefore difficult to solve by conventional maximization algorithms.

Nevertheless this study was essential for the continuation of our development efforts in structural optimization, because it finally led to the convex linearization method that is now employed for sizing and for shape optimal design problems. Indeed one important conclusion of reference [9] was that further research was needed to make the explicit constraints (4) convex, for example by partially linearizing them with respect to the reciprocal variables. When carefully examining the dual problem, it can be seen that the terms responsible for nondifferentiability are those which correspond to negative coefficients in the linear constraints (4). Hence, to obtain a sound dual formulation, it is sufficient to replace those negative terms with positive approximate ones.

The convex linearization method (CONLIN) is based on generalizing this idea to all the functions describing the mathematical programming problem to be solved. In Problem E4 the objective function is linearized with respect to the direct variables x_i while the constraints are linearized with respect to the reciprocal variables z_i. In CONLIN each function defining the optimum design problem is linearized with respect to a properly selected mix of variables so that a convex and separable subproblem is generated. The selection of direct and reciprocal variables is made on the basis of the signs of the first partial derivatives. It is easily proven that, considering any differentiable function $g(\mathbf{x})$, the following linearization scheme yields a convex approximation (hence the appellation "convex linearization") [10]:

$$g(\mathbf{x}) \simeq g(\mathbf{x}^0) + \sum_{+} g_i^0 (x_i - x_i^0) - \sum_{-} (x_i^0)^2 g_i^0 (z_i - z_i^0)$$

References pp. 319–320

where g_i denote the first derivatives of $g(\mathbf{x})$ with respect to the original variables x_i. The symbol $\sum_{+}$ ($\sum_{-}$) means summation over the terms for which g_i is positive (negative). One of the most interesting features of the convex linearization scheme is that it also leads to the most conservative approximation amongst all the possible combinations of mixed direct and reciprocal variables. This property was initially demonstrated in reference [11], in which conservative approximation was employed to handle difficult buckling constraints.

The CONLIN algorithm proceeds by applying the foregoing convex linearization scheme to the objective function and to all the constraint functions, including the linear constraints (4). The resulting explicit approximations take on a simpler form if the design variables are normalized so that they become equal to unity at the current point $\mathbf{x}^0$ where the problem is linearized. The following subproblem is then generated:

CONLIN: Minimize

$$\sum_{+} f_i x_i - \sum_{-} f_i / x_i$$

subject to

$$\sum_{+} u_{ij} x_i - \sum_{-} u_{ij} / x_i \leq \bar{u}_j$$

$$\underline{x}_i \leq x_i \leq \bar{x}_i$$

where f_i and u_{ij} denote the first derivatives of the objective and constraint functions, respectively, evaluated at the current point $\mathbf{x}^0$. These derivatives must be taken with respect to the same set of variables for all the functions. The behavior constraints c_j and the linear constraints d_j from equations (2) and (4) have been collected in a single set of constraints u_j. Note that the upper bounds $\bar{u}_j$ have been modified to contain the zero-order contributions in the Taylor series expansion.

Therefore, in CONLIN, the initial problem is transformed into a sequence of explicit subproblems having a simple algebraic structure. Furthermore this subproblem is convex and separable. These remarkable properties make it attractive to solve the subproblem by using dual algorithms. In the dual approach, the constrained primal minimization problem is replaced by maximizing a quasi-unconstrained dual function depending only on the Lagrange multipliers associated with the linearized constraints. These multipliers are the dual variables subject to simple non-negativity constraints. The efficiency of this dual formulation is due to the fact that maximization is performed in the dual space, whose dimensionality is relatively low and depends on the number of active constraints at each design iteration. The CONLIN approach can be viewed as a generalization of well-established approaches to pure sizing structural optimization problems, namely, approximation concepts and optimality criteria techniques [12]; as such, it is capable of addressing a broader class of problems with considerable facility.

As previously mentioned, the convex linearization scheme yields explicit approximations that are locally conservative; that is, they tend to overestimate the values of the true functions. In other words the approximate feasible domain corresponding to the explicit subproblem is generally located inside the true feasible domain corresponding to the primary problem (1–4). As a result the CONLIN optimizer has a tendency to generate a sequence of design points that "funnel down the middle" of the feasible region. Despite the use of a dual solution scheme, a primal philosophy is maintained; that is, the method often produces a sequence of steadily improving feasible designs. This represents an attractive property from an engineering point of view, since the designer may stop the optimization process at any stage and still get an acceptable noncritical design, better than its initial estimate.

Although conservativeness is a desirable property most of the time, it is not desirable when the initial starting point is seriously infeasible. In such a case it can happen that the approximate feasible domain is empty, so that the CONLIN method can no longer be applied. To cope with this difficulty, it is convenient to work in an expanded design space by increasing the upper bounds assigned to the constraints. Denoting by v_j the maximal increments that the user accepts to add to the bounds $\bar{u}_j$, and introducing an additional "relaxation" variable r, the following explicit subproblem must now be solved by the CONLIN optimizer:

Minimize

$$\tilde{f}(\mathbf{x}) + r f(\mathbf{x}^0)$$

subject to

$$\tilde{u}_j(\mathbf{x}) \leq \bar{u}_j + v_j(1 - 1/r)$$
$$\underline{x}_i \leq x_i \leq \bar{x}_i$$
$$r \geq 1$$

where $\tilde{f}(\mathbf{x})$ and $\tilde{u}_j(\mathbf{x})$ are the convex approximation of $f(\mathbf{x})$ and $u_j(\mathbf{x})$ appearing in the CONLIN subproblem statement. Note that separability is maintained with regard to the added variable r.

Clearly if the relaxation variable r hits its lower bound ($r = 1$), nothing is changed in the problem statement, which will usually happen when the starting point $\mathbf{x}^0$ is feasible or nearly feasible. On the other hand, if the starting point $\mathbf{x}^0$ is seriously infeasible, the algorithm will find a value of r greater than unity, which means that the approximate feasible domain is artificially enlarged. If the shape optimal design problem to be treated has a solution, this relaxation technique will usually be activated only once or twice. The subsequent iterations will then yield a unit relaxation variable. However if the feasible domain corresponding to the primary problem (1–4) is empty (for example because some of the requirements are incompatible) the optimization process will converge to the best possible design point. Obviously an infeasible design will be generated, although it should represent a good compromise between the conflicting requirements. Note that the user can act on the CONLIN results by entering appropriate values for each increment v_j. For example if the kth constraint must be exactly satisfied because it corresponds to a particularly severe requirement, it is sufficient to input $v_k = 0$.

Because of its many attractive features, the CONLIN algorithm now forms the basis of all the optimization capabilities to be used in our research projects. At each successive iteration point, the CONLIN method only requires evaluation of the objective and constraint functions and their first derivatives with respect to the design variables. These quantities are provided by the finite element analysis and sensitivity analysis results. The CONLIN optimizer then selects by itself an appropriate approximation scheme on the basis of the sign of the derivatives. CONLIN benefits from many interesting features that make it especially well suited to structural synthesis problems.

- The CONLIN approach is very general, requiring only values and derivatives of the functions describing the optimization problem to be solved; therefore it permits direct interfacing to the finite element software.
- Because it is based on conservative approximation concepts, CONLIN does not demand a high level of accuracy for the sensitivity analysis results, which can thus be obtained through a simple finite difference technique.
- CONLIN usually generates a nearly optimal design within less than ten finite element analyses.
- CONLIN has an inherent tendency to produce a sequence of steadily improving feasible designs.
- The CONLIN method is simple enough to lead to a relatively small computer code and is well organized enough to avoid high core requirements. This feature is useful for interactive optimization.

APPLICATIONS

In this section two examples of application of the convex linearization method to shape optimal design are offered. Although the two structures under consideration are simple from an analysis point of view, their geometric description is fairly complex when considering the design aspects. Therefore the present paper will use these examples to emphasize the fundamental concept of creating a design model well suited to shape optimization. Both examples are extracted from reference [13], which describes in detail all the steps followed to generate appropriate design models.

Axisymmetric Disk—The first example consists of designing an axisymmetric turbine disk. The meridian cross-section of the disk is made up of four parts, as indicated in Figure 6(a): the hub (uniform thickness), the disk itself, the rim and the blades. The shape optimal design problem is posed as finding the shape of the disk and the thickness of the hub that yield minimum weight while satisfying upper limits on radial and circumferential stresses under thermal and centrifugal loads. The thickness of the rim and attached blades are considered to be fixed to predetermined values. Note that because of symmetry only one half of the structure needs to be studied.

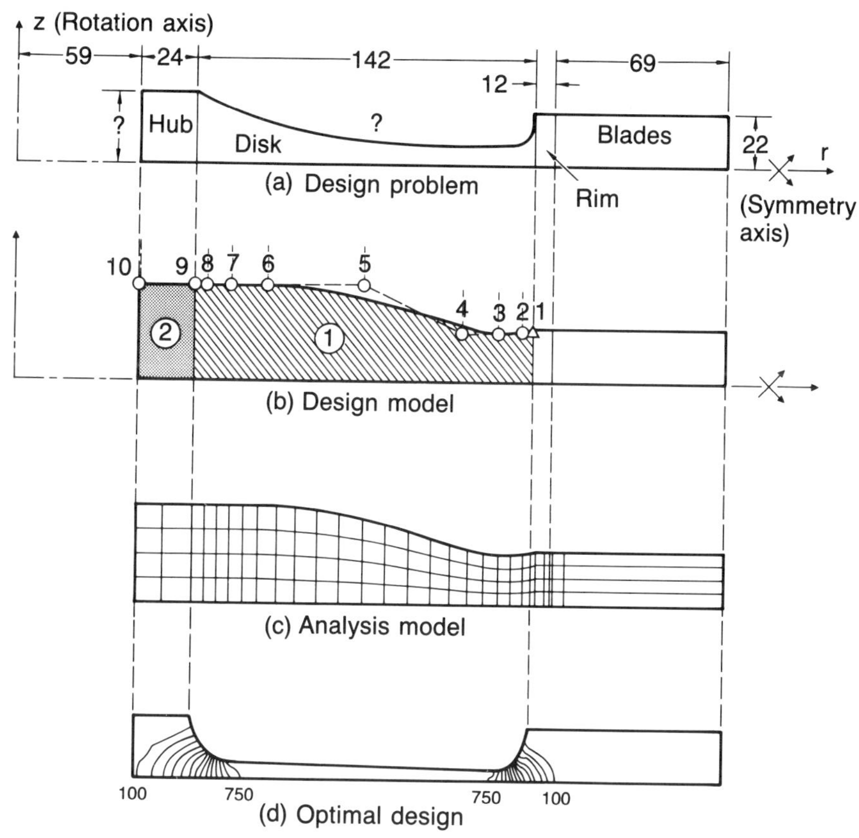

Figure 6. Shape optimal design of a turbine disk.

The geometric description of the design model begins with defining all its boundaries and selecting appropriate parametric curves. In the present case many parts of the boundary are required to be straight segments. Each segment can be considered, for example, as a particular case of a Bézier curve. The only curved part of the boundary will be represented by a B-spline with sixth-order continuity (i.e. degree five). To control it, nine master nodes are employed, as indicated in Figure 6(b). The second step is to subdivide the structure into design elements, which is quite obvious for this problem. As shown in Figure 6(b), two design elements are sufficient to fully describe the moving boundaries, corresponding to the disk and its hub. The two other subregions, representing the rim and the blades, have a fixed geometry. Next the design variables must be selected in order to monitor acceptable changes in shape of the two design elements. As a reminder, the design variables provide the positions of the master nodes describing the boundary curves. Here the hub thickness and the disk shape have to be determined while keeping the rim and blade thickness constant. Therefore only nodes numbered 2 through 10 in Figure 6(b) will be permitted to move. Finally the fourth step consists of expressing constraints restricting the control node displacements. For example, the structure may not move into the negative side of the Z axis. Also all the design elements must keep reasonable geometries.

To facilitate the introduction of these requirements, the design variables are defined as the distances separating each moving node from its corresponding reference node. In addition the move direction of each control node is kept constant. In the present case the control nodes are required to move in the Z direction. With this definition of the design variables, the geometric requirements can be easily stated and treated by the optimization algorithm.

- The hub must have a uniform thickness. It is sufficient to impose constraints such that the displacements of nodes 9 and 10 be the same. This is a simple equality constraint between two variables which can be eliminated before entering the optimizer (design variable linking).
- In order to prevent the moving nodes from penetrating the negative Z space, the design variables are required to remain positive (side constraints).

Having constructed a proper design model that involves only eight independent variables, an analysis model can now be created. The finite element mesh is shown in Figure 6(c). An optimal design is generated in five structural analyses, each followed by an optimization step using the CONLIN algorithm. The final design is given in Figure 6(d), together with representation of typical stress contours.

Bracket—Many shape optimal design capabilities are restricted to small changes in the geometry for the reason that very often the design model definition is such that the possible shape modifications depend largely upon the initial geometry. In other words the subdivision into design elements requires an a priori estimate of the optimum shape, because many of the design element boundaries are fixed in advance. In order to avoid this important limitation, a more flexible design model must be envisioned. The purpose of the following example is to illustrate how Coons patches [3] can be employed to permit larger changes in the geometry. Design elements defined by Coons patches become essentially characterized by their boundaries. As a result there is no longer a need to update the positions of the internal master nodes. Another important consequence is that all the boundaries of each design element are allowed to move.

The bracket shown in Figure 7(a) is clamped at its two lower ends to a rigid foundation, and acted upon by a concentrated load transmitted through a rigid axle. The objective is to minimize the structural weight while assigning maximum allowable values to the von Mises stress. All the boundaries can be modified except the internal circular hole (the radius of the axle is fixed). The design model is made up of the six subregions shown in Figure 7(b). The corresponding analysis model (i.e., the finite element mesh) is represented in Figure 7(c) for the initial design, and Figure 7(d) shows the optimal design generated by CONLIN.

For this particular example, it is helpful to provide a somewhat detailed explanation of the procedure followed to create a meaningful optimization model. The six design elements (referred to as A-F) are depicted in Figure 8(a) and the

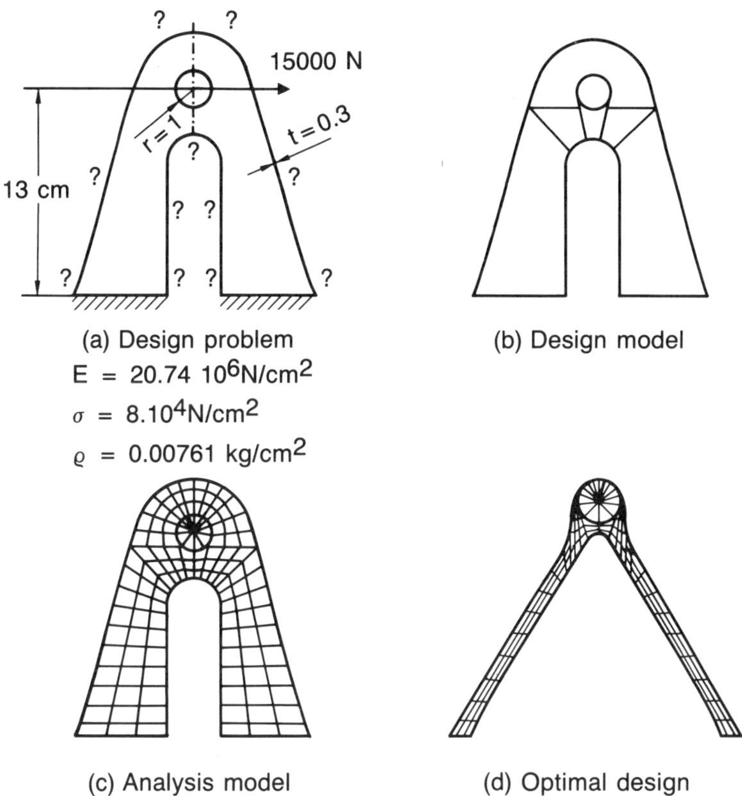

Figure 7. Shape optimal design of a bracket.

associated design variables are shown schematically in Figure 8(b). Design element A has to be defined through a Cartesian product of two families of curves, so that connection to elements B and D is possible. The first family of curves is described by cubic B-splines with seven nodes. They represent, in particular, the external and internal curved boundaries (A_1-A_2 and A_3-A_4). The second family of curves is also cubic, and the curves define, in particular, the straight boundaries A_1-A_4 and A_2-A_3. As usual, the design variables measure the distances between each fixed control node (i.e., attached to the internal hole) and the corresponding moving control node (i.e., attached to the external unknown boundary). Because of its definition by the Cartesian product, design element A possesses internal master nodes, which will be moved homothetically with respect to the external control nodes. Note that this will have to be reflected in the definition of the corresponding design variables associated with elements B and D. So seven design variables fully describe the element. However, because a symmetrical design is sought, only four independent variables are retained (see Figure 8(b)).

On the other hand, the remaining design elements B through F are defined by Coons patches, which means that all their control nodes can be considered as design variables. For these elements, two opposed edges of each patch are characterized

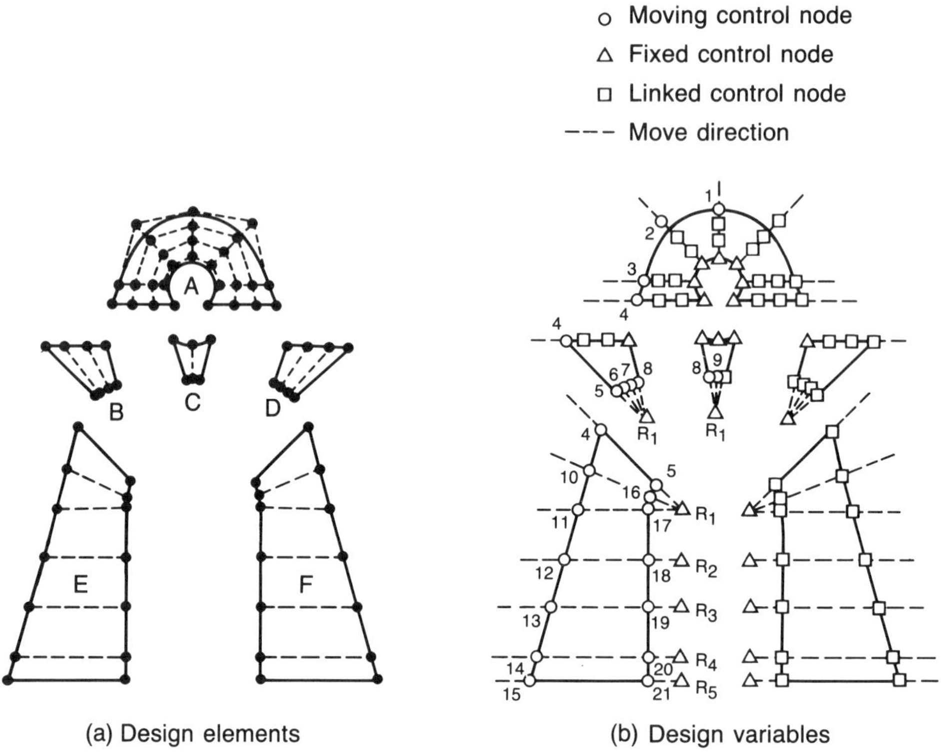

(a) Design elements (b) Design variables

Figure 8. Detailed design model.

by Bézier curves, the two other opposed edges being straight segments. Design variable linking is again employed to impose symmetry. Elements B and D are each described by eight control nodes and the curved boundaries are governed by cubic functions. However only four additional design variables are introduced because the upper parts also belong to the previously considered element A. For the same reason, element C involves quadratic curves but only one new design variable. Note that one single master node, denoted R_1 in Figure 8(b), serves as a reference node to define the design variables associated with elements B, C and D. Finally, element E and its symmetrical element F are described by 14 control nodes; the boundaries are Bézier curves with 7 control nodes. All the control nodes are permitted to move along the directions shown in Figure 8(b). These move directions are defined by introducing fixed reference nodes, denoted R_1 through R_5. Elements E and F involve 12 additional design variables, since variables 4 and 5 have been previously considered.

Therefore, by taking into account this sophisticated linking scheme, only 21 independent design variables are sufficient to properly characterize a design model suitable for optimization. To complete the description of the model, it should be mentioned that linear constraints (see equation (4)) must be added to the problem statement in order to impose tangential continuity of the curved boundaries at the junctions between design elements (points denoted 4, 5 and 8 in Figure 8(b)). For example, the control nodes corresponding to the design variables 3, 4 and 10 are assigned to reside on a straight line.

The optimal design produced by CONLIN in a few iterations is represented in Figure 7(d). It can be seen that the finite element mesh remains quite reasonable, despite the large modifications in shape produced by the optimizer. Note that the final design looks very much like a two-bar truss, which brings us close to topological design.

CONCLUDING REMARKS

The convex linearization method applied in this paper to shape optimal design problems has proven to be a highly efficient and reliable optimization tool. The CONLIN algorithm offers many attractive features that make it ideal for most of our research projects. CONLIN is especially adapted to structural synthesis problems, and it has the ability of solving fairly large-scale optimization problems with hundreds of design variables and constraints. The computational time needed is moderate, which is useful for interactive use. It has a built-in constraint relaxation capability that allows the user to start from any infeasible initial design, and even to find a solution to infeasible problems (in the form of minimal relaxation). Last but not least, because we know our algorithm well, we can correct it or improve it whenever necessary. All this explains why our future research goals shall mostly rely on the CONLIN optimizer.

CONLIN is not, however, the only possible choice, and many other good optimizers have now become commercially available. Therefore the critical issue is no longer which optimizer to select, but rather how to state properly the shape optimal design problems and how to implement optimization concepts in the context of an industrial environment.

An important task in the future will be to develop a new type of preprocessor especially devised to facilitate the introduction of design optimization concepts into the finite element system. To this end it is convenient, as proposed in this paper, to distinguish between a design model and an analysis model. The design model must be simple enough to be easily defined by a nonspecialist in the finite element discretization processes. All the necessary information about the design geometry, the materials to be used, the applied loads and boundary conditions should be introduced interactively. The design model must of course be represented in an appropriate data base, in which the subdivision of the model parameters between prescribed quantities and design variables is clearly identified according to the user's instructions. These inputs are transmitted to the finite element system, which will be capable of retrieving the design variable definition in order to perform the sensitivity analysis. The concept of a design model is one level above that of an analysis model. Ultimately, when the discretization and analysis schemes will have become sufficiently reliable, it can be expected that only the design model will remain accessible to the user whereas most of the analysis models will become

References pp. 319–320

completely transparent.

The formulation employed in this paper offers great flexibility in building up a design model. By entering a relatively small number of design elements it is possible to create a compact model that describes well the two-dimensional structure to be optimized. Because of the internal parametric representation, the analysis model can be directly derived from the design model at any stage of the iterative optimization process. Provided that appropriate mapping rules are adopted to update the coordinates of the internal nodes, the integrity and validity of the resulting finite element mesh can be ensured at any step.

If optimization concepts are to become fully introduced into the design cycle, another substantial effort must be devoted to develop appropriate tools to facilitate the task of the designer. Indeed, an important aspect of finite element analysis and optimization capabilities should be their ability to really help the user in accelerating the design process. In addition to tabular outputs, which are useful in many circumstances, it is necessary to produce extensive graphical output and to devise expressive ways of displaying meaningful results. The interactive engineering design system of the future should contain most of the conventional graphics displays usually found in preprocessing and postprocessing programs (e.g., undeformed and deformed geometry plots, contours of specific stress components, representation of applied forces and boundary conditions). In addition innovative graphics capabilities should be introduced in connection with the new optimization concepts that are now becoming highly reliable. For instance, drawing the new shape generated after each iteration represents an extremely valuable tool to verify the validity of the results produced by the optimizer. Figure 9 shows an example related to the turbine disk problem previously discussed. Furthermore the following functions deserve consideration.

- Visualization of the design model, with clear indication of the design variable locations.
- Visualization of the analysis model, with special features to facilitate verification of the validity of the finite element mesh.
- User interaction with the system, for example by being able to modify the positions of some control nodes and examine the effect on the design model as well as on the analysis model.
- Evolution plots giving the objective function and the constraint functions in terms of the number of iterations.
- Visualization of slices in the design space: the user selects two significant design variables and the program plots the corresponding two-dimensional design space (contours of objective function constraint surfaces defining the feasible domain, location of optimum design, etc.). This representation of the design space could be made on the basis of the convex linearization scheme which constitutes an excellent explicit approximation.

These graphic display capabilities should be organized for interactive use. For

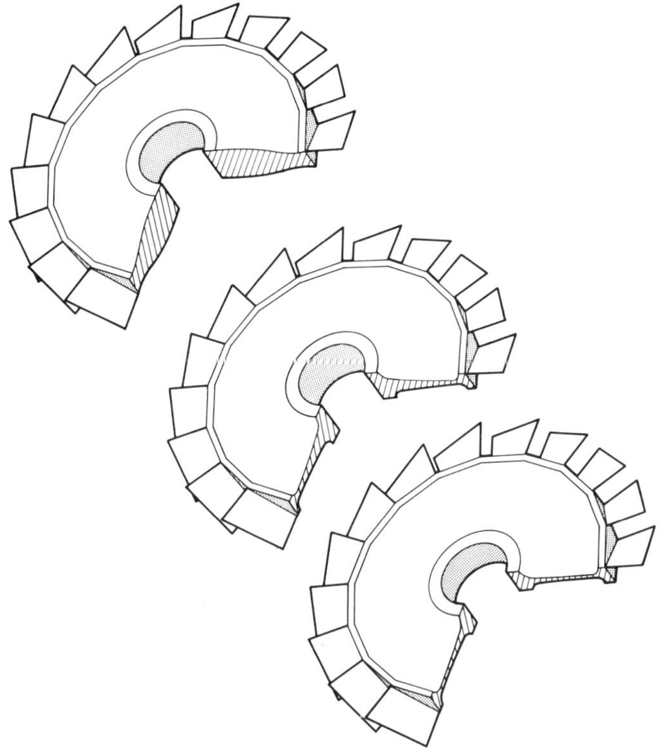

Figure 9. Visualization of iteration history (turbine disk).

example, it might be interesting to produce simultaneously a plot of the design model and a plot of a design subspace. The user would select on the screen the design variables to form the design subspace. He would have the opportunity of requiring a particular constraint to be highlighted both in the design model plot (location) and in the design space plot (restraint surface).

In summary, then, reliable optimization algorithms such as CONLIN have now become available to solve shape optimal design problems. Of course further research is needed to improve the optimizers and to expand the class of problems they can now address, such as increasing the number of design variables and constraints, the efficient treatment of nonlinear equality constraints, discrete variables, etc. However, to facilitate the introduction of optimization concepts within the design cycle, the main development efforts in the future will likely be devoted to devising appropriate user interfaces.

REFERENCES

1. V. Braibant and C. Fleury, Shape optimal design using b-splines. *Comput. Meth. Appl. Mech. Eng.* **44**, 247–267 (1984).

2. W. M. Newman and R. F. Sproull, *Principles of Interactive Computer Graphics.* McGraw-Hill, New York (1979).

3. A. R. Forrest, On Coons' and other methods for the representation of curved surfaces. *Comput. Graphics Image Process.* **1**, 341–359 (1972).

4. J. A. Bennett and M. E. Botkin, Structural shape optimization with geometric description and adaptive mesh refinement. *AIAA J.* **33** (3), 458–464 (1985).

5. G. N. Vanderplaats, CONMIN - A FORTRAN program for constrained function minimization: User's manual. NASA-TM-X 62.282 (1973).

6. C. Fleury and L. A. Schmit, Dual methods and approximation concepts in structural synthesis. NASA-R 3226 (1980).

7. V. Braibant and C. Fleury, Aspects théoriques de l'optimisation de forme par variation de noeuds de contrôle in *Conception Optimale de Forme.* INRIA, Rocquencourt (France) (1983).

8. C. Fleury, Structural weight optimization by dual methods of convex programming. *Int. J. Numer. Meth. Eng.* **14** (12), 1761–1783 (1979).

9. C. Fleury and V. Braibant, Prise en compte de constraints lineaires dans les methodes duales d'optimisation structurale. Rapport LTAS SF-107, University of Liège, Belgium (1982).

10. C. Fleury and V. Braibant, Structural optimization - A new dual method using mixed variables. *Int. J. Numer. Meth. Eng.* (1985), to appear.

11. J. H. Starnes and R. T. Haftka, Preliminary design of composite wings for buckling, stress and displacement constraints. *J. Aircr.* **16**, 564–570 (1979).

12. C. Fleury, Reconciliation of mathematical programming and optimality criteria approaches to structural optimization. Chapter 10 in *Foundations of Structural Optimization: A Unified Approach* (Edited by A. J. Morris). John Wiley & Sons, New York (1982).

13. V. Braibant, Optimisation de forme des structures en vue de la conception assistée par ordinateur. Doctoral Dissertation, University of Liège, Belgium (1985).

DISCUSSION

S. D. Rajan *(Arizona State University)*

You were able to optimize, I presume, truss and constant strain triangle or membrane-like structures because the design variables in sizing optimization are the area and the thickness where you can pull these design variables out of the stiffness matrix and calculate your expressions for the objective function, and the constraints as the design variable times a certain constant. Can you extend that idea to beams and shells where it is not easy to do that, and then use CONLIN-type methods?

Fleury

You can do it for shells. It has already been done for plates. For beams it is more complex because you usually have several design variables to describe the beam, so you need some way of describing the cross-section. You must know which type of cross-section you are using, which makes the problem somewhat more complex.

It is also feasible just to use a finite difference at the element level to get the sensitivity of the stiffness matrix. Having this, you will get the gradients of the constraints in the usual semianalytical approach, and you do not have to worry about how to get the gradients. You get them and you go to CONLIN or ADS or any other optimizer, and it should work—provided you have a good model, of course. You can take a beam with an appropriate way of defining the design variables. If not, you could have a big surprise—you might obtain zero area, or something like that.

C. Fleury

Rajan

My question is not how difficult or easy it is to compute the gradients; it is whether it would work with an approximation technique. You use the Taylor series approximation. Would this work with elements like beams and plates?

Fleury

I know that it works with plates. With beams, I repeat, it depends on how you do it. It works if you take a simple beam with a single design variable. For example, you can take a rectangular cross-section and define only the height or the width. Then it should work. However, if you take a more complex problem, I don't know. It depends on how you define your design variables. Maybe we need a new theoretical approach or type of approach for beams.

R. Lust *(University of California–Los Angeles)*

We have successfully used the CONLIN-type approximations to solve beam problems, and we have recently presented the results of this work in a paper. The major difficulty in this case is that you must be careful about how you solve the dual maximization problem.

Rajan

In CONLIN do you use the active set strategy or do you consider all the constraints?

Fleury

It is a special type of algorithm based on the dual methods. In fact the algorithm is a second-order maximization method which works on the dual. What is nice about the algorithm is that you start from a single active constraint and then progressively expand the dimensionality of the dual space. This means you inspect the number of active behavior constraints so that the critical number is the number of active constraints, not the total number of constraints (from the computational point of view and from the point of view of the optimizer's efficiency).

Rajan

Suppose you start with an infeasible design, with multiple cases, which is very common in structural optimization. You will have a tremendous number of active constraints. How do you go from an infeasible point to a feasible point without resorting to computing the gradients of violated constraints?

Fleury

The number of violated constraints has no relationship with the number of constraints that are active at the optimum of the subproblem.

Rajan

I was referring to your initial guess, not the optimum.

Fleury

Wherever. If you start from a high degree of infeasibility with a lot of violated constraints, this doesn't mean that the number of active constraints will be high for the subproblem. There is no relationship between the two.

V.B. Venkayya *(Wright-Patterson AFB)*

How do you make a decision on which constraints are active at the optimum?

Fleury

This is the subject of another paper. I have actual experience in solving the dual problem.

Venkayya

It is not a question of subjective choices. How do you make a decision on this, in principle?

Fleury

By solving the dual problem. If a dual variable is positive, you have an active constraint. If the dual variable is zero, you have an inactive constraint.

Venkayya

That could be, but that is not necessarily at the optimum. It could be that point where you made the analysis.

Fleury

No, not at that point. For this particular point, it approximates the problem.

Venkayya

Not at the optimum?

Fleury

I will know what is the set of active constraints when it will not change between two stages; in other words, when you will have obtained convergence.

Venkayya

That's not the point you made. How do you decide at the optimum what your optimum active constraint is?

Fleury

It will come out naturally from the dual solution scheme or from any mathematical programming method. Let's forget the analysis using dual methods.

If CONMIN is used as the optimizer rather than the dual methods, even CONMIN will tell you which are the active constraints.

Venkayya

Which are the active constraints at the optimum?

Fleury

Any optimizer will tell you which are the active constraints. One purpose of an optimizer is to select a set of active constraints.

Venkayya

At the optimum? You don't know where the optimum was?

Fleury

You know if it converges. If there is no answer, you have no convergence.

B. Prasad *(Electronic Data Systems)*

Coming back to the previous question of whether you can handle the beam problem with the approximation techniques, I have some experience working with that problem. I think you can do it by using a power form of approximation such as X^{p} in which the p could be real functions. You can use X^{p} or $1/x^{\mathrm{p}}$ and then form a Taylor's series. You can limit the domain to be conservative and at the same time you can get better approximations. Have you implemented anything similar in CONLIN?

Fleury

No, because the direct and reciprocal variables were OK for shape problems. But we are doing it another way. It is no longer a direct and reciprocal variable. This is an idea from Mr. Esping in Sweden, who wrote some good papers—some while he was in Arizona. The idea is similar to another type of change of variable in which you use asymptotes in your approximation. By moving the asymptotes you can control the degree of convexity and curvature.

Prasad

You said that you linearized the geometrical constraints. Do you really linearize them, or do you consider that to be part of the generic model?

Fleury

I have to linearize them just to get a convex, conservative approximation.

Prasad

But do you include that in your generic model? You don't have to include that, I think.

Fleury

I don't know. For most of them, I do not have any geometric constraints. However, in some cases, I need to impose linear equality constraints or inequality constraints, from one control node to the next, to ensure that the two successive Bézier or B-splines will have first-order continuity.

Prasad

Does the generic model have the linking of variables built into it? Will it satisfy the compatibility requirement automatically so you don't have to worry later?

Fleury

Most of the time, that is what we do. Most of the time, we implement this into generic models so that we do not have to impose these constraints. But in some cases we have to do it because there is no easy way to do it in the generic models. This happens, for example, when the geometric constraint must be stated as an inequality.

J. A. Swanson *(Swanson Analysis Systems, Inc.)*

First, I wonder which major software suppliers are missing, since MSC, Swanson Analysis Systems, SDRC, EMRC and PDA are all here. Second, the general statement to be made is that if the approximate functions do not adequately represent the precise solution, then you will not converge. Conversely, if your approximating functions do in fact represent bh^3 in the case of beams, then the approximation procedure works quite well. Our experience has been that if you have a quadratic approximation to a cubic function, you get a minimum but it is not right, not quite what you would like it to be.

Fleury

It's true that this can happen in cases where the approximation is not good enough or not convex enough. But moving asymptotes is like moving limits: you can modify them to have the type of convex approximation that you want. The initial CONLIN method has been developed for simple sizing problems and then applied to shape. For shape it works perfectly, so we did not have to improve it; but now we are applying it to the configuration optimization of trusses, and there it does not work well. We need some control parameters, such as moving asymptotes.

R. Levy *(Jet Propulsion Laboratory)*

From the description you gave of using CONLIN, there doesn't seem to be anything in it that would limit the use of multiple loading conditions. Have you had any experience with the shape optimization where you had more than one load and the loads were very different?

Fleury

Yes. In fact the disk problem that I showed has multiple loading. I didn't give the details, but we have at least three different loading conditions: thermal, regular and centrifugal loads. We have studied several occasions in which each of the three loads act simultaneously, or when the loads act independently. It doesn't change anything—you just increase the number of constraints.

Levy

But don't you get strange shapes?

Fleury

Yes.

R. T. Haftka *(Virginia Polytechnic Institute and State University)*

To amplify on an earlier question, when you start with a highly infeasible domain or design, you have to calculate the derivatives of all the constraints just to get the initial approximation problem, if my understanding was correct. That would be extremely expensive, not because of CONLIN, but if you start with an initial design for any approximation problem where the constraints are violated, then the sensitivity part would be extremely expensive, wouldn't it?

Fleury

Yes. What usually happens is that infeasibility is not due to starting from a very bad design for which all the constraints are violated, but rather to starting from a design in which many constraints are satisfied but two or three constraints are really in conflict. You usually have a problem with only two or three constraints, not much more, so it is not such a large problem. But it is always better to start from an initial feasible design.

SESSION IV
NEW FRONTIERS IN SHAPE OPTIMIZATION

Session Chairman
J. SOBIESKI
NASA Langley Research Center
Hampton, Virginia

A NUMERICAL METHOD FOR SHAPE DESIGN SENSITIVITY ANALYSIS AND OPTIMIZATION OF BUILT-UP STRUCTURES

K. K. CHOI and H. G. SEONG

Department of Mechanical Engineering
and
Center for Computer Aided Design
The University of Iowa
Iowa City, Iowa

Abstract

A general purpose design sensitivity analysis and optimization method for built-up structures is proposed and software implementation of the method is considered. Both conventional (sizing) and shape design variables for components of built-up structures are studied. For conventional design sensitivity analysis, distributed parameter structural design sensitivity analysis theory is used. For shape design sensitivity analysis, the material derivative concept of continuum mechanics is employed. A domain method of shape design sensitivity analysis is presented as the foundation of a design component method that leads to systematic organization of computations for design sensitivity analysis, which can be implemented with existing finite element codes. A sparse matrix symbolic factorization technique for iterative structural optimization is used for efficient repeated solution of matrix finite element equilibrium and eigenvalue equations. For optimization, Pshenichny's linearization method is employed. Feasibility of the proposed method is demonstrated through design optimization of a truss-beam-plate built-up structure.

INTRODUCTION

Design sensitivity analysis of distributed parameter built-up structures whose component shapes are defined by conventional design variables such as cross-sectional area and thickness has been treated in reference [1]. In such problems, design variables that specify the shape of structural components are defined on fixed physical domains. Shape design sensitivity analysis, where the geometric shapes of structural components are regarded as design variables, has been treated in references [1], [2] and [3] using the material derivative idea of continuum mechanics. Shape design sensitivity information is explicitly expressed as boundary integrals, using integration by parts and boundary and/or interface boundary conditions. For accurate numerical calculation of design sensitivity information, accurate evaluation of stresses, strains and/or normal derivatives of the state and adjoint variables on the boundary is crucial. However, when the finite element method is used for analysis of built-up structures, the accuracy of numerical results for state and adjoint variables on interface boundaries may not be good [4].

To overcome this difficulty, a domain method of shape design sensitivity analysis [5] is developed, in which design sensitivity information is expressed as domain integrals instead of boundary integrals. The domain and the boundary methods are analytically equivalent. However, when one uses an approximate numerical method such as finite element analysis, the resulting design sensitivity approximations may give quite different numerical values. Moreover, the domain method offers a remarkable simplification in derivation of shape design sensitivity formulas for built-up structures. Using the domain method and the results of conventional design sensitivity analysis theory [1], a design component method [6] is developed for unified and systematic organization of design sensitivity analysis for built-up structures, with both conventional and shape design variables. That is, conventional and shape design sensitivity formulas for each standard component type can be derived. The result is standard formulas that can be used for design sensitivity analysis of built-up structures by simply adding contributions from each component. The method gives a systematic organization of computations for design sensitivity analysis that is similar to the way in which computations are organized within a finite element code.

A numerical method has been developed [7] to implement the results of the design component method, using the versatility and convenience of existing finite element codes. It is shown in reference [7] that calculations can be carried out outside existing finite element codes, using postprocessing data only. Thus, design sensitivity analysis software does not have to be imbedded in an existing finite element code.

The purpose of this paper is to combine these developments, to propose a general purpose method of design sensitivity analysis for built-up structures and to suggest software implementation of the method with existing finite element codes. Even though only static response is considered here, the method is also applicable for eigenvalue design sensitivity analysis [1].

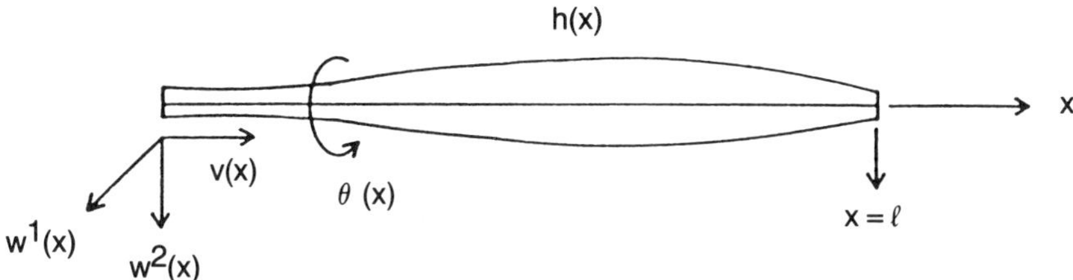

Figure 1. Beam-truss component.

Using design sensitivity information, the designer can optimize structural systems by using iterative methods of nonlinear programming or optimality criteria. At a more practical level, the designer can carry out interactive trade-off analysis [1] and improve the design.

To demonstrate feasibility of the proposed method, a truss-beam-plate built-up structure is optimized using 292 design parameters and 543 inequality constraints. The finite element model has 480 elements and 1,281 degrees of freedom. For iterative optimization, a sparse matrix symbolic factorization technique of reference [8] is used for efficient repeated solution of matrix finite element equilibrium and eigenvalue equations. The technique is an efficient method for iterative structural optimization, since each iteration involves only one numerical factorization and $1 + k$ forward and backward substitutions where k is the number of active constraints for which design sensitivity information is to be computed. Symbolic factorization, which requires large storage, is executed just once, as long as connectivity of the finite elements is preserved. For optimization problems in which design changes affect only a few elements, the designer can further save computing time and storage space [8]. For nonlinear programming, Pshenichny's linearization method [9] is used. A Cray-1S supercomputer was used to solve this large scale structural design problem

VARIATIONAL EQUATIONS OF BUILT-UP STRUCTURES

A built-up structure is made up of combinations of a variety of structural components that are interconnected by kinematic constraints at their interfaces. While a substantial library of structural components is required to represent a large class of built-up structures, the library developed in this paper is limited to truss, beam, and plane elastic solid and plate components. In the formulation, truss and beam components are incorporated into a single component because they have the same physical domain. Likewise, plane elastic solid and plate components are combined.

Before constructing variational equations for built-up structures, consider the energy bilinear and load linear forms of each structural component, which are characterized by both conventional and shape design variables.

Consider the beam-truss component of Figure 1. The energy bilinear form (internal virtual work)[1] of the component is

References p. 349

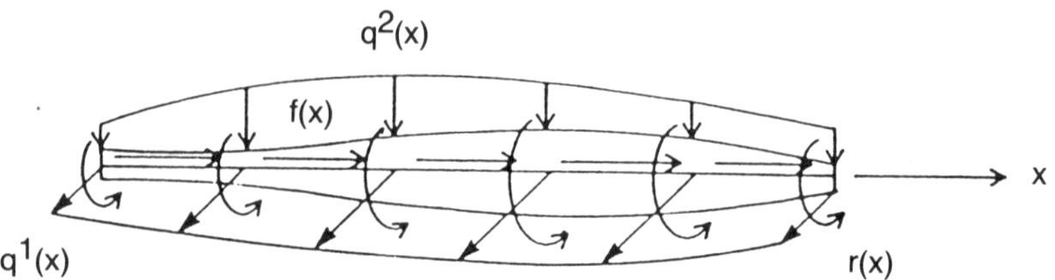

Figure 2. External loads for beam-truss.

$$a_{u,\Omega}(\mathbf{z},\bar{\mathbf{z}}) = \int_0^\ell EI^1 w^1_{xx}\bar{w}^1_{xx}\,dx + \int_0^\ell EI^2 w^2_{xx}\bar{w}^2_{xx}\,dx + \int_0^\ell GJ\theta_x\bar{\theta}_x\,dx + \int_0^\ell hEv_x\bar{v}_x\,dx \tag{1}$$

where w^1, w^2, θ and v are two orthogonal lateral displacements, the angle of twist and the axial displacement, respectively, and $\mathbf{z} = [w^1, w^2, \theta, v]^T$. Throughout this paper an overbar, e.g., $\bar{\mathbf{z}}$, denotes a virtual displacement. The subscript x in equation (1) denotes a derivative with respect to x. In equation (1), E, G, I^1, I^2, J and h are Young's modulus, shear modulus, two moments of inertia, the torsional moment of inertia and the cross-sectional area of the component, respectively. The conventional design variable is $u = h(x)$ and the shape design variable is the length of the domain $\Omega = [0, \ell]$. The load linear form (external virtual work) [1] of the component is

$$\ell_{u,\Omega}(\bar{\mathbf{z}}) = \int_0^\ell q^1\bar{w}^1\,dx + \int_0^\ell q^2\bar{w}^2\,dx + \int_0^\ell r\bar{\theta}\,dx + \int_0^\ell f\bar{v}\,dx \tag{2}$$

where q^1, q^2, r and f are two orthogonal lateral loads, twisting moment and axial load, respectively, as shown in Figure 2 [1]. If there are point loads, a Dirac delta measure can be used for q^1, q^2, r and f in equation (2) [1].

Next, consider the plate-plane elastic solid component of Figure 3. The energy bilinear form [1] of the component is

$$a_{u,\Omega}(\mathbf{z},\bar{\mathbf{z}}) = \iint_\Omega \hat{D}(t)\,[(w_{11}+\nu w_{22})\,\bar{w}_{11} + (w_{22}+\nu w_{11})\,\bar{w}_{22} + 2(1-\nu)w_{12}\bar{w}_{12}]\,d\Omega + \iint_\Omega t\left[\sum_{i,j=1}^{2}\sigma^{ij}(v)\varepsilon^{ij}(\bar{v})\right]d\Omega \tag{3}$$

where $\mathbf{z} = [w, v^1, v^2]^T$ is the displacement field and subscript $i (i = 1, 2)$ denotes the derivative with respect to variable x_i. In equation (3) $\hat{D}(t) = Et^3/[12(1-\nu^2)]$

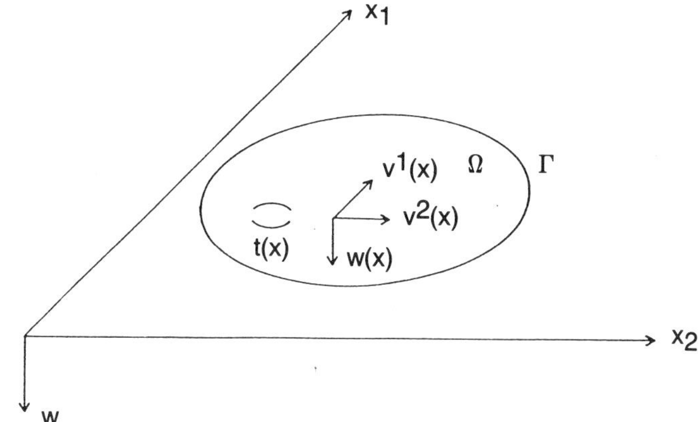

Figure 3. Plate-plane elastic solid component.

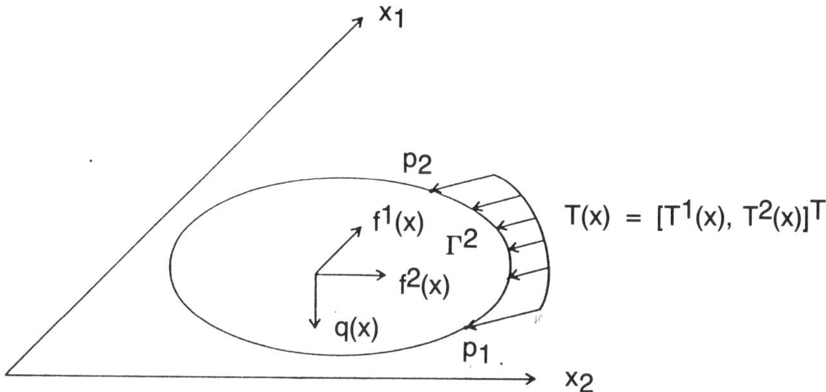

Figure 4. External loads for plate-plane elastic solid.

is flexural rigidity, ν is Poisson's ratio and t is the thickness of the component. Also, $\sigma^{ij}(\mathbf{v})$ and $\varepsilon^{ij}(\mathbf{v})$ are respectively the stress and strain due to an in-plane displacement field $\mathbf{v} = [v^1, v^2]^T$. For this component, the conventional design variable is $u = t(x)$ and the shape design variable is the shape of the domain Ω. The load linear form [1] of the component is

$$\ell_{u,\Omega}(\mathbf{z}) = \int\int_{\Omega} qw \, d\Omega + \int\int_{\Omega} \sum_{i=1}^{2} f^i v^i \, d\Omega + \int_{\Gamma^2} \sum_{i=1}^{2} T^i v^i \, d\Gamma \tag{4}$$

where q, f and T are lateral load, body force and traction force, respectively, as shown in Figure 4. As in the beam-truss component, if there are point loads, a Dirac delta measure [1] can be used.

Consider a built-up structure that is made up of $m \geq 1$ structural components that are interconnected by kinematic constraints at their interfaces. Using the principle of virtual work for built-up structures [1], one can obtain the variational

References p. 349

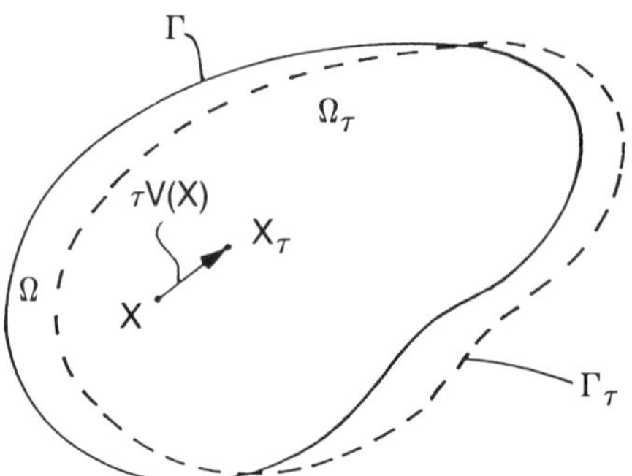

Figure 5. Variation of domain.

formulation of the governing equations:

$$a_{u,\Omega}(\mathbf{z}, \bar{\mathbf{z}}) = \ell_{u,\Omega}(\bar{\mathbf{z}}), \qquad \text{for all } \bar{\mathbf{z}} \in Z \tag{5}$$

where

$$a_{u,\Omega}(\mathbf{z}, \bar{\mathbf{z}}) = \sum_{i=1}^{m} a_{u^i,\Omega^i}(\mathbf{z}, \bar{\mathbf{z}}), \tag{6}$$

$$\ell_{u,\Omega}(\bar{\mathbf{z}}) = \sum_{i=1}^{m} \ell_{u^i,\Omega^i}(\bar{\mathbf{z}}) \tag{7}$$

and Z is the space of kinematically admissible displacements [1, 2], which is defined as the set of displacement fields that satisfy homogeneous boundary conditions and kinematic interface conditions between components. In equations (6) and (7), $a_{u^i,\Omega^i}(\mathbf{z}, \bar{\mathbf{z}})$ and and $\ell_{u^i,\Omega^i}(\bar{\mathbf{z}})$ are energy bilinear and load linear forms of component i with domain Ω^i. Note that in equations (6) and (7) the energy bilinear and load linear forms of equation (5) are simply summations of corresponding terms from each component. Thus, as will be seen later, the design sensitivity analysis of the built-up structure is a simple additive process.

DESIGN SENSITIVITY ANALYSIS OF BUILT-UP STRUCTURES

In this section, design sensitivity information for displacement and locally averaged stress functionals is derived for general built-up structures. Consider a domain Ω shown schematically in Figure 5. Suppose that an initial point $\mathbf{x} \in \Omega$ moves to $\mathbf{x}_\tau = \mathbf{x} + \tau\mathbf{V}(\mathbf{x})$ due to a change in shape of the domain Ω to Ω_τ, where $\mathbf{V}(\mathbf{x})$ is a design velocity field.

Define $\dot{\mathbf{z}}$ as the total variation of $\mathbf{z}$, due to both conventional and shape design changes [1, 2]:

$$\begin{aligned}\dot{\mathbf{z}} &= \frac{d}{d\tau}\mathbf{z}_\tau(\mathbf{x} + \tau\mathbf{V}(\mathbf{x}), u + \tau\delta u)\big|_{\tau=0} \\ &= \frac{d}{d\tau}\mathbf{z}(\mathbf{x}, u + \tau\delta u)\big|_{\tau=0} + \frac{d}{d\tau}\mathbf{z}_\tau(\mathbf{x} + \tau\mathbf{V}(\mathbf{x}), u)\big|_{\tau=0}.\end{aligned} \tag{8}$$

The first variation of equation (5) is [1, 2]

$$a'_{\delta u,\Omega}(\mathbf{z},\bar{\mathbf{z}}) + a'_{u,\mathbf{V}}(\mathbf{z},\bar{\mathbf{z}}) + a_{u,\Omega}(\dot{\mathbf{z}},\bar{\mathbf{z}}) = \ell'_{\delta u,\Omega}(\bar{\mathbf{z}}) + \ell'_{u,}\mathbf{V}(\bar{\mathbf{z}}), \qquad \text{for all } \bar{\mathbf{z}} \in Z \quad (9)$$

where $\dot{\mathbf{z}} = [\dot{w}^1, \dot{w}^2, \dot{\theta}, \dot{v}]^T$ for beam-truss components and $\dot{\mathbf{z}} = [\dot{w}, \dot{v}^1, \dot{v}^2]^T$ for plate-plane elastic solid components. The notation of equation (9) is chosen to clearly display which variables are held fixed and which vary.

Consider first a displacement functional that defines the displacement $\mathbf{z}$ at nodal point $\mathbf{x}^k \in \Omega^r$:

$$\psi_k = \int\!\!\int_{\Omega^r} \hat{\delta}(\mathbf{x} - \mathbf{x}^k)\mathbf{z}(\mathbf{x})\, d\Omega \tag{10}$$

where $\hat{\delta}(\mathbf{x})$ is the Dirac measure at the origin. Taking the first variation of equation (10), one obtains [1, 2]

$$\psi'_k = \int\!\!\int_{\Omega^r} \hat{\delta}(\mathbf{x} - \mathbf{x}^k)\dot{\mathbf{z}}(\mathbf{x})\, d\Omega. \tag{11}$$

Define a variational adjoint equation by replacing $\dot{\mathbf{z}}$ in the term on the right of equation (11) by a virtual displacement $\bar{\boldsymbol{\lambda}}$, and equate the result to the energy bilinear form evaluated at the adjoint variable $\boldsymbol{\lambda}$; i.e.,

$$a_{u,\Omega}(\boldsymbol{\lambda},\bar{\boldsymbol{\lambda}}) = \int\!\!\int_{\Omega^r} \hat{\delta}(\mathbf{x} - \mathbf{x}^k)\bar{\boldsymbol{\lambda}}(\mathbf{x})\, d\Omega, \qquad \text{for all } \bar{\boldsymbol{\lambda}} \in Z. \tag{12}$$

Denote the solution of equation (12) as $\boldsymbol{\lambda}^{(k)}$. Since $\dot{\mathbf{z}}$ satisfied kinematic boundary and interface conditions [1, 2], equation (12) can be evaluated at $\bar{\boldsymbol{\lambda}} = \dot{\mathbf{z}}$ and equation (9) can be evaluated at $\bar{\mathbf{z}} = \boldsymbol{\lambda}^{(k)}$ to obtain

$$\begin{aligned} \psi'_k &= a_{u,\Omega}(\dot{\mathbf{z}},\boldsymbol{\lambda}^{(k)}) \\ &= \sum_{i=1}^{m} \left[\ell'_{\delta u^i,\Omega^i}(\boldsymbol{\lambda}^{(k)}) - a'_{\delta u^i,\Omega^i}(\mathbf{z},\boldsymbol{\lambda}^{(k)})\right] \\ &\quad + \sum_{i=1}^{m} \left[\ell'_{u^i,}\mathbf{V}^i(\boldsymbol{\lambda}^{(k)}) - a'_{u^i,}\mathbf{V}^i(\mathbf{z},\boldsymbol{\lambda}^{(k)})\right] \end{aligned} \tag{13}$$

where the first term on the right is due to conventional design variation and the second term is due to shape design variation. Note that equation (13) is valid for general built-up structures that are composed of $m \geq 1$ structural components. To calculate equation (13) numerically, one must construct a library of explicit expressions of the terms in equation (13) for each standard structural component.

Next consider a locally averaged stress functional over a finite element Ω_p in a plate-plane elastic solid component Ω^r, $\Omega_p \subset \Omega^r$:

$$\psi_p = \int\!\!\int_{\Omega^r} \left[g^1(t, w_{ij}) + g^2(\sigma(\mathbf{v}))\right] m_p\, d\Omega \tag{14}$$

References p. 349

where $g^1(t, w_{ij})$ and $g^2(\sigma(\mathbf{v}))$ are principal stress, von Mises yield stress, or some other stress measures due to lateral displacement w and in-plane displacement field $\mathbf{v}$, respectively. Here, $g^1(t, w_{ij})$ is measured at the extreme fiber and m_p is a characteristic function on Ω_p, defined as

$$m_p = \begin{cases} 1/\int\int_{\Omega_p} d\Omega, & \mathbf{x} \in \Omega_p \\ 0 \quad\quad\quad\quad , & \mathbf{x} \notin \Omega_p \end{cases}. \tag{15}$$

Taking the first variation of equation (14) and using the domain method for shape design variation, one obtains [1, 3, 5]

$$\begin{aligned} \psi_p = & \int\int_{\Omega^r} g_t^1 m_p \, \delta t \, d\Omega + \int\int_{\Omega^r} \sum_{i,j=1}^{2} g^1_{w_{ij}} \left[\dot{w}_{ij} - (\nabla w^T \mathbf{V})_{ij}\right] m_p \, d\Omega \\ & + \int\int_{\Omega^r} \left[\sum_{i,j=1}^{2} g^2_{\sigma^{ij}} \sigma^{ij}(\dot{\mathbf{v}})\right] m_p \, d\Omega + \int\int_{\Omega^r} \operatorname{div}\,(g^1 \mathbf{V}) m_p \, d\Omega \\ & - \int\int_{\Omega^r} \sum_{i,j=1}^{2} \left[\sum_{k,\ell=1}^{2} g^2_{\sigma^{ij}} D^{ijk\ell} (\nabla v^{k^T} \mathbf{V}_\ell)\right] m_p \, d\Omega \\ & + \int\int_{\Omega^r} g^2 \operatorname{div} \mathbf{V} m_p \, d\Omega - \int\int_{\Omega^r} (g^1 + g^2) m_p \, d\Omega \int\int_{\Omega^r} m_p \operatorname{div} \mathbf{V} d\Omega \end{aligned} \tag{16}$$

where $D^{ijk\ell}$ is the elastic modulus tensor which satisfies [1, 2]

$$\sigma^{ij}(\mathbf{v}) = \sum_{k,\ell=1}^{2} D^{ijk\ell} \varepsilon^{ij}(\mathbf{v}), \qquad i, j, k, \ell = 1, 2. \tag{17}$$

Define an adjoint equation by replacing $\dot{w}$ and $\dot{\mathbf{v}}$ in equation (16) with virtual displacements $\bar{\eta}$ and $\bar{\boldsymbol{\xi}}$, respectively, and equating terms involving $\bar{\eta}$ and $\bar{\boldsymbol{\xi}}$ in equation (16) to the energy bilinear form $a_{u,\Omega}(\boldsymbol{\lambda}, \bar{\boldsymbol{\lambda}})$ of the built-up structure; thus

$$\begin{aligned} a_{u,\Omega}(\boldsymbol{\lambda}, \bar{\boldsymbol{\lambda}}) = & \int\int_{\Omega^r} \left[\sum_{i,j=1}^{2} g^1_{w_{ij}} \bar{\eta}_{ij}\right] m_p \, d\Omega \\ & + \int\int_{\Omega^r} \left[\sum_{i,j=1}^{2} g^2_{\sigma^{ij}} \sigma^{ij}(\bar{\boldsymbol{\xi}})\right] m_p \, d\Omega, \qquad \text{for all } \bar{\boldsymbol{\lambda}} \in Z \end{aligned} \tag{18}$$

where $\boldsymbol{\lambda} = [\eta^1, \eta^2, \omega, \xi]^T$ for the beam-truss components and $\boldsymbol{\lambda} = [\eta, \xi^1, \xi^2]^T$ for plate-plane elastic solid components are adjoint variables. Following the procedure to derive equation (13) from equation (11), one obtains

$$\begin{aligned}
\psi_p' = &\sum_{i=1}^{m} \left[\ell'_{\delta u^i,\Omega^i}(\boldsymbol{\lambda}^{(p)}) - a'_{\delta u^i,\Omega^i}(\mathbf{z},\boldsymbol{\lambda}^{(p)})\right] + \iint_{\Omega^r} g_t^1 m_p \delta t \, d\Omega \\
&+ \sum_{i=1}^{m} \left[\ell'_{u^i,\mathbf{V}^i}(\boldsymbol{\lambda}^{(p)}) - a'_{u^i,\mathbf{V}^i}(\mathbf{z},\boldsymbol{\lambda}^{(p)})\right] \\
&- \iint_{\Omega^r} \left[\sum_{i,j=1}^{2} g^1_{w_{ij}} (\boldsymbol{\nabla} w^t \mathbf{V})_{ij}\right] m_p \, d\Omega \\
&+ \iint_{\Omega^r} \text{div } (g^1 \mathbf{V}) m_p \, d\Omega - \iint_{\Omega^r} \sum_{i,j=1}^{2} \left[\sum_{k,\ell=1}^{2} g^2_{\sigma ij} D^{ijk\ell} (\boldsymbol{\nabla} v^{k^T} \mathbf{V}_\ell)\right] m_p \, d\Omega \\
&+ \iint_{\Omega^r} g^2 \text{ div } \mathbf{V} m_p \, d\Omega - \iint_{\Omega^r} (g^1 + g^2) m_p \, d\Omega \iint_{\Omega^r} m_p \text{ div } \mathbf{V} \, d\Omega
\end{aligned} \tag{19}$$

where the first two terms on the right of equation (19) are due to conventional design variation, and the remaining six terms are due to shape design variation. As in the displacement functional case, to calculate equation (19) numerically one must find explicit expressions of the first and third terms in the equation. Note that these terms have the same form as those of equation (13). The difference is that terms in equation (13) are evaluated at $\lambda^{(k)}$ and terms in equation (19) are evaluated at $\lambda^{(p)}$; that is, once the expressions for terms in equation (13) are derived, they can be used for different built-up structures and different constraints. Design sensitivity information for locally averaged stress functionals over finite elements in a beam-truss component can be derived using the same procedure.

DESIGN COMPONENTS

In this section, explicit expressions for terms in equation (13) are derived to construct a library of design components by taking conventional and shape design variations of the energy bilinear and load linear forms of each component. This results in standard expressions that can be used for design sensitivity analysis of built-up structures by simply adding contributions from each component. For shape design variations, the domain method of reference [5] is employed.

Conventional Design Variations — First, assume that shape design variables are fixed, and consider a structure made of a single beam-truss component. Then $m = 1$ and the governing variational equation is equation (5), with $a_{u,\Omega}(\mathbf{z},\bar{\mathbf{z}})$ and $\ell_{u,\Omega}(\bar{\mathbf{z}})$ given by equations (1) and (2), respectively. By taking the first variation of equation (5) one obtains [1]

$$a'_{\delta u,\Omega}(\mathbf{z},\bar{\mathbf{z}}) + a_{u,\Omega}(\dot{\mathbf{z}},\bar{\mathbf{z}}) = \ell'_{\delta u,\Omega}(\bar{\mathbf{z}}), \qquad \text{for all } \bar{\mathbf{z}} \in Z \tag{20}$$

where

$$a'_{\delta u,\Omega}(\mathbf{z},\bar{\mathbf{z}}) = \int_0^\ell EI_h^1 w_{xx}^1 \bar{w}_{xx}^1 \delta h \, dx + \int_0^\ell EI_h^2 w_{xx}^2 \bar{w}_{xx}^2 \delta h \, dx$$

References p. 349

$$+ \int_0^{\ell} GJ_h \theta_x \bar{\theta}_x \delta h \, dx + \int_0^{\ell} E v_x \bar{v}_x \delta h \, dx \tag{21}$$

and

$$\ell'_{\delta u,\Omega}(\bar{\mathbf{z}}) = 0. \tag{22}$$

Subscript h in equation (21) denotes the derivative with respect to h, and $\dot{\mathbf{z}}$ in equation (20) is equal to the first term on the right of equation (8) for the beam-truss component.

If the structure is made of a single plate-plane elastic solid component, then equation (5) is valid where $m = 1$ and one uses equations (3) and (4) for the energy bilinear and load linear forms. By taking the first variation of (5), one obtains equation (20), where [1]

$$\begin{aligned} a'_{\delta u,\Omega}(\mathbf{z}, \bar{\mathbf{z}}) &= \int\!\!\int_{\Omega} \frac{Et^2}{4(l-\nu^2)} [w_{11}\bar{w}_{11} + \nu(w_{11}\bar{w}_{22} + w_{22}\bar{w}_{11}) + w_{22}\bar{w}_{22} \\ &\quad + 2(l-\nu) w_{12}\bar{w}_{12}] \delta t \, d\Omega + \int\!\!\int_{\Omega} \left[\sum_{i,j=1}^{2} \sigma^{ij}(\mathbf{v}) \varepsilon^{ij}(\bar{\mathbf{v}}) \right] \delta t \, d\Omega \\ &= \int\!\!\int_{\Omega} \left[\sum_{i,j=1}^{2} \sigma^{ij}(w) \varepsilon^{ij}(\bar{w}) \right] \delta t \, d\Omega \\ &\quad + \int\!\!\int_{\Omega} \left[\sum_{i,j=1}^{2} \sigma^{ij}(\mathbf{v}) \varepsilon^{ij}(\bar{\mathbf{v}}) \right] \delta t \, d\Omega \end{aligned} \tag{23}$$

and

$$\ell'_{\delta u,\Omega}(\bar{\mathbf{z}}) = 0. \tag{24}$$

In equation (23) $\sigma^{ij}(w)$ and $\varepsilon^{ij}(w)$ are stress and strain at the extreme fibers, due to the lateral displacement w [7].

Shape Design Variations — Next, assume that conventional design variables are fixed and consider a structure made of a single beam-truss component. By taking the first variation of equation (5), using the domain method and equations (1) and (2), one obtains [1, 5]

$$a'_{u,\mathbf{V}}(\mathbf{z}, \bar{\mathbf{z}}) + a_{u,\Omega}(\dot{\mathbf{z}}, \bar{\mathbf{z}}) = \ell'_{u,\mathbf{V}}(\bar{\mathbf{z}}), \qquad \text{for all } \bar{\mathbf{z}} \in Z \tag{25}$$

where

$$\begin{aligned} a'_{u,\mathbf{V}}(\mathbf{z}, \bar{\mathbf{z}}) &= \int_0^{\ell} \{ -EI^1 [3 w^1_{xx} \bar{w}^1_{xx} V_x + (w^1_{xx} \bar{w}^1_x + w^1_x \bar{w}^1_{xx}) V_{xx}] \\ &\quad + EI^1_x w^1_{xx} \bar{w}^1_{xx} V \} \, dx + \int_0^{\ell} \{ -EI^2 [3 w^2_{xx} \bar{w}^2_{xx} V_x \end{aligned}$$

$$
\begin{aligned}
&+ (w_{xx}^2 \bar{w}_x^2 + w_x^2 \bar{w}_{xx}^2) V_{xx}] + EI_x^2 w_{xx}^2 \bar{w}_{xx}^2 V \} \, dx \\
&+ \int_0^\ell (-GJ\theta_x \bar{\theta}_x V_x + GJ_x \theta_x \bar{\theta}_x V) \, dx \\
&+ \int_0^\ell (-hEv_x \bar{v}_x V_x + h_x E v_x \bar{v}_x V) \, dx
\end{aligned} \tag{26}
$$

and

$$
\begin{aligned}
\ell'_{u,\mathbf{V}}(\bar{\mathbf{z}}) = &\int_0^\ell (q_x^1 \bar{w}^1 V + q^1 \bar{w}^1 V_x) \, dx + \int_0^\ell (q_x^2 \bar{w}^2 V + q^2 \bar{w}^2 V_x) \, dx \\
&+ \int_0^\ell (r_x \bar{\theta} V + r\bar{\theta} V_x) \, dx + \int_0^\ell (f_x \bar{v} V + f\bar{v} V_x) \, dx.
\end{aligned} \tag{27}
$$

In equation (25), $\dot{\mathbf{z}}$ is equal to the second term on the right of equation (8) for the beam-truss component.

As in the conventional design case, if the structure is made of a single plate-plane elastic solid component, by taking the first variation of (5) and by using the domain method and equations (3) and (4), one obtains equation (25), where [1, 5]

$$
\begin{aligned}
a'_{u,\mathbf{V}}(\mathbf{z}, \bar{\mathbf{z}}) = \int\!\!\int_\Omega \Big[&-\hat{D}(t) \{ 4(w_{11} \bar{w}_{11} V_1^1 + w_{22} \bar{w}_{22} V_2^2) + [\nu(w_{11} \bar{w}_{22} + w_{22} \bar{w}_{11}) \\
&- (w_{11} \bar{w}_{11} + w_{22} \bar{w}_{22}) + 2(1-\nu) w_{12} \bar{w}_{12}] \text{ div } \mathbf{V} \\
&+ 2(w_{11} \bar{w}_{11} + w_{12} \bar{w}_{11} + w_{22} \bar{w}_{12} + w_{12} \bar{w}_{22})(V_2^1 + V_1^2) \\
&+ [w_1 \bar{w}_{11} + w_{11} \bar{w}_1 + \nu(w_1 \bar{w}_{22} + w_{22} \bar{w}_1)] V_{11}^1 \\
&+ [w_1 \bar{w}_{22} + w_{22} \bar{w}_1 + \nu(w_1 \bar{w}_{11} + w_{11} \bar{w}_1)] V_{22}^1 \\
&+ [w_2 \bar{w}_{11} + w_{11} \bar{w}_2 + \nu(w_2 \bar{w}_{22} + w_{22} \bar{w}_2)] V_{11}^2 \\
&+ [w_2 \bar{w}_{22} + w_{22} \bar{w}_2 + \nu(w_2 \bar{w}_{11} + w_{11} \bar{w}_2)] V_{22}^2 \\
&+ 2(1-\nu)(w_1 \bar{w}_{12} + w_{12} \bar{w}_1)(V_{12}^1 + V_{12}^2) \} \\
&+ \frac{Et^2}{4(1-\nu^2)} [w_{11} \bar{w}_{11} + \nu(w_{11} \bar{w}_{22} + w_{22} \bar{w}_{11}) + w_{22} \bar{w}_{22} \\
&+ 2(1-\nu) w_{12} \bar{w}_{12}] \boldsymbol{\nabla} t^T \mathbf{V} \Big] \, d\Omega \\
&+ \int\!\!\int_\Omega \Bigg\{ -t \sum_{i,j=1}^{2} [\sigma^{ij}(\mathbf{v})(\boldsymbol{\nabla} \bar{v}^{i^T} \mathbf{V}_j) + \sigma^{ij}(\bar{\mathbf{v}})(\boldsymbol{\nabla} v^{i^T} \mathbf{V}_j) \\
&- \sigma^{ij}(\mathbf{v}) \varepsilon^{ij}(\bar{\mathbf{v}}) \text{ div } \mathbf{V}] + \sum_{i,j=1}^{2} \sigma^{ij}(\mathbf{v}) \varepsilon^{ij}(\bar{\mathbf{v}}) \boldsymbol{\nabla} t^T \mathbf{V} \Bigg\} \, d\Omega,
\end{aligned} \tag{28}
$$

and

$$
\ell'_{u,\mathbf{V}}(\bar{\mathbf{z}}) = \int\!\!\int_\Omega [\bar{w}(\boldsymbol{\nabla} q^T \mathbf{V}) + q\bar{w} \text{ div } \mathbf{V}] \, d\Omega
$$

References p. 349

$$+\iint_{\Omega}\sum_{i=1}^{2}[\bar{v}^i(\nabla f^{i^T}\mathbf{V})+f^i\bar{v}^i \text{ div } \mathbf{V}]\,d\Omega$$

$$+\int_{\Gamma}\sum_{i=1}^{2}\{-T^i(\nabla\bar{v}^{i^T}\mathbf{V})+[\nabla(T^i\bar{v}^i)^T\mathbf{n}+HT^i\bar{v}^i](\mathbf{V}^T\mathbf{n})\}\,dT$$

$$+\sum_{i=1}^{2}\left[T^i\bar{v}^iV_{T}\Big|_{p_2}-T^i\bar{v}^iV_{T}\Big|_{p_1}\right]. \tag{29}$$

In equation (29), H is the curvature of the loaded boundary Γ^2, and the last two terms on the right account for corner effects due to movement of points p_1 and p_2 in Figure 4 [10]. In these two terms, the notation $\mid p_i$ indicates that the terms are evaluated at point p_i, and V_T is the component of velocity $\mathbf{V}$ tangent to Γ which is positive if it is in a counterclockwise direction in Figure 4.

Results given in equations (21)–(24) and (26)–(29) can be used in equations (13) and (19) for each structural component. This allows one to systematically organize design sensitivity expressions for built-up structures. Moreover, one can develop a modular computer program that will carry out numerical integrations of terms in equations (21)–(24) and (26)–(29), using the same shape functions that are employed in finite element analysis codes. The result will then be a general algorithm and numerical method for design sensitivity analysis that can be implemented with existing finite element codes [7].

For numerical implementation of shape design sensitivity analysis, one must parameterize the boundary Γ of the domain Ω. For this purpose, one may use B-spline curves and surfaces [11]. The next step is to develop a general method of defining and computing a velocity field in the domain, in terms of the perturbation of the boundary Γ. Moreover, the velocity field must satisfy certain regularity conditions. It is shown in references [1] and [2] that C^1-regular and C^2-regular velocity fields are sufficient for shape design sensitivity analysis of truss and plane elastic solid problems and beam and plate problems, respectively. However, observing equations (26)–(29), one may relax these regularity conditions. That is, for truss and plane elastic solid problems, the highest-order derivative of the velocity field which appears in equations (26)–(29) is one. Thus one may use a C^0-regular velocity field with an integrable first derivative. Similarly, one may use a C^1-regular velocity field with an integrable second derivative for beam and plate problems. Therefore, regularity of the velocity field must be at least at the level of regularity of the displacement field of the structural component considered. This suggests the use of displacement shape functions to systematically define the velocity field in the domain. Moreover, one can select a velocity field that obeys the governing equation of the structure; that is, the perturbation of the boundary can be considered as a displacement at the boundary. With no external forces and a displacement on the boundary, one can use the finite element code to find the displacement (domain velocity) field that satisfies the required regularity conditions.

IMPLEMENTATION OF DESIGN SENSITIVITY ANALYSIS WITH EXISTING FINITE ELEMENT CODES

To obtain design sensitivity information, equations (5) and (12) must be solved for displacement functionals, and equations (5) and (18) must be solved for stress functionals. Once the original and adjoint structures are solved, one can integrate equations (13) and (19) numerically to obtain the desired sensitivity information. The finite element method can be viewed as an application of the Galerkin method to equations (5), (12) and (18) for an approximate solution of the boundary value problem. Note that the energy bilinear forms for equations (5), (12) and (18) are the same; therefore, the adjoint structures of equations (12) and (18) are the same as that of (5), with different adjoint loads. The adjoint load of equation (12) is a simple unit load at the point $\mathbf{x}^k$ in the positive direction of $\mathbf{z}(\mathbf{x}^k)$. To calculate the adjoint load using the load functional on the right side of equation (18), one should use the same shape functions that are used in the finite element code. Since m_p is a characteristic function defined on finite element Ω_p, numerical integration of the load functional is done on Ω_p only and the adjoint equivalent nodal force acts only on the nodal points of Ω_p.

For numerical implementation with existing finite element codes, one can proceed as in the flow chart of Figure 6. In the beginning, the model is defined by identifying the finite element model, original structural load, design variables and constraint functionals. In the next step, an existing finite element code is used to obtain structural response. With the structural response obtained, one calculates an adjoint load, external to the finite element code, using the shape functions of the code. The adjoint load is then input to the finite element code to obtain an adjoint response for each constraint functional. For adjoint analysis, one can use the multiloading (restart) option of the finite element code, so that only forward and backward substitutions are performed to obtain each adjoint response. Using the original and adjoint structural responses, design sensitivity information is calculated for each constraint functional by carrying out only numerical integration. This procedure allows one to perform calculations outside finite element codes using only postprocessing data. That is, the design sensitivity software does not have to be imbedded into finite element codes. Moreover, the method does not require differentiation of stiffness and mass matrices, and the uncertain numerical accuracy associated with selection of a finite difference perturbation can be eliminated.

AN OPTIMIZATION PROBLEM

To demonstrate the feasibility of the general purpose design sensitivity analysis method for built-up structures in structural design optimization, a truss-beam-plate built-up structure is optimized using a sparse matrix symbolic factorization technique for iterative structural optimization [8] and Pshenichny's linearization method [9].

Consider the truss-beam-plate built-up structure shown in Figure 7. A distributed vertical load $f(\mathbf{x})$ is applied to the plates. The points supported by the

References p. 349

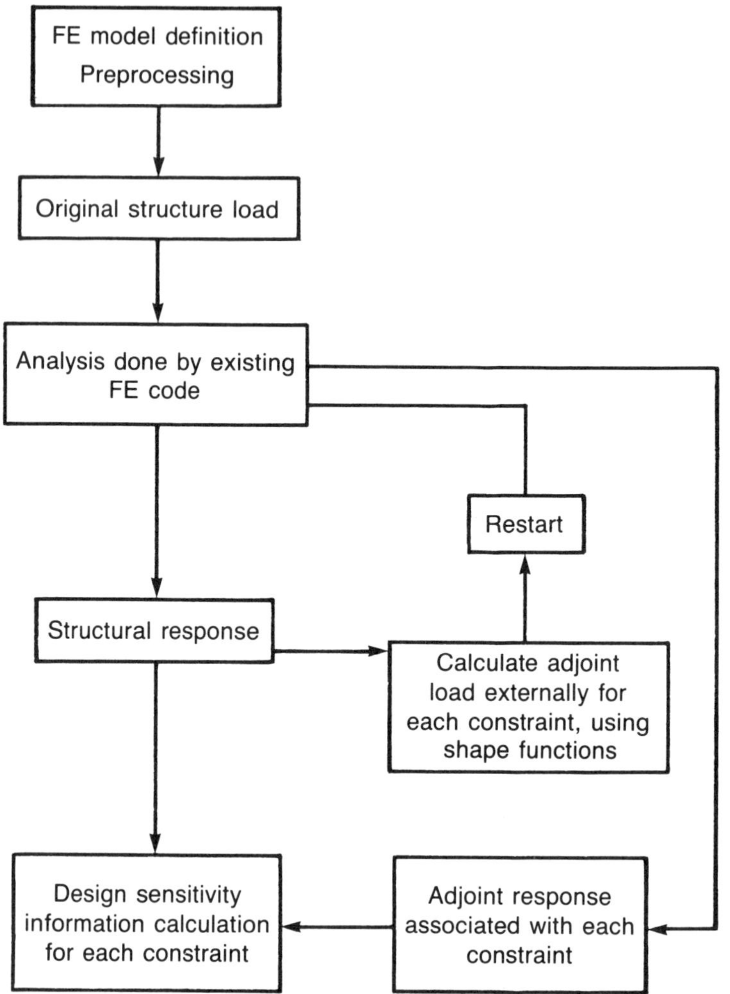

Figure 6. Flow chart of design sensitivity calculation procedure.

trusses are at the intersections of two crossing beams nearest to the free edges of the structure. No external loads are applied to the truss and beam components. The plates and beams are assumed to be welded together. Coordinates of intersection points of beams and plates are supposed to be in the midplanes of the plates and neutral axes of the beams. Beam components have rectangular cross-sections.

The design variables for this built-up structure are thickness $t^{ij}(\mathbf{x})$ of each plate component, width $\tilde{d}^{ij}(x_1)$ and height $\tilde{b}^{ij}(x_2)$ of each longitudinal beam component, width $\hat{d}^{ij}(x_2)$ and height $\hat{b}^{ij}(x_1)$ of each transverse beam component, and positions $\alpha_i(i = 1,4)$ and $\beta_j(j = 1,4)$ of transverse and longitudinal beam components, respectively. The lengths and cross-sectional areas of the trusses are

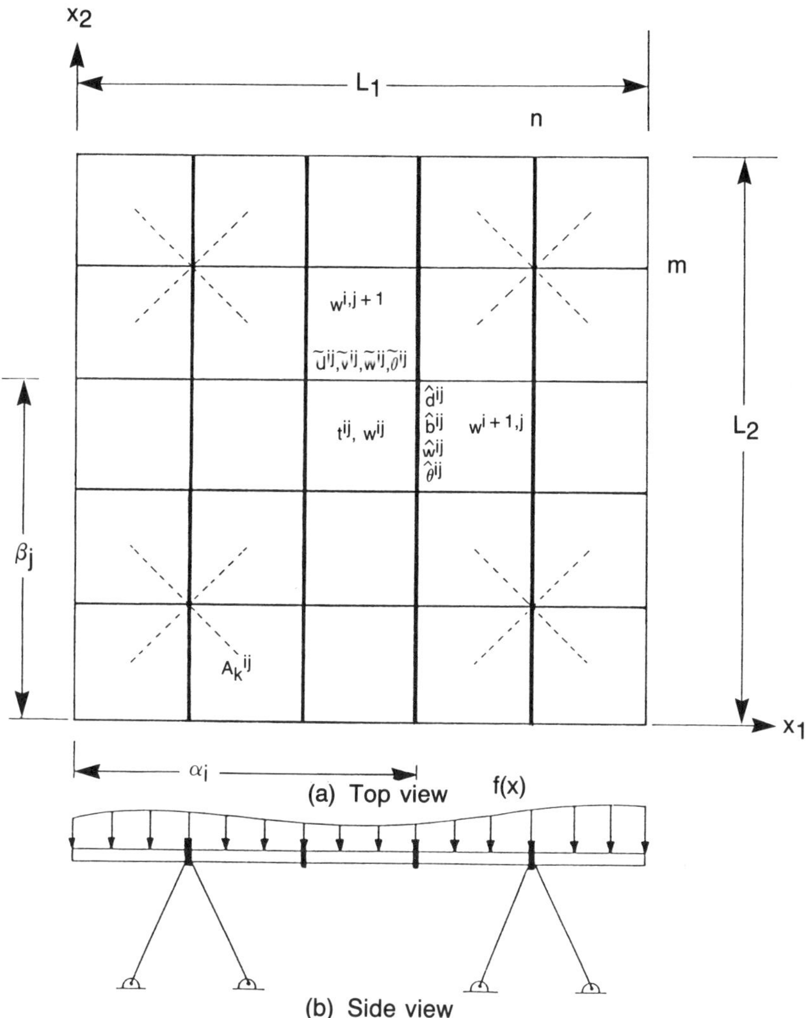

Figure 7. Truss-beam-plate built-up structure.

fixed, but they may change their ground positions. The outside boundary of the entire structure is fixed; i.e., only the locations α_i and $\beta_j (i, j = 1, 4)$ of beams are shape design variables. Dimensions of the structure and the numbering and spacing of beams in both directions are shown in Figure 7.

For plate components, 12 degree-of-freedom nonconforming rectangular elements [12] are used. For beam components, Hermite cubic shape functions are used. The finite element model used for design sensitivity analysis and optimization is shown in Figure 8. Only one-quarter of the entire structure is analyzed because of symmetry. A total of 484 elements with 1,281 degrees of freedom are used to model one quarter of the built-up structure, including 400 rectangular plate elements, 80 beam elements and 4 truss elements.

References p. 349

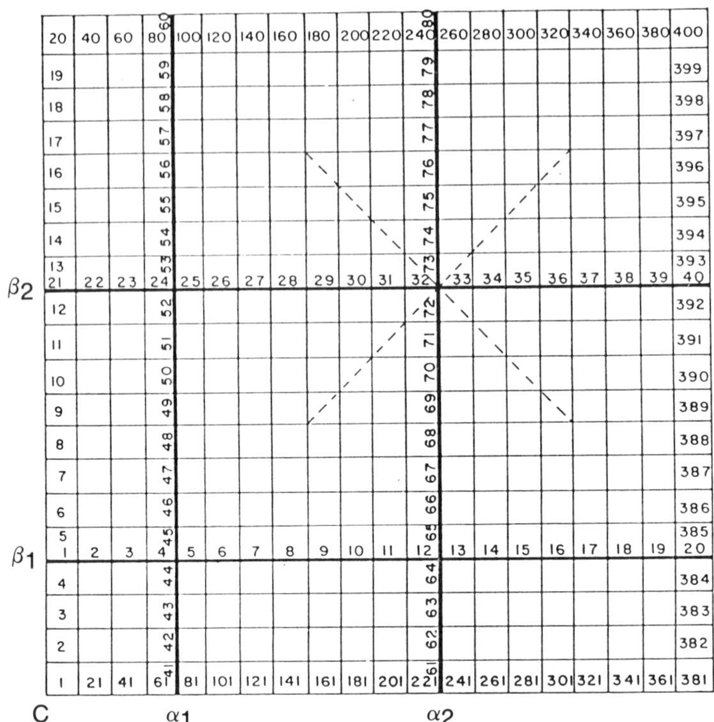

Figure 8. Finite element model of a truss-beam-plate built-up structure.

For numerical design sensitivity analysis and optimization, design variables are discretized; that is, each plate element has constant thickness and each beam element has constant width and height. Since the built-up structure is symmetric with respect to the center C, thickness t_i $(i = 1, 210)$, width d_i and height b_i $(i = 1, 40)$, and the locations α_i $(i = 1, 2)$ of transverse and longitudinal beams (measured from the center C), are taken as design parameters Thus the total number of design parameters is 292.

The optimal design problem of the built-up structure is to minimize weight of the structure subject to the following constraints:

Displacement at C: $\psi_1 = z(C) \leq 0.105$ in.

Plate element von Mises stress: $\psi_i \leq 17{,}500$ psi, $i = 2, 211$

Beam element bending stress: $-70{,}000$ psi $\leq \psi_i \leq 70{,}000$ psi, $i = 212, 251$

Plate thickness: 0.05 in. $\leq t_i \leq 0.25$ in., $i = 1, 210$

Beam width: 0.075 in. $\leq d_i \leq 0.30$ in., $i = 1, 40$

Beam height: 0.25 in. $\leq b_i \leq 1.00$ in., $i = 1, 40$

Beam position: $0 \leq \alpha_1 < \alpha_2 \leq L_1/L_2$.

Thus, the total number of inequality constraints is 543.

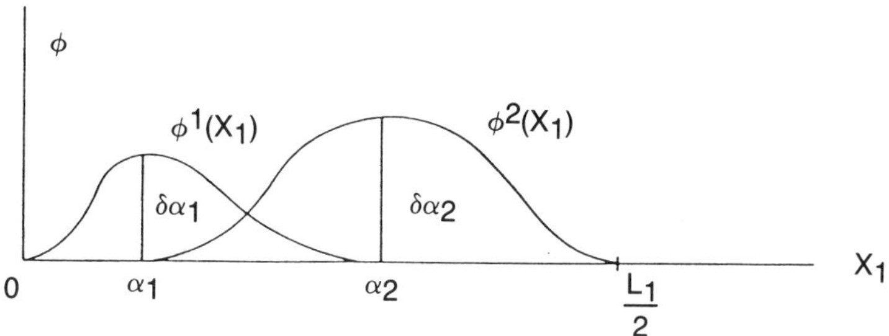

Figure 9. Shape functions for the velocity $V^1(x_1)$.

For numerical data, Young's modulus and Poisson's ratio are 3×10^7 psi and 0.3, respectively. The overall dimensions are $L_1 \times L_2 = 15$ in. × 15 in. and weight density is 0.1 lb./in.^{3} For truss components, the length is 5.364 in. and the cross-sectional area is 0.1 in.^{2} A uniformly distributed load $f = 17.5$ lb./in.^{2} is applied to the plate components.

As mentioned before, for shape design sensitivity calculations one must define a velocity field that has C^1 regularity, with an integrable second derivative. The beam components are allowed to move in transverse directions only. Hence V^1 is a function of x_1 only, and V^2 is a function of x_2 only. The velocity fields in both plate components $V^1(x_1)$ and $V^2(x_2)$ are represented by Hermite cubic functions in each direction. To see the velocity field representation graphically, consider Figure 9, in which the shape functions for $V^1(x_1)$ are plotted and $\delta\alpha_1$ and $\delta\alpha_2$ denote perturbations of locations of transverse beams. From Figure 9 one obtains $V^1(x_1) = \phi^1(x_1) + \phi^2(x_1)$:

$$V^1(x_1) = \begin{cases} -\frac{2x_1^2}{\alpha_1^3}(x_1 \frac{3\alpha_1}{2})\delta\alpha_1, \quad 0 \leq x_1 \leq \alpha_1 \\ \frac{2(x_1 - \alpha_1)^2}{(\alpha_2 - \alpha_1)^3}\left[(x_1 - \alpha_1) - \frac{3(\alpha_2 - \alpha_1)}{2}\right](\delta\alpha_1 - \delta\alpha_2) + \delta\alpha_1, \\ \qquad \alpha_1 \leq \alpha_2 \\ \frac{2(x_1 - \alpha_2)^2}{(\frac{L_1}{2} - \alpha_2)^3}\left[(x - \alpha_2) - \frac{3(\frac{L_1}{2} - \alpha_2)}{2}\right]\delta\alpha_2 + \delta\alpha_2, \\ \qquad \alpha_2 \leq x_1 \leq \frac{L_1}{2}. \end{cases} \quad (30)$$

A similar expression is obtained for $V^2(x_2)$.

For numerical computation of the shape design sensitivity information, derivatives I_x, J_x, and h_x in equation (26) for the beam component and ∇t in equation (28) for the plate component must be computed. Since each plate element has constant thickness and each beam element has constant width and height, these derivatives are Dirac delta measures, so computations of the shape design sensitivity information become complicated.

References p. 349

To avoid this difficulty, the design process is divided into two phases. In Phase 1 each plate and beam component (not element) has constant thickness and constant width and height, respectively. Hence in Phase 1, the design parameter set includes six plate thicknesses, six beam heights and widths, and two beam locations with a total of 20 design parameters. To assign the same design parameter to elements in a component, design variable linking is used in Phase 1. Once an optimum point is reached in Phase 1, the design process is switched to Phase 2 where shape design parameters are fixed and each plate and beam element is allowed to have different design parameters. Hence the total number of design parameters is 290 in Phase 2.

For numerical computation, PRIME-750 and Cray-1S computers are used for design Phases 1 and 2, respectively. The initial and final designs of Phase 1 are given in Table 1. The initial cost is 0.7875 pound and the final cost of Phase 1 is 0.5894 pound. There are 12 design iterations in Phase 1 with average CPU time 15,986 seconds per iteration on a PRIME-750 computer. It is observed in design Phase 1 that inner beam stiffners move inward (α_1 = 1.4211 pound) and outer beam stiffners move outward (α_2 = 4.5872 in.). The number of active stress and displacement constraints at the final design of Phase 1 is 31. Thus, it is necessary to calculate sensitivity information for 31 constraints out of 251.

There are nine design iterations in Phase 2 with average CPU time 25.84 seconds per iteration on a Cray-1S computer. The cost function history of Phases 1 and 2 is shown in Figure 10. The final cost of Phase 2 is 0.5388 pound with 54 active stress and displacement constraints at the final design. A profile of the upper half of the final design is shown in Figure 11.

CONCLUSIONS

In this paper, a general purpose design sensitivity analysis and optimization method for both conventional and shape design of built-up structures is proposed. Results presented in this paper indicate that software implementation of the results of the design component method with existing finite element codes is feasible by assembling a modular computer program to carry out calculations outside such codes using only postprocessing data. An important result is that the method does not require differentiation of the stiffness and mass matrices with respect to conventional and shape design parameters.

The capability of the proposed method for large-scale structural design problems is demonstrated with the design of a truss-beam-plate built-up structure using the Cray-1S supercomputer.

Table 1

Initial and final designs of Phase 1

	Design parameter	Initial design	Final design
	t_1		0.0874
Plate	t_2		0.0500
component	t_3	0.1	0.0518
thickness	t_4		0.0501
	t_5		0.0815
	t_6		0.0910
	b_1		0.4234
Beam	b_2		0.6902
component	b_3	0.5	0.4990
height	b_4		0.3974
	b_5		0.4266
	b_6		0.4326
	d_1		0.1087
Beam	d_2		0.2214
component	d_3	0.15	0.2499
width	d_4		0.0754
	d_5		0.0838
	d_6		0.1286
Beam	α_1	1.50	1.4221
position	α_2	4.50	4.5872

References p. 349

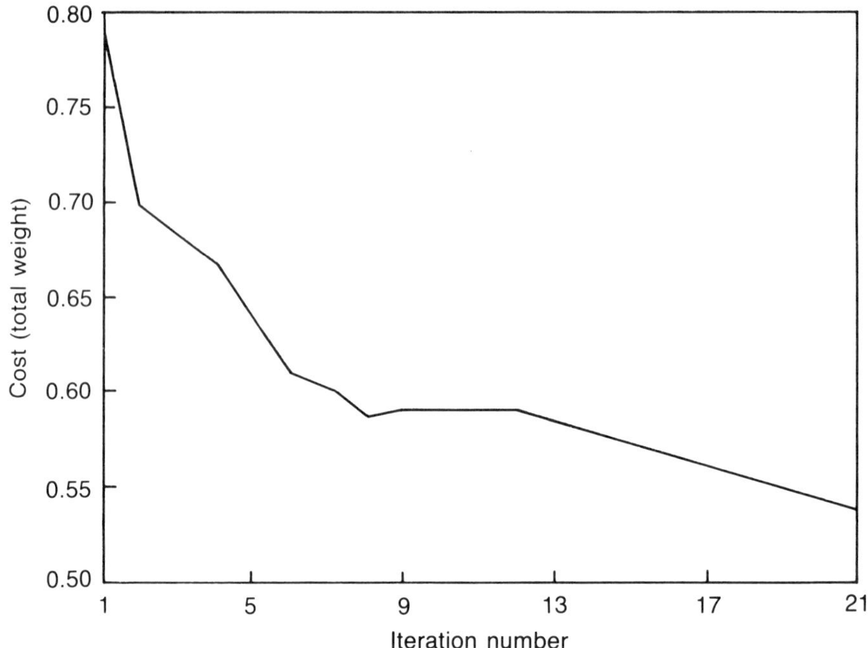

Figure 10. Cost function history.

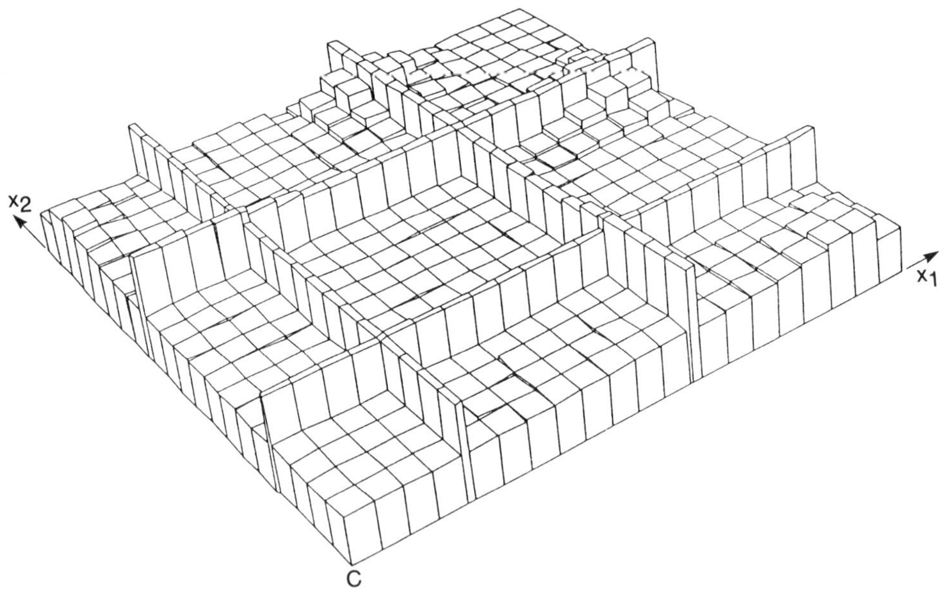

Figure 11. A profile of the final design.

ACKNOWLEDGEMENT

Research was supported by NSF Grant No. CEE 83-19871.

REFERENCES

1. E. J. Haug, K. K. Choi and V. Komkov, *Design Sensitivity Analysis of Structural Systems.* Academic Press, New York (1986).
2. K. K. Choi and E. J. Haug, Shape design sensitivity analysis of elastic structures. *J. Struct. Mech.* **11** (2), 231–269 (1983).
3. K. K. Choi, Shape design sensitivity analysis of displacement and stress constraints. *J. Struct. Mech.* **13** (1), 27–41 (1985).
4. I. Babuška and A. K. Aziz, Survey lectures on the mathematical foundations of the finite element method, pp. 1–359 in *The Mathematical Foundations of the Finite Element Method with Applications to Partial Differential Equations* (Edited by A. K. Aziz). Academic Press, New York (1972).
5. K. K. Choi and H. G. Seong, Domain method for shape design sensitivity analysis of built-up structures. *Computer Methods Appl. Mech. Eng.* (1986), to appear.
6. K. K. Choi and H. G. Seong, Design component method for sensitivity analysis of built-up structures. *J. Struct. Mech.* 14 (3), (1986).
7. K. K. Choi, J. L. T. Santos and M. C. Frederick, Implementation of design sensitivity analysis with existing finite element codes. *J. Mech. Transm. Autom. Des.* 85-DET-77.
8. H. L. Lam, K. K. Choi and E. J. Haug, A sparse matrix finite element technique for iterative structure optimization. *Comput. Struct.* **16** (1–4), 289–295 (1983).
9. K. K. Choi, E. J. Haug, J. W. Hou and V. N. Sohoni, Pshenichny's linearization method for mechanical system optimization. *J. Mech. Transm. Autom. Des.* **105** (1), 97–103 (1983).
10. J. P. Zolesio, Gradient des coûts governés par des problèms de Neumann posês sur des ouverts angeleux en optimization de domain. CRMA-Report 1116, University of Montreal, Canada (1982).
11. D. F. Rogers and J. A. Adams, *Mathematical Elements for Computer Graphics.* McGraw-Hill, New York (1976).
12. R. D. Cook, *Concepts and Applications of Finite Element Analysis.* John Wiley & Sons, New York (1981).

DISCUSSION

J. Taylor *(The University of Michigan)*

There is some indication from studies, particularly in Denmark by Olhoff and Cheng and Bendsøe and by workers in the U.S.S.R., that these sorts of solutions cannot converge if you look at what happens with increasing refinement of the grid. Could you comment on that?

Choi

If you broaden your design space by increasing the number of design variables, then you may have introduced a lot of thin fibers running across the structure. What I am talking about here is the optimum in this specific design space.

K. K. Choi

K. Izadpanah *(MacNeal-Schwendler Corporation)*

I have two comments about the method that has been presented several times at this symposium. It seems you dismissed the auxiliary or the adjoint problem as a simple case for computation by the finite element method or some other methods. In my experience, by using the finite element method you might run into trouble in solving that kind of adjoint problem if you are looking at positions very close to the boundary or if you are looking at very thin strips. This is because the elasticity solution by itself is not very nice. I simply want to comment that in your methodology it is essential to have a good solution coming from the adjoint problem, and this is not always that easy to achieve.

With regard to the plate elements that you used (it seemed to me they were Kirchhoff plate elements), with the variation of thickness you allowed your program to have, and with all the computation involved, I question the final result coming out in the finite element code because I do not think it would give you good enough accuracy. I doubt that finite element analysis is in a state in which we can deliver very accurate results in the problem you have posed.

Choi

We are not saying that this result is optimal for the real structure. You are never going to have a better design than your analysis can handle. Certainly if you can only handle a limited design space with the current capability, then you have to limit your design space. We are not saying that ours would be the true solution for the real physical problem; but given the finite element capability and our design variable concept, this is the optimum.

Izadpanah

To follow up on that, I would recommend that if you are going to use plate elements in future, that you use the Reissner-Mindlin type of plate formulation. That will probably improve both your iterations and the final results.

D. Vasilopoulos *(General Motors Research Laboratories)*

Do you assume that the upper surface of the plate is always straight and therefore any variation in the thickness will be underneath?

Choi

No, it is symmetric with respect to the mid plane, so both surfaces will move. I was just showing the upper half.

L. Schmit *(University of California - Los Angeles)*

One of the main points you made is the transparency or the independence of the finite element code, and I think that is a very persuasive and powerful argument for the sensitivity analysis method presented in your paper. Is it true that in effect you've managed to stay outside the existing finite element code? In many respects you have had to repeat much of the work which is done inside the finite element code, which is in the integrations over the elements using Gauss point integration, and in the assembly process because you assemble this element by element. Now my recollection of how finite element analysis code works is that it does the same thing, only the integrand is different and it does not require the adjoint solution.

Choi

I am not sure whether I understand your question. The first reason we could do the sensitivity computations outside of the analysis program is that we do not need the explicit expression for the stiffness matrix. If you do find explicit expressions, fine; but you have to take derivatives with respect to design parameters. In our method, all we need for the sensitivity calculation are stress, strain and displacement, which will be given by any finite element code. I'm sure a well-written finite element code would give you information on the shape function.

Of course, we are doing the calculation using exactly the same numerical integration approach used in the finite element analysis. But the basic reason that this method is independent of the finite element code is that we are starting from the variational equation of elasticity and not from a matrix finite element equation. If you start with the matrix finite element equation of a finite element code, you are obliged to use that code through the whole process. The philosophy of the method starting from the matrix finite element equation is quite different from that using the variational equation, although the computational procedures look very similar.

C. Fleury *(University of California - Los Angeles)*

You mentioned that you can get the shape function in any finite element code, Do you think this is really true?

Choi

From any *well-written* finite element code, yes. We have been working with ANSYS, and it worked really well. We have also worked with EAL (a hybrid method) and it worked fine. It is interesting that we can apply this method to both the displacement method and the hybrid method. It is a medium size program and it worked almost as well as ANSYS. So we are sure that as long as you are given the shape functions, you can use the finite element code.

Even if you are *not* given the shape functions, you are certainly given nodal information for displacement and stress. With careful judgment, you can apply some outside shape function and interpolate these values at the Gauss points. This works because integration is a smoothing process. Even though you may not have exactly the right value at the Gauss point, integration gives you an averaged value.

H. Miura *(NASA Ames Research Center)*

What is the motivation to separate Phase 1 and 2?

Choi

When we look at the sensitivity expression, we see that it requires a gradient of thickness. That means we should limit ourselves to continuous and differentiable thicknesses. Since we are doing analysis with constant thickness for each element, there is a discontinuity in thickness—so we have to separate into two phases.

If we do not want two phases, we have to use differentiable shape functions for thickness. That is, if we use smooth shape functions for thickness, then we can use a single phase. There is no problem in that case.

ANOMALIES ARISING IN ANALYSIS AND COMPUTATIONAL PROCEDURES FOR THE PREDICTION OF OPTIMAL SHAPE

J. E. TAYLOR
Aerospace Engineering Department
The University of Michigan
Ann Arbor, Michigan

Abstract

Methods presently available for the treatment of shape optimal design problems fail to perform adequately in certain applications. Cases where difficulties may be identified with limitations inherent in the discretization models are examined. Other anomalies that sometimes arise in connection with shape optimization are examined as well.

INTRODUCTION

Analytical formulations for problems in shape optimal design are reasonably well established, particularly for the classical mechanics of two- and three-dimensional objects. Thus necessary conditions and sensitivity analysis are readily available for shape design within a broad array of objectives and criteria. Methods for the computational treatment of shape design problems, sometimes based on the results from analysis or otherwise as applications of available techniques from the field of nonlinear programming also are well developed. However, it is possible to identify a number of important and challenging problem areas that call for additional study. The objective of this paper is to elaborate on several such areas and to speculate on the prospects for their effective treatement. We consider certain difficulties that may be identified with limitations in the discretization methods, and also anomalies that may arise in the prediction of shape where connectivity of the design domain is not fixed.

DIFFICULTIES ASSOCIATED WITH COMPUTATIONAL ERROR

Procedures for the prediction of shape design are susceptible to all of the pitfalls ordinarily identified with the use of discretization models for the computational solution of continuum problems, as well as to those difficulties that are peculiar to the treatment of variable domain problems. While it is possible to consider such issues separately, e.g., the instabilities that may occur in standard finite element procedures for analysis versus the independent potential for instability associated with the prediction of domain boundary, all these factors tend to be coupled in shape design problems. In fact, procedures for the prediction of optimal shape are sometimes exceptionally sensitive to discretization error. Here we examine a number of sample problems as a means to consider these issues in some detail.

There is ample precedent in the practice of solving shape design problems for the kind of unsatisfactory performance of a solution procedure where the results define the design boundary as an irregular shape, while the expected solution should have the form of a smooth curve. A mild difficulty of this kind is exemplified in the treatment by Braibant and Fleury [1] for design of a hole in a sheet. Results obtained there with the use of a coarse finite element mesh predict the optimum hole to have a star shape, as indicated in Figure 1. The correct shape for the optimal hole is simulated successfully by results obtained in the same study using splines. The introduction of splines to model the design boundary in fact limits the space of feasible designs to a set of smooth curves, i.e., the irregular shape is effectively avoided in this case by the imposition of constraints.

It is not uncommon to find that the fault can be remedied in cases where a procedure produces poor results or where a problem is "ill-posed," simply by the addition of constraints. On the other hand, this is not to suggest that poor performance of a computational procedure implies that the problem is not well-posed. For the hole-in-a-sheet problem described above, it has been demonstrated in a recent study by Chung [2] that the incorporation of appropriate means to refine the finite element mesh makes it possible to predict the optimal hole without imposing constraints on the shape of the design boundary. This confirms that the problem is well-posed in its original form, i.e., without smoothing or other additional constraints, and that the difficulty in predicting shape is associated in some way with properties of the discretization model.

The design of a fillet provides an example of a shape design problem where the difficulty described above is compounded by the presence of a form of instability. Not only does an ordinary procedure fail to predict a smooth shape for the design (see Figure 2), but it tends at some point in the solution process to oscillate between dissimilar irregular shapes. This result, which is also cited in the study by Chung [2] where the objective in design is to minimize the maximum stress, suggests that the design criterion becomes relatively insensitive to design change. It was also established for this example that the difficulty is resolved by use of an appropriate scheme for systematic refinement of the finite element grid. The scheme is based on a separate optimization problem, namely, design of the grid to minimize a local

Figure 1. Finite element method design of a hole in a plate using a coarse grid.

measure of error in the finite element computation. For practical reasons, the grid refinement and the adjustment in shape design are performed in separate steps rather than at once.

Difficulties such as those cited for the fillet problem generally arise in cases where the design to be predicted has sharply varying shape. The determination of optimum shape in problems where the system includes mechanical contact interfaces tends to be troublesome as well. The design of the base part of a turbine blade involves both contact interfaces and sharply varying shape along the design boundaries. This problem also was treated in the study by Chung. An example of his result for the refined grid and shape of the blade root design is shown in Figure 3. The use of optimal grid adaptation in various other examples of such problems is reported in the presentation given earlier in this symposium by Kikuchi. Choi [3] has reported on developments for a substantially different approach to the management of difficulties in shape design problems.

As an adjunct to the discussion of this section, the combined shape and finite element grid optimization mentioned in the context of the above-described examples is interpreted here as a *multicriteria design problem.* It is supposed for present purposes that the shape design problem alone corresponds to the statement

$$\min_{y_i}[\max_{x_\alpha} \mathbf{f}_d].$$

Here y_i symbolizes the set of values (points) that defines the design boundary; x_α represents the set of points in the discretized domain for which the state variable

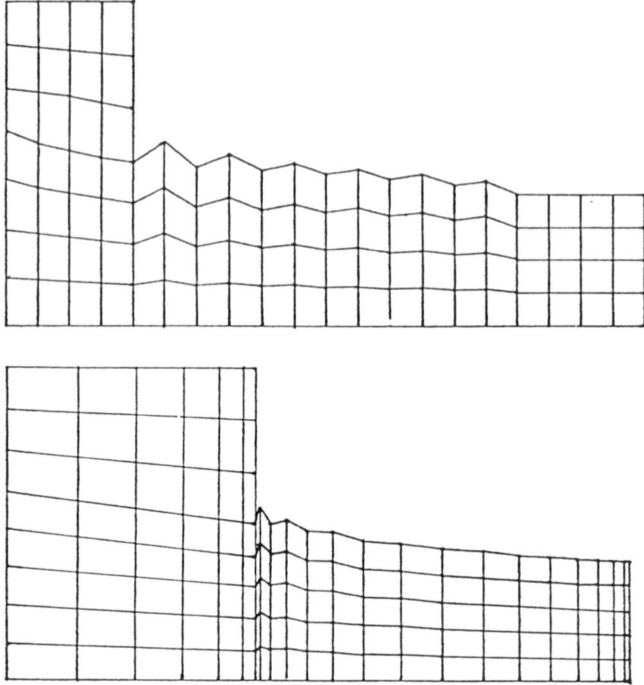

Figure 2. Fillet designs.

and the argument $\mathbf{f}_d$ are defined; and $\mathbf{f}_d$ stands for the local measure of the design criterion. Similarly the grid optimization problem by itself might be stated as

$$\min_{z_k}[\max_{x_\alpha} \mathbf{E}_h]$$

where z_k represents values (nodal coordinates) that serve to define the finite element grid, and $\mathbf{E}_h$ is the symbol for the local measure of discretization error. Details on this form for grid optimization are provided in Diaz et al. [4].

The combined problem corresponding to these separate statements may be characterized in the form

$$\min_{y_i, z_k}[\max_{x_\alpha}\{\mathbf{f}_d, \mathbf{E}_h\}].$$

Elements $\mathbf{f}_d$ and $\mathbf{E}_h$ of the vector argument must be nonnegative and bounded on the design set for which (y_i, z_k) is defined. This minmax problem with a multicriteria argument is interpreted into scalar form as

$$\min_{y_i, z_k} \beta$$

$$\text{within} \qquad \left.\begin{array}{r} \mathbf{f} - \beta \leq 0 \\ \alpha\mathbf{E}_h - \beta \leq 0 \end{array}\right\} \forall x_\alpha$$

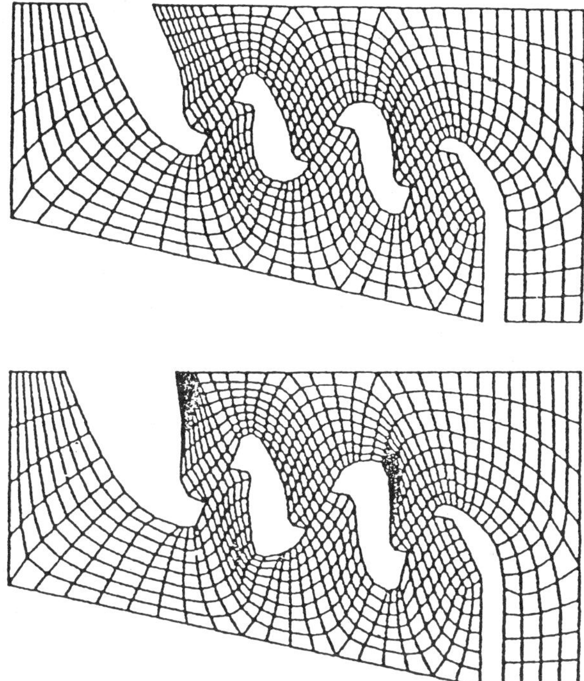

Figure 3. Turbine blade root: design with account of friction on contact surfaces.

where weights α are specified, and the bound value β is an additional variable. Details on the interpretation of minmax problems via the bound formulation are provided in reference [5], and applications to multicriteria optimization are discussed in [6]. The accomplishment of this reduction of the combined optimization problem to a simple scalar form suggests that means might be available for the practical treatment of the original problem. The merit of addressing problems in shape design in a way that incorporates explicit means for mesh optimization is established simply from the argument that the capability to predict authentic results for shape is enhanced by any facility that provides for greater precision with respect to discretization error. Of course the incorporation into the design procedure of other devices to reduce the discretization error may be justified on the same basis.

CONNECTIVITY

In the treatment of shape design problems, the degree of connectivity associated with the design domain may itself be unknown. Clearly for such cases some means are needed to identify appropriate connectivity as a part of the procedure to predict optimum shape. With certain examples, at least, the determination of whether a design would be improved in a given step of the design process by the

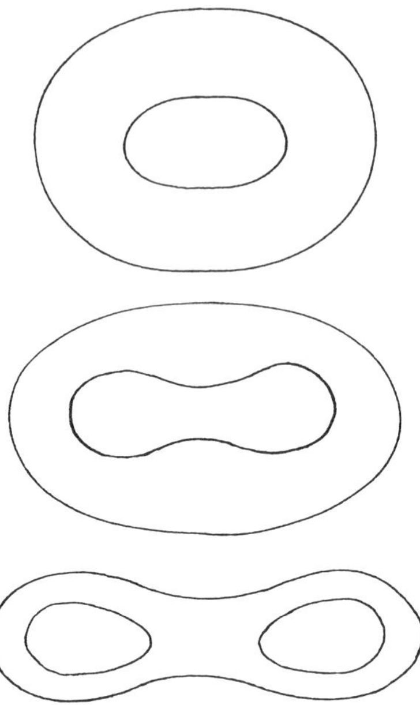

Figure 4. Design of cross-section for torsion bar, depending on aspect ratio.

introduction (or elimination) of holes can be made by monitoring an index of sensitivity. An indication of how this might be done is provided by Cea [7] in conjunction with his description of the speed method. Techniques for handling design problems where connectivity may be an issue are discussed in this section through examples of the design of sheets and the design of cross-sectional shape for torsion members.

A technique for the prediction of shape including holes was demonstrated by Na [8, 9] on several examples in the design of torsion bars. In contrast to the speed method and other approaches based on sensitivity analysis, Na's algorithm to predict design change at a given stage of the computational procedure is based on necessary conditions for the optimal modification of shape. In the case of design for maximum elastic stiffness, these conditions provide that the domain of a hole is identified by monitoring a measure of unit elastic energy (more generally the measure has the form of unit mutual elastic energy). Explicitly, for a prescribed reduction in cross-sectional area, the hole shape is identified with the set of area elements for which the measure of unit energy has value below a threshold value.

A set of solutions obtained by applying this method is shown in schematic form in Figure 4. The results substantiate a capability to identify the topological change from one to two holes (as well as hole shape) which occurs in this example as the overall shape of the cross-section becomes more oblong. In a different problem, the design of a sheet to transmit load to prescribed support curves [8], the method was applied to predict the shapes of both outer boundaries and the hole (see Figure 5).

In a distinctly different approach, it is possible to predict the optimum shape

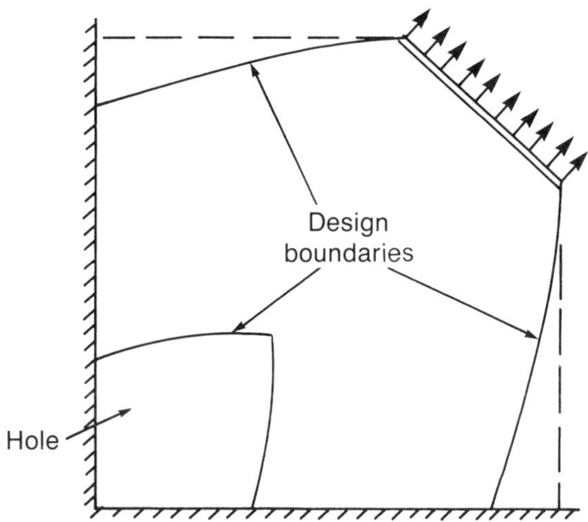

Figure 5. Design of a sheet using procedure for optimal modification of shape.

for the sheet by exercising in an appropriate way a procedure for the determination of optimal material distribution within the fixed domain. (A procedure for doing this was described in the thesis by Rossow [10, 11].). Consider the latter problem, expressed for the discretized sheet with element thickness t_i and characterized by the problem statement

$$\min_{t_i}[\max_{x_a} f_d]$$

within

$$t_i - \bar{t} \leq 0$$

$$\underline{t} - t_i \leq 0$$

$$SA_i t_i - R \leq 0.$$

The definitions given above for x_a and f_d still apply, and A_i represents element area. Upper and lower bounds $\bar{t}$ and $\underline{t}$ on thickness and the bound R on volume of material are specified.

Clearly the element boundaries that separate those elements for which $t_i = \underline{t}$ from the rest of the elements in the set comprising the sheet will be established as part of the solution in the material distribution problem. Note in particular that for the solution associated with the limits as $\underline{t} \to 0$, the cited element boundaries effectively define holes in the sheet. In the upper one of the example results [10] for this method shown in Figure 6, only the outer boundaries are determined. Both the shape of the hole and the outer boundaries are predicted for the other examples in this figure.

References pp. 361–362

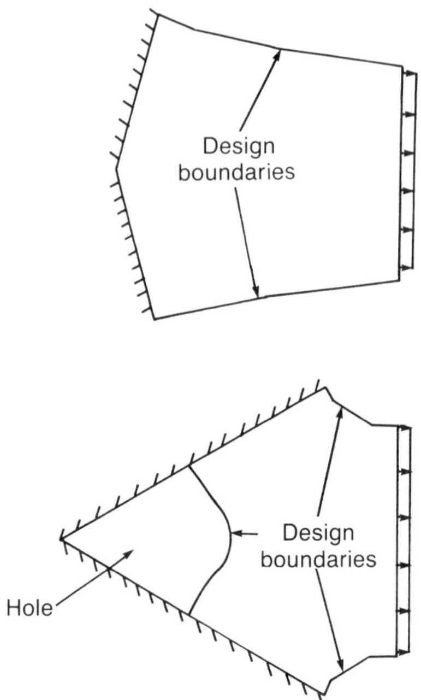

Figure 6. Design of a sheet, obtained with the procedure for optimal material distribution.

DISCUSSION

Substantial additional development is needed in the area of combined grid plus shape optimization. Experience of the kind described above shows that at least for the more sensitive design problems, a capability to achieve exceptional precision in computational results is essential to the prediction of shape. For practical reasons, computational procedures for the combined problem should be adaptive, i.e., they should have means automatically to produce systematic improvement in precision. Automatic adaptation is similar in a sense to adaptive control.

On another topic, it seems likely that consideration of both *skeletal structure* and *microstructure* will enter into the treatment of shape design. A useful discussion of microstructure in the context of other design problems is given in [18]. This refers to situations where characterization of the physical structure by means of models for mechanics of the simple continuum are not adequate for proper formulation of the design problem (the difficulty is identified in analysis with the property that the problem lacks G closure).

Results from contemporary developments in plate optimization [12–14], design for elastic-plastic response [15] and optimization of composites [16, 17], for example, provide some indication of how the appropriate modifications in problem formulation are to be made. However, conclusions from the most recent studies indicate that even for these areas the issues are not fully settled. In plate design, a study performed this year establishes that two or more levels of microstructure are needed to model the general optimum plate [18]. For the composite sheet, it has been confirmed that precisely six levels of microstructure are required [19]. Also, while the skeletal forms associated with some other systems are known, e.g., Michell trusses for the sheet, grillages for plates [20] and archgrids for membrane shells [21], apparently such general limiting forms as might arise in conjunction with shape design have not been identified. However, it is possible that certain computational results such as the final shape obtained for the design of a suspension bracket [22] are suggestive of a skeletal form.

REFERENCES

1. V. Braibant and C. Fleury, Shape optimal design using b-splines. *Comp. Meth. Appl. Mech. Eng.* **44**, 247–267 (1984).
2. K. Y. Chung, Shape optimization and free boundary problems. Ph.D. Thesis, University of Michigan (1985).
3. K. K. Choi and E. J. Haug, Shape optimization of elastic structure. Soc. Engr. Sci., 21st Annual Mtg. VPI & SU, Blacksburg, VA (1984).
4. A. Diaz, N. Kikuchi and J. E. Taylor, Optimal design formulations for finite element grid adaption, in *Sensitivity of Functionals with Applications to Engineering Sciences* (Edited by V. Komkov), Lecture Notes in Math No. 1086. Springer-Verlag, Berlin (1984).
5. J. E. Taylor and M. P. Bendsøe, An interpretation for minmax structural design problems including a method for relaxing constraints. *Int. J. Solids Struct.* **20** (4), 301–314 (1984).
6. M. P. Bendsøe, N. Olhoff and J. E. Taylor, A variational formulation for multicriteria structural optimization. *J. Struct. Mech.* **11** (4), 523–544 (1983-84).
7. J. Cea, Problems of shape optimal design, in *Optimization of Distributed Parameter Structures* Vol. II (Edited by E. J. Haug and J. Cea). Sijthoff and Noordhoff, The Netherlands (1981).
8. M. S. Na, Optimal shape remodeling of elastic bodies by the finite element method. Ph.D. Thesis, University of Michigan (1982).
9. M. S. Na, N. Kikuchi and J. E. Taylor, Optimal modification of shape for two-dimensional elastic bodies. *J. Struct. Mech.* **11** 111–135 (1983).
10. M. P. Rossow, A finite element approach to optimal structural design. Ph.D. Thesis, University of Michigan (1973).
11. M. P. Rossow and J. E. Taylor, An optimal structural design algorithm using optimality criteria. Soc. Engrg. Sci., 13th Annual Mtg, Hampton, VA. (Nov. 1976).
12. K.-T. Cheng and N. Olhoff, An investigation concerning optimal design of solid elastic plates. *Int. J. Solids Struct.* **17**, 305–332 (1981).
13. M. P. Bendsøe, Generalized plate models and optimal design. *Proc. IMA Workshop on Homogenization and Effective Moduli.* University of Minnesota (Oct. 1984) to appear.
14. G. I. N. Rozvany, N. Olhoff, K.-T. Cheng and J. E. Taylor. On the solid plate paradox in structural optimization. *J. Struct. Mech.* **10** 1–32 (1982).

15. R. V. Kohn and G. Strang, Optimal design for torsional rigidity, in *Hybrid and Mixed Finite Element Methods* (Edited by S. N. Atluri, R. H. Gallagher, and O. C. Zienkiewicz). Wiley and Sons (1983).
16. K. A. Lurie, A. V. Fedorov and A. V. Cherkaev, Regularization of optimal design problems for bars and plates, I and II. *J. Optim. Theory Appl.* **37**, 499–521, 523–543 (1982).
17. L. Tartar, Estimations de coefficients homogénéisés, pp. 364–373 in *Computer Methods in Applied Sciences and Engineering, Proc. 1977 IRIA Paris* (Edited by R. Glowinski and J. L. Lions). Lecture Notes in Mathematics No. 704. Springer-Verlag, Berlin (1979).
18. G. I. N. Rozvany, T. G. Ong, N. Olhoff, M. P. Bendsøe and W. T. Szeto, Least-weight design of perforated elastic plates, I and II. *Int. J. Solids Struct.*, submitted.
19. G. A. Francfort and F. Murat, Homogenization and optimal bounds in linear elasticity, Paper No. 85009. Laboratoire d'Analyse Numerique, Universite Pierre et Marie Curie (1985).
20. G. I. N. Rozvany, *Optimal Design of Flexural Systems*. Pergamon Press, Oxford (1976).
21. G. I. N. Rozvany and W. Prager, A new class of optimization problems: Optimal archgrids. *Comput. Meth. Appl. Mech. Eng.* **19**, 127–150 (1979).
22. J. A. Bennett and M. E. Botkin, Structural shape optimization with geometric descriptions and adaptive mesh refinement. *AIAA J.* **23** (3), 458–464 (1985).

DISCUSSION

D. J. Wilde *(Stanford University)*

Do you see a lot of problems where there are many shapes that have just about the same minimum? Are these minima flat with respect to shape parameters?

J. E. Taylor

Taylor

When the measure of the criterion function is flat in the neighborhood of the optimum, there is usually some difficulty with convergence—and this happens often. The complement to that observation is that frequently in the structural optimization problem, the greatest part of the advantage is achieved in the early part of moving toward an optimal design.

Wilde

Is there anything done on establishing lower bounds on these optimums so that you know that you are close enough? This is being done in other fields. Do you know of any work that is being done with respect to shape?

Taylor

The only work I know of that would demonstrate useful measures of a lower bound come from academic examples, where explicit results are available in analytical form or very nearly so. I do not know of any expression in general form for a lower bound to the value of the criterion at the optimum.

C. Fleury *(University of California - Los Angeles)*

As a comment on Professor Wilde's question, there is one work by Alan Morris from England. This numerical work uses some type of dual formulation to obtain lower bounds on the optimum.

GEOMETRIC MODELING FOR STRUCTURAL AND MATERIAL SHAPE OPTIMIZATION

E. L. STANTON
PDA Engineering
Santa Ana, California

Abstract

A challenge many structural designers must soon face is shape optimization for components made from new, advanced materials. Designs reinforced with graphitic or ceramic materials offer mechanical and thermal performance gains that can give a competitive edge to an automotive product. Unfortunately, these advanced materials introduce material geometry into an already difficult problem as well as introducing new manufacturing constraints. In this paper, an interactive modeling system for both structural and material geometry is applied to the design of a connecting rod to explore modeling requirements for shape optimization of composites. These include new geometric and material data exchange requirements necessary for communication among design analysis and manufacturing activities for composite structures. Extensions to the current IGES system for material shape definition are recommended for components fabricated from these new materials.

Shape optimization is also explored for basic geometric design parameters such as the ratio of surface area to volume (which in large measure establish material processing requirements). Length, area volume, and point shape sensitivity derivatives are then presented for parametric cu bic geometric design variables used to construct finite element analysis models by the PATRAN solid modeling system. In this way, a natural and significant reduction of the dimension of the mathematical programming problem for shape optimization results following the work of Fleury and his colleagues. Design sensitivity derivatives are also presented for global composite design variables used to define the construction of certain laminated composite shell structures. Unlike laminate point design variables based on local thickness, these variables directly define the global shape of the structural components.

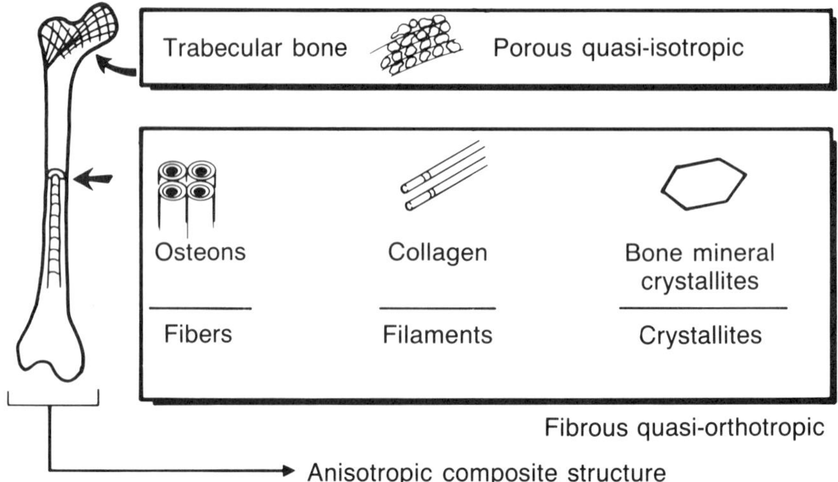

Figure 1. The femur—a natural continuum composite.

INTRODUCTION

Shape optimization in structural mechanics usually refers to structural geometry. This is quite natural for components that are homogeneous if not isotropic in their mechanical properties. Cost and performance are largely determined by their external shape for a given material. Today, an increasing number of mechanical designs are being manufactured using composite materials. These materials introduce additional design variables, especially material geometry which defines the internal shape of the microstructure of the component. A very common example of a composite design taken from nature appears in Figure 1 and illustrates the way in which material geometry can be used to optimize performance. To introduce the notion of shape optimization for material geometry, consider the external shape of the femur as fixed and ask what distribution and orientation of bone mineral crystallites best perform the mechanical and system functions of the femur in the skeletal system. In asking the question, we are led to think about functions the femur must perform in the context of a larger system that performs an enormous variety of functions. The system is dynamic and self-propelled so that weight must be kept to a minimum. The femur is a primary structural member that transmits large axial loads between joints that are virtually moment free. The two ends must be hardened against the contact pressures among bones in these two pin joints. The overall system contains a producing chemical plant with diffusion-controlled processes that require some porosity in our structural member.

To carry axial load efficiently the design uses unidirectional fibers called osteons, made from filaments that spiral around and down the centerline to form the fiber. Twisted yarn construction is very common in fiber-reinforced composites that must have some flexibility in addition to fiber direction stiffness and strength; tire cord is one example. At the two ends where bearing loads are high and constantly change direction, the design uses a quasi-isotropic construction called trabecular or spongy bone permeated with liquids. The basic material is the same; only the distribution and orientation of the short filaments here is different. At the external bearing surfaces, there is a thin layer of high-density bone similar to a case-hardened metal. When examined in structural coordinates, the mechanical properties are found to be anisotropic by virtue of the changing orientation of the orthotropic filaments. No algorithm is capable of material shape optimization at this level—only natural selection.

In this paper we begin to develop the geometric models necessary to represent material orientation and distribution in continuum composite structures—a first step in the evolution of material shape optimization for continuum composites. They are applied to a fiber-reinforced connecting rod to illustrate modeling of fiber orientation and distribution in a three-dimensional finite element model. New material geometry data exchange requirements are identified to facilitate communication among analysts, designers and fabricators of composite structures. These are necessary because the current Initial Graphics Exchange Specification (IGES) includes only point composite design data for essentially thin laminates. The paper concludes with the development of shape sensitivity derivatives for parametric cubic models and design sensitivity derivatives for global composite design variables. The mathematical tools are similar to those for external shape optimization, but developments in material shape optimization are really just beginning.

SOLID MODELING FOR CONTINUUM COMPOSITES

Geometric modeling with finite line, surface and volume models based on Hermite parametric cubic basis functions is referred to as analytic solid modeling [1]. This area has matured in recent years, and there are now textbooks [2] and major computer aided engineering software systems such as PATRAN that are based on the use of analytic solid modeling. Briefly, the representations of a line, a surface and a solid in canonical form are:

Parametric cubic line	$\mathbf{Z}(\xi) = S_i^\alpha F_i(\xi)\mathbf{e}_\alpha; \quad F_i(\xi) = \xi^{4-i}$
Bicubic surface patch	$\mathbf{Z}(\xi_1, \xi_2) = S_{ij}^\alpha F_i(\xi_1)F_j(\xi_2)\mathbf{e}_\alpha$
Tricubic solid or hyperpatch	$\mathbf{Z}(\xi_1, \xi_2, \xi_3) = S_{ijk}^\alpha F_i(\xi_1)F_j(\xi_2)F_k(\xi_3)\mathbf{e}_\alpha$

where $i, j, k = 1, 2, 3, 4; \alpha = 1, 2, 3$; and the summation convention for repeated indices is adopted. Greek subscripts and superscripts refer to coordinate directions.

The utility of Hermite parametric cubics for shape representation stems from their ability to ensure continuity between surfaces or solids, including tangent vector continuity. These properties hold uniformly over the entire edge or surface interface and not just at isolated points. However, the designer is able to control shape using tangent vectors only at the corners, with simple transformations available between the canonical or algebraic coefficients and the tangent vector control point data.

References pp. 378–379

$$S_i^\alpha = M_{ij} P_j^\alpha$$
$$S_{ij}^\alpha = M_{i\ell} M_{jm} P_{\ell m}^\alpha$$
$$S_{ijk}^\alpha = M_{i\ell} M_{jm} M_{km} P_{\ell mn}^\alpha$$

There are other representations of the same shape, including Bézier control points, that are especially useful for interactive graphics work. Each representation has its own shape-independent transformation matrix **M**. Recently similar parametric models have been used as master control points for shape optimization in the work of Fleury [3] and Wang, Sun and Gallagher [4]. Control points are also used with other primitives, as in Botkin and Bennett [5], but in the present system we use Hermite parametric cubics for shape definition and finite element analysis.

As in the case of the femur, continuum structural composites are often the synthesis of crystallites into filaments, fibers and fabrics or other higher-dimensional structures (see Figure 2). Here the designer needs to control the orientation as well as distribution of fibers on the interior of a solid object, and to do that he needs a systematic way of representing material geometry. At each material point, orientation of the material coordinate frame must be defined (Figure 3), and in general that requires three angles. Note the difference with thin composite shells for which one angle will define orientation relative to the local surface coordinate frame. Another basic difference is the sensitivity of critical stresses to material angle modeling in a finite element analysis [6]. In continuum composite analyses, the matrix-dominated stresses are often more important than the larger fiber-dominated stresses, because they determine delamination and other matrix-initiated failure modes. Processing failures are a common problem that occurs when designs are evaluated considering only the fiber-dominated stresses. These observations are introduced here to help explain why it is necessary to add the complexity of three-dimensional material geometry modeling to the structural design process for continuum composites.

Consider the design of a fiber-reinforced connecting rod made using a metal matrix material and DuPont FP ceramic fibers. In order to evaluate a metal matrix design, it is important to analyze all transverse stresses as well as the fiber stresses, and that analysis requires modeling fiber orientation in a three-dimensional finite element model, shown in Figure 4. Finite elements in the model have their shape defined by structural geometry and stress response so that elements are not aligned with local material directions (Figure 5). To define analytically the fiber directions in these elements, we use PATRAN vector field models that later can be evaluated over any set of finite elements (see Figure 6) to create up to three material angles at every analysis point. These models are defined by a small number of parameters that could be used as design variables for material shape optimization. At any space point **Z**, a vector field defines a vector **V** originating at **Z**,

$$\mathbf{V} = f_\alpha(\mathbf{Z})\mathbf{e}_\alpha; \quad f_\alpha(\mathbf{Z}) = C_{\alpha\beta} Z_\beta^{E_{\alpha\beta}} + C_{\alpha 0}$$

	Graphite crystals	Crystallites	Filament	Yarn	Graphite fabric
Constituents	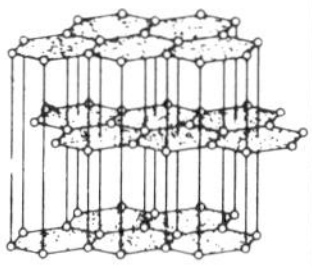	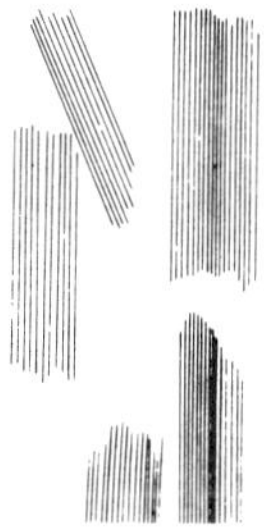			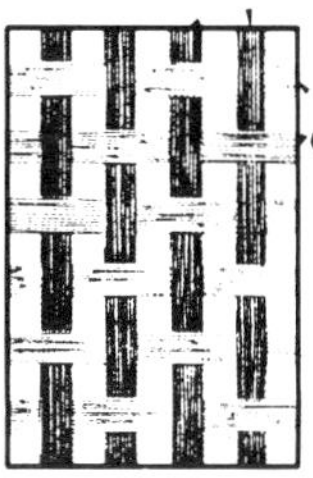 Warp ends Fill ends PLAIN WEAVE
Scale	0.000342μ	0.005μ	5μ	175μ	625μ
Modulus	154 MSI	~	80 MSI	33 MSI	16 MSI warp 5 MSI fill

Figure 2. Elements of a fibrous composite structure.

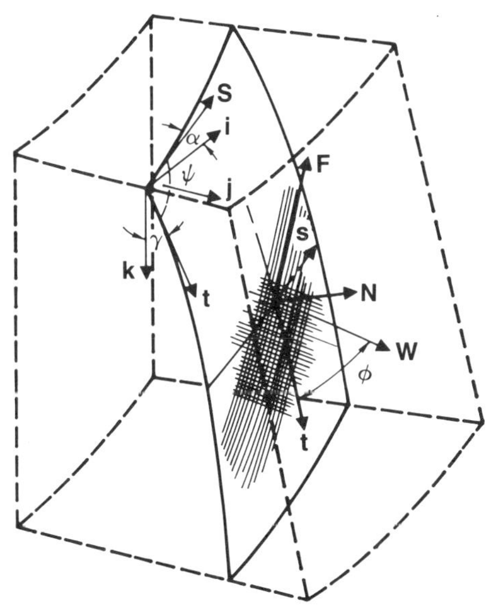

Structural coordinates

$$\mathbf{z} = \theta\mathbf{i} + r\mathbf{j} + z\mathbf{k}$$

Material coordinates

$$\mathbf{i}, \mathbf{j}, \mathbf{k} \xrightarrow{R\,(\alpha,\gamma,\phi)} \mathbf{W}, \mathbf{F}, \mathbf{N}$$

W = Warp direction
F = Fill direction
N = Normal direction

In general, three rotations (α, γ, ϕ) are required to define material orientation at every material point.

α = Arc angle
γ = Tilt angle
ϕ = Helix or bias angle

Figure 3. Material geometry modeling and definitions.

References pp. 378–379

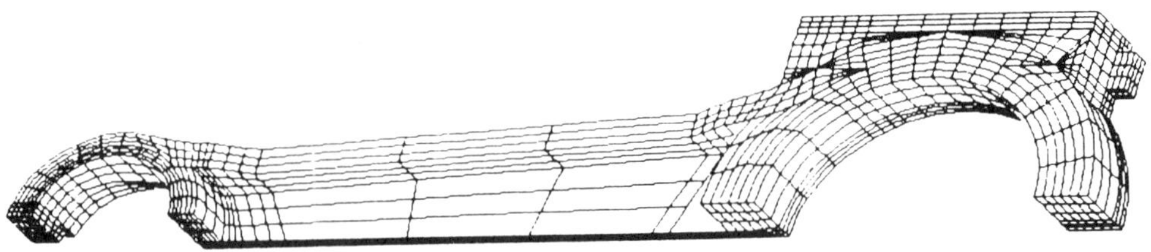

Figure 4. Connecting rod finite element model.

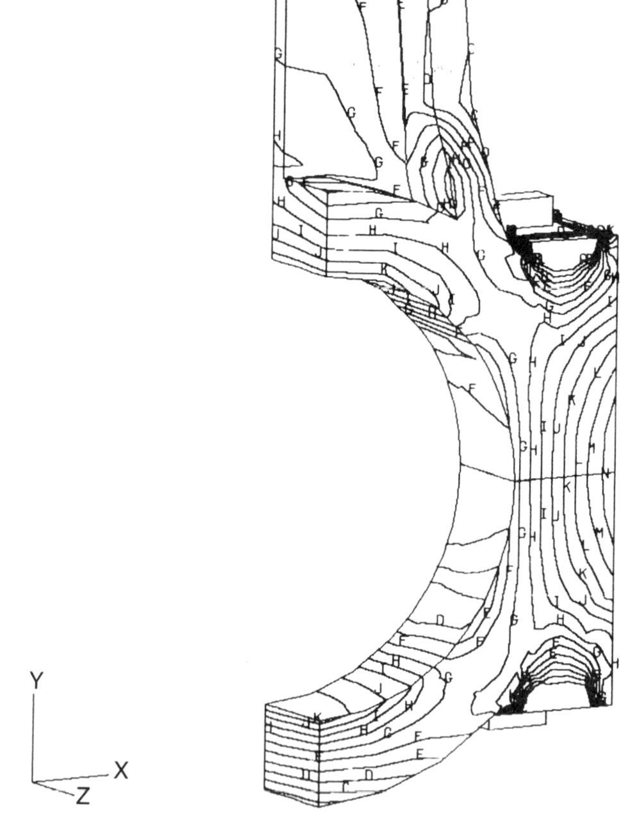

Figure 5. Connecting rod fiber stresses.

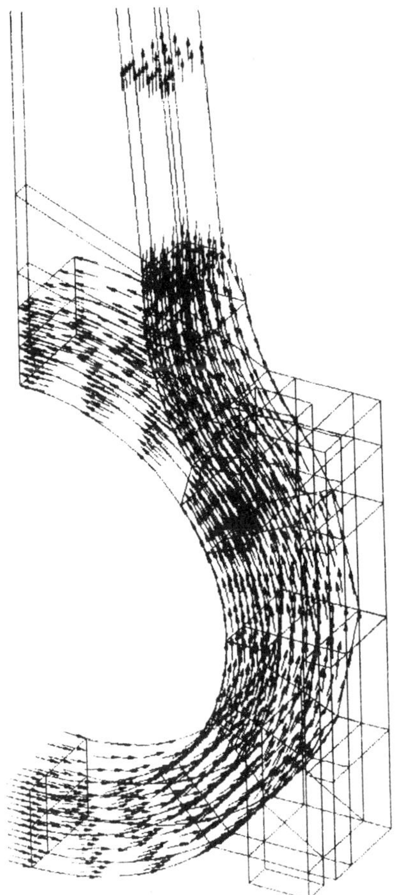

Figure 6. Connecting rod fiber distribution model.

and having components in PATRAN that are exponential functions of coordinates in the frame. This frame can be any rectangular, cylindrical or spherical coordinate frame, and several will be required to model material geometry for most components. The connecting rod shown in Figure 7 required four, and the definition of the coordinate frame itself could be used as material shape design parameters. In many cases the material angles can be defined directly using scalar fields

$$\phi = f(\mathbf{Z}); \quad f = C_\beta Z_\beta^{E_\beta} + C_0$$

that define the angle at a space point $\mathbf{Z}$ in terms of the coordinates of the point in a convenient local coordinate frame. To illustrate this point, consider the arc angle at a point in a thick cylinder laminated using involute construction [7]. Assuming small angle approximations for the sine function,

$$f(\mathbf{Z}) = -57.3(CR^{-1} + \theta)$$

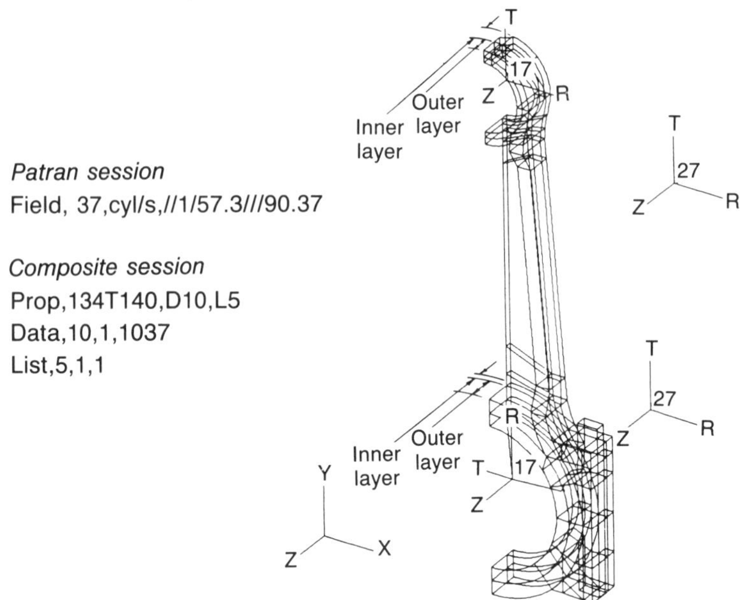

Figure 7. Modeling material geometry with vector fields.

where R and θ are coordinates of $\mathbf{Z}$ in a cylindrical frame

$$\mathbf{Z} = R\mathbf{e}_R + \theta\mathbf{e}_\theta + Z\mathbf{e}_Z.$$

Here the material angle is independent of the axial coordinate and the design variable C is directly related to the number and thickness of the plies used to manufacture the cylinder using involute construction:

$$C = Nt/2\pi.$$

The definition of material geometry in terms of global composite design variables is a critical step in moving from simple point designs typical of optimization today to component material geometry shape optimization.

Experiences at PDA Engineering in the design, analysis and manufacture of continuum composites over the last decade indicate there is a serious communication gap among the line engineering organizations required to produce a composite structure. Basic information necessary to effectively control processing risks and service life are not in the exchange system used for product definition by most CAD/CAM programs. IGES Version 2.0 includes basic finite element data: the current release, Version 3.0, includes point material data for laminates but no material geometry data [8]. The manufacture of a composite requires controlling not only the external shape but also the internal shape. Tape winding, braiding, weaving, molding and a variety of other constructions produce this internal geometry, and the designer analyst needs to understand these processes at least well enough to include them in a comprehensive product definition. Then it becomes possible to optimize material geometry including processing risks and realistic manufacturing constraints. At the very least, the next generation computer-aided engineering exchange system, Product Design Exchange Specification (PDES), should include parameters that describe the internal distribution and orientation of fibers in a continuum composite.

SENSITIVITY ANALYSIS FOR ANALYTIC SOLID MODELS

Sensitivity derivatives are a key component of shape optimization that are useful for geometric design as well as structural design. In this section we develop these derivatives for basic geometric parameters and relate them to classic problems such as isoparametric area minimization, the Brachistochrone problem, and to a problem suggested by Casale [9] that illustrates the use of parametric variables in geometric design. Consider first the sensitivity of the length of a line to changes in parametric cubic coefficients defining the line.

$$L = \int_0^1 \sqrt{E} d\xi; \quad E = \mathbf{Z}_{,\xi} \mathbf{Z}_{,\xi}$$

$$\partial L / \partial S_i^\alpha = \int_0^1 S_k^\alpha \dot{F}_k \dot{F}_i / \sqrt{E}\, d\xi$$

where

$$\dot{F}_i = \partial(\xi^{4-i}) / \partial\xi$$

and the comma notation indicates partial derivatives of a vector's coordinate functions. If we define the integrals in this expression symbolically

$$E_k^i = (4-k)(4-i) \int_0^1 \xi^{6-k-i} / \sqrt{E}\, d\xi,$$

then the sensitivity derivatives are defined by the quasi-linear transformation

$$\partial L / \partial S_i^\alpha = S_k^\alpha E_k^i.$$

The transformation coefficients are not independent of shape but they could be considered constant for all line coefficients in a small region sufficiently close to those used to compute the integrals.

This explicit formulation is used when changing control points for a given line parameterization. One of the reasons parametric models are so useful in shape optimization is that even when all four control points are fixed, an infinite number of parametric cubic lines still can be found that pass through the points by simply changing parametrization. Casale analyzed a line minimization problem of the type in which the four space points are fixed and the parametric coordinates of the two interior points are varied to find the minimum length line. His result was obtained

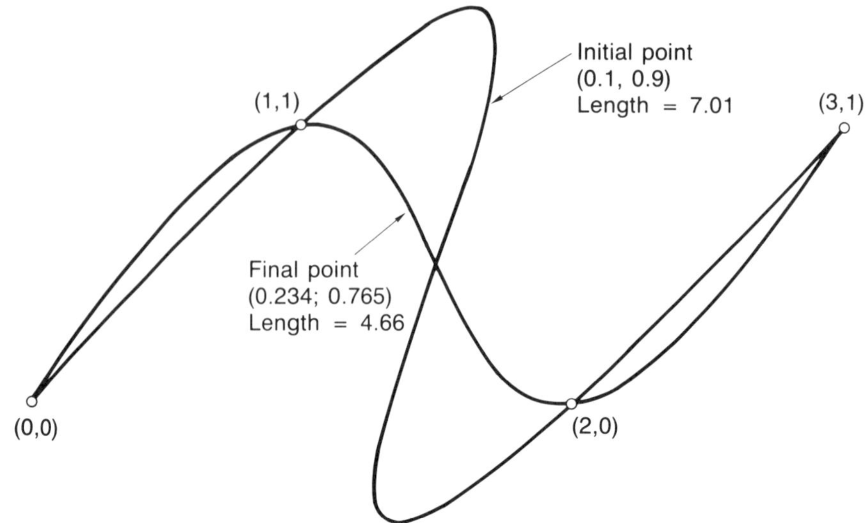

Figure 8. Length minimization in parametric coordinates.

Table 1
Parametric line minimization results

Length	Coordinate 1	Coordinate 2	Gradient 1	Gradient 2	Cycle/Method	
5.68370	0.3	0.6	6.773	−8.929	0	
4.66303	0.2300	0.7640	−0.203	−0.039	113	F.D.
4.66239	0.2346	0.7654	−0.006	0.007	70	Analytic
7.01937	0.1	0.9	−23.221	23.221		
4.66239	0.2344	0.7654	−.014	.006	29	F.D.
4.66239	0.2349	0.7651	.009	−.009	11	Analytic

using Vanderplaats' ADS code with finite difference derivatives (see Figure 8). After deriving analytic sensitivity derivatives, ADS was used again and a slightly better result was obtained in far fewer iterations (Table 1). These derivatives were obtained implicitly by expressing the variation in parametric coordinates in terms of space coordinates.

$$\begin{Bmatrix} P(o) \\ P(u) \\ P(v) \\ P(1) \end{Bmatrix}^{\alpha} = \begin{bmatrix} o & o & o & 1 \\ u^3 & u^2 & u & 1 \\ v^3 & v^2 & v & 1 \\ 1 & 1 & 1 & 1 \end{bmatrix} \begin{Bmatrix} S_1 \\ S_2 \\ S_3 \\ S_4 \end{Bmatrix}^{\alpha}$$

Differentiating

$$\begin{Bmatrix} o \\ o \\ o \end{Bmatrix} = \begin{bmatrix} u^3 & u^2 & 1 \\ v^3 & v^2 & 1 \\ 1 & 1 & 1 \end{bmatrix} \begin{Bmatrix} ds_1 \\ ds_2 \\ ds_3 \end{Bmatrix}^\alpha + \begin{bmatrix} \bar{u}^\alpha & o \\ o & \bar{v}^\alpha \\ o & o \end{bmatrix} \begin{Bmatrix} du \\ dv \end{Bmatrix}$$

where

$$\bar{u}^\alpha = 3u^2 S_1^\alpha + 2u S_2^\alpha + S_3^\alpha$$
$$\bar{v}^\alpha = 3v^2 S_1^\alpha + 2v S_2^\alpha + S_3^\alpha,$$

then

$$dL = (\partial L / \partial S_i^\alpha) dS_i^\alpha$$

gives

$$dL = -(du, dv) \begin{bmatrix} \bar{u}^\alpha & o & o \\ o & \bar{v}^\alpha & o \end{bmatrix} \begin{bmatrix} u^3 & u^2 & 1 \\ v^3 & v^2 & 1 \\ 1 & 1 & 1 \end{bmatrix}^{-1} \begin{Bmatrix} \partial L/\partial S_1 \\ \partial L/\partial S_2 \\ \partial L/\partial S_3 \end{Bmatrix}^\alpha$$

which implicitly defines the sensitivity of L to the two parametric coordinates

$$dL = (\partial L/\partial u) du + (\partial L/\partial v) dv.$$

In general, of course, the interior points are not known and must be kept in most minimization problems. The minimum surface area of revolution between two points, a catenary, can be closely approximated with a parametric cubic line. If the Y axis is the axis of rotation, then find the line that minimizes

$$A = 2\pi \int_0^1 x\sqrt{E}\, d\xi$$

where the line is approximated by a parametric cubic with the control point sensitivity derivatives

$$\partial A/\partial S_i^\alpha = 2\pi \int_0^1 [(x\sqrt{E}) S_k^\alpha \dot{F}_k \dot{F}_i + \sqrt{E}\delta_i^\alpha F_i]\, d\xi.$$

Similarly the minimum fall time between two points, the Brachistochrone problem, requires finding the line that minimizes

$$T = 1/\sqrt{2g} \int_0^1 \sqrt{E/y}\, d\xi$$

and here the sensitivity derivatives are

$$\partial T/\partial S_i^\alpha = 1/\sqrt{2g} \int_0^1 [(1/\sqrt{Ey}) S_k^\alpha \dot{F}_k \dot{F}_i - (1/2y)\sqrt{E/y}\, \delta_2^\alpha F_i]\, d\xi.$$

References pp. 378–379

A number of classic shape optimization problems have been solved using the shapes representable by parametric cubics as a way of checking our software. Consider next the area of a general surface patch

$$A = \int_0^1 \int_0^1 \sqrt{EG - F^2}\, d\xi_1\, d\xi_2$$

where

$$\begin{aligned} E &= \mathbf{Z}_{,\xi_1} \mathbf{Z}_{,\xi_1} \\ G &= \mathbf{Z}_{,\xi_2} \mathbf{Z}_{,\xi_2} \\ F &= \mathbf{Z}_{,\xi_1} \mathbf{Z}_{,\xi_2}\,; \end{aligned}$$

then

$$\begin{aligned} \partial A/\partial S_{ij}^{\alpha} = \int_0^1 \int_0^1 \frac{1}{\sqrt{EG - F^2}} \Big\{ & E S_{k\ell}^{\alpha} F_k(\xi_1) \dot{F}_\ell(\xi_2) F_i(\xi_1) \dot{F}_j(\xi_2) \\ & + G S_{k\ell}^{\alpha} \dot{F}_k(\xi_1) F_\ell(\xi_2) \dot{F}_i(\xi_1) F_j(\xi_2) \\ & - 2F S_{k\ell}^{\alpha} F_k(\xi_1) \dot{F}_\ell(\xi_2) \dot{F}_i(\xi_1) F_j(\xi_2) \Big\} d\xi_1 d\xi_2. \end{aligned}$$

Again defining the integrals

$$\begin{aligned} E_{ij}^{k\ell} &= (4-\ell)(4-j) \int_0^1 \int_0^1 (E/\sqrt{EG - F^2})\, \xi_1^{8-i-k} \xi_2^{6-\ell-j}\, d\xi_1 d\xi_2 \\ G_{ij}^{k\ell} &= (4-i)(4-k) \int_0^1 \int_0^1 (G/\sqrt{EG - F^2})\, \xi_1^{6-i-k} \xi_2^{8-\ell-j}\, d\xi_1\, d\xi_2 \\ F_{ij}^{k\ell} &= (4-\ell)(4-i) \int_0^1 \int_0^1 (F/\sqrt{EG - F^2})\, \xi_1^{7-i-k} \xi_2^{7-\ell-j}\, d\xi_1\, d\xi_2 \end{aligned}$$

the sensitivity derivatives can be reduced to the form of a quasi-linear transformation

$$\partial A/\partial S_{ij}^{\alpha} = S_{k\ell}^{\alpha} \left[E_{k\ell}^{ij} + G_{k\ell}^{ij} - 2F_{k\ell}^{ij} \right]$$

which is a convenient form for computational work. The expression for volume sensitivity derivatives are too cumbersome to include here because they have not been reduced to transformation form.

The shape analyses presented to this point have been for structural geometry or external shape definition. In the manufacture of many composites there are billet geometry constraints associated with cure cycle deformations so that external shape cannot be optimized for service stress conditions exclusively. To analyze and control these risks requires that we model the sensitivity of the structural response to material geometry as well as external geometry. The state of the art in this area

PROP, ID, INVSL, D10, L5

DATA, 10, DTID, NPLY, TPLY, PHI_0

DTPC, DTID, OPTION, Z1, R1, Z2, R2, ..., ZN, RN

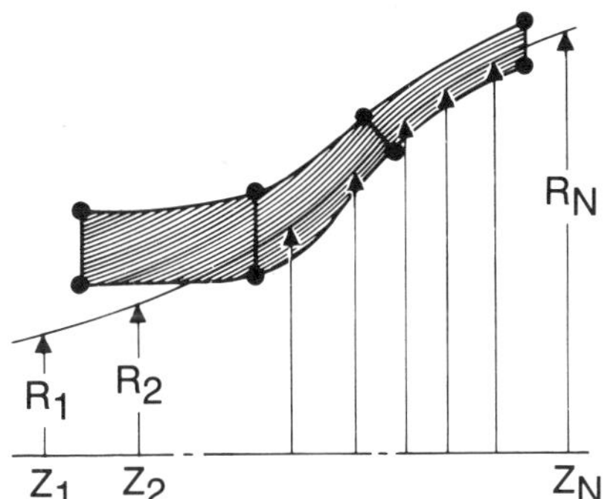

$(R, Z)_1$ = Startline coordinates
NPLY = Number of plies
TPLY = Ply thickness
PHI_0 = Ply orientation on broadgoods

Figure 9. Global composite design variables.

is in its infancy, just beginning to move from point material geometry models to billet or component models. To indicate the basic ideas involved, consider the case of involute composites studied extensively by Pagano [7] and others over the last few years. Here the material geometry and structural geometry are both defined by a ply pattern that is in turn defined by a single line in space called a start line (Figure 9). A complete sequence of relations has been developed from basic designer data to global material geometry models that can be evaluated at any space point to create a finite element model of any detail. Using these relations we can develop analytic sensitivity derivatives beginning with the designers start line $R(Z)$

$$C^2 = (1 + R_{,z}^2)/[(2\pi/Nt)^2 + (R_{,z}/R)^2]$$

and the number of plies N and their thickness t. The material angles in Figure 3 are defined analytically by Pagano in terms of these design variables, and structural orientation is defined by strain transformations using direction cosine products. The change in local stiffness at any material point in terms of a global design variable change can be evaluated using the chain rule

$$dK = \frac{\partial K}{\partial \theta}\frac{\partial \theta}{\partial c}\frac{\partial c}{\partial t}\,dt$$

where in this case the sequence begins with the sensitivity of the involute constant to a change in ply thickness:

$$\frac{\partial c}{\partial t} = \frac{2(1 + m^2)4\pi^2/N^2t^3}{[(2\pi/Nt)^2 + (m/R)^2]^2}.$$

Note that this derivative depends on the location where $m = R_{,z}$ is the slope of the start line at an axial station. A more interesting design change which changes

References pp. 378–379

ply pattern shape very substantially is one in which the slope of the start line m is varied. The point to be made here is that material geometry sensitivity derivatives are defined for this construction over the entire composite structure in terms of a very small number of design parameters. Also these are the parameters used to define tooling for fabrication of the composite. The sensitivity of the overall structural response to these changes can be determined using the finite element analyses as described by Wang, Sun and Gallagher [4] and several others. The only difference is that stiffness matrix changes are now defined for all finite elements in a region from global material geometry models for the composite construction used to manufacture the component in that region.

CONCLUSIONS

Shape optimization for structures fabricated from composite materials introduces material or internal geometry to the set of design variables that must be considered. This adds three angles to the design space at each material point for continuum composites which must be included in any comprehensive definition of a component. Material geometry models based on vector fields offer one efficient method of representing these data independent of any particular analysis model. Experiences to date with a fiber-reinforced connecting rod and several laminate constructions have been positive, and they represent a first step in the evolution of material shape optimization for continuum composites. Here the designer analyst needs to know the distribution and orientation of reinforcing materials in a mechanical component to evaluate processing constraints as well as service stress constraints and mechanical performance. This same information is required after manufacture to inspect composite structures for internal geometry quality.

Another important ingredient in composite shape optimization is the availability of sensitivity derivatives for length, area, volume and internal material geometery. Cure cycle reactions are diffusion-controlled processes that are sensitive to surface area, to volume ratios and to other external shape parameters. Parametric cubic models for lines, surfaces and solids lend themselves to such analyses, and sensitivity derivatives have been obtained for control point variations and for parametric coordinate variations. The limited applications to date have significantly improved convergence of ADS algorithms over finite difference approximations. Also, analytic sensitivity derivatives for global composite design variables were introduced for involute composites that will form the basis for design sensitivity analyses using finite element models. Internal or material geometry shape optimization is in its infancy, and its future development will require similar models for many other constructions.

REFERENCES

1. M. S. Casale and E. L. Stanton, An overview of analytic solid modeling. *IEEE Comp. Graph. Appl.* **5** (2), 45–56 (1985).

2. M. E. Mortenson, *Geometric Modeling.* John Wiley & Sons (1985).

3. C. Fleury and V. Braibant, Shape optimal design: A performing CAD formulation. *Proc. AIAA/ASME/ASCE/AHS 25th SDM Conf.*, pp. 50–57 (1984).
4. S. Wang, Y. Sun and R. H. Gallagher, Sensitivity analysis in shape optimization of continuum structures. *Comput. Struct.* **20** (5), 855–867 (1985).
5. M. E. Botkin and J. A. Bennett, Shape optimization for three-dimensional folded plate structures. *Proc. AIAA/ASME/ASCE/AHS 25th SDM Conf.* pp. 41–49 (1984).
6. E. L. Stanton, Material modelling for accurate stresses in involute composites. *Int. J. Numer. Meth. Eng.* **21**, 1925–1934 (1985).
7. N. J. Pagano, General relations for exact and inexact involute bodies of revolution, pp. 129–137 in *Advances in Aerospace Structures, Materials and Dynamics, AD-03* (Edited by R. M. Laurenson and U. Yuceoglu). ASME (1982).
8. R. L. Ivey, IGES and the exchange of finite element modeling data. *Proc. ASME Pressure Vessels and Piping Conf.* PVP Vol. 98, no. 5, pp. 31–35 (1985).
9. M. S. Casale, Numerical methods for shape optimization. Unpublished notes (1985).

DISCUSSION

Z. Mróz *(Polish Academy of Science)*

On this composite surface you have three parameters: one normal to the surface and two angles within the surface, with two fiber families in the surface. It is not clear how you formulate these in your program. What was the mechanical model of the fiber and the matrix?

Stanton

We used a three-dimensional finite element analysis based on experimental data. For material characterization, we laminated many parts. Some parts were made up in blocks 3 x 3 x 6 inches so that we could characterize interlaminar properties. All of the materials we worked with were different in tension and compression. All of the mechanical constants—the three Poisson ratios, the three G's, the three E's—and all six stress-strain diagrams were obtained.

Mróz

Were they brittle or ductile materials?

Stanton

They were quasi-ductile. They do have some brittle matrix failure modes in them; but because the matrix phase is relatively tough, they tend to be (depending on how you view ductility) fairly ductile for this class of material in which the matrix is carbon.

Mróz

Do you design against fiber breaking or against matrix breaking or against the fibers pulling out from the matrix?

Stanton

Fiber pullout in a woven material is relatively unlikely. In some of the three-dimensional pre-forms, fiber pullout can cause failure because of shear lag. In woven two-dimensional pre-forms, we use one criterion for fiber failure and another criterion to indicate interlaminar failure. For the interaction between tension and interlaminar shear, we follow more or less the kinds of models that Hahn has suggested. We try to do as much experimental work as we can, because this is one area where it is risky to make up property numbers for analysis. We spend about a million dollars a year on testing, just for material characterization.

E. L. Stanton

W. J. Stroud *(NASA Langley Research Center)*

Have you done any work on tailoring composite structures for prosthetic devices? Could you comment on that?

Stanton

Prosthetic devices designed for total hip replacement is one interesting area. Hip prostheses have been around for a while, and replacement is one of the more common procedures done in this country and in Europe. Historically, most of the devices have been made of steel, which doesn't tend to work too well over a long period of time in the body. You have to bond the steel to the bone with a fairly reactive material, polymethylmethacrylate, which is difficult for the surgeon to work with. It also tends to loosen up over time, so you may have to go back in and repair the joint.

There are people working with composite materials, both ceramics and carbon composites, who are trying to tailor both the stiffness and to some extent the open porosity of the material. Then when a replacement part goes into the top end of the femur that nests into the acetabulum, you promote some circulation between the femur and implant and get some knitting with the device itself. This results in a better bond. In so many of the composite applications, the key issue is the fiber-matrix adhesion—the ability of things to stay together over a long period.

J. A. Bennett *(General Motors Research Laboratories)*

Some speakers earlier in the symposium have started from a shape optimization problem and almost resolved that into a discrete fiber solution. You started from the other direction, from the discrete fiber approach. Is there any thought that discrete fibers might be a more basic variable to work with than a parametric description of the part?

Stanton

I suspect it might be more suitable for the class of materials that are orientable and have microstructure. On the other hand, for materials that are truly homogenous and isotropic, working with the external boundary type of information is probably the correct place to start work for design. You may be right if we are trying to design a system using fibrous composite materials where material geometry and structural geometry are coupled, but we are obviously incompetent to come up with a design such as the femur. However, even in the optimum design there must have been some trade-offs in terms of the shape of the part and the material orientation. This may be a "chicken and the egg" situation in which the structural design needs mechanical properties and the material design needs a structural shape.

R. Kohn *(Courant Institute of Mathematics)*

The structure of the femur may have more to do with optimal redesign than with optimal design. In fact, much of the work that I know in the field of bone behavior is oriented toward bone growth and self-repair. Perhaps this occurs so as to redesign the femur optimally for some criterion. Do you know of any work in this direction?

Stanton

Wolff's law is often referred to by people in bone mechanics. Nature tends to order things to avoid bending. If you put bending into bone, the bone will redistribute to reduce bending. The way in which it does this is described as Wolff's law. In terms of redesign, you might try to match porosity. For example, jaw bone replacements tend to use a porous graphite. Sometimes you might even dope the porosity with a polymeric material that body chemistry likes to consume. It will absorb some of this polymeric material and the carbon implant will knit with the bone in some cases when everything works out right. Promoting circulation between the replacement part and the natural repairing material is also one of the things biomechanics engineers like to consider.

W. J. Anderson *(The University of Michigan)*

On this topic, I am acquainted with a young man at the University of Michigan who is doing the redesign problem. He is performing in vivo experiments with dog femurs, as I understand—putting them under compression and then checking out the restrengthening of the bone. What he is doing is interesting because it is like an identification problem: he is trying to figure out how nature reinforces the bone when you fracture it. There are a number of people at the University of Michigan working in biomedical engineering.

Stanton

There is a tremendous amount of interesting work going on. Unfortunately, as a structural designer you do not have an active mechanism inside your design to repair things when it doesn't work too well. Composites can be relatively unforgiving compared to a metal; so if you have made a mistake, you are more at risk in this kind of design than you are using some of the more ductile and isotropic kinds of materials.

J. Taylor *(The University of Michigan)*

One other point in response to Bob Kohn's question. I think one of the better medical treatments for the degradation of bone that comes with aging is to recommend that people keep exercising. In particular, older people need to do exercises that put the bone under a heavy load. The idea is to inspire the natural mechanism to reinforce the bone by redepositing material in areas that have become weakened.

H. A. Kamel *(University of Arizona)*

One of the negative aspects of natural reinforcement that I came across is that when a bone cracks (usually these are minute cracks), it may be stiffened to the point at which it becomes too stiff. The cartilage in the knee, for example, will then absorb more energy than it used to, or it absorbs too much energy in relation to the bone, causing its deterioration. This is the basis for one theory of the development of arthritis in joints.

Stanton

Bone damage could lead to damage in the ligaments.

R. L. Benedict *(Goodyear Tire and Rubber Company)*

I recall that there is fortunately a bound for loads. If the load is too low, the bone dissolves away; and if the load is too high, it creeps away. This could be an extremely difficult design problem because you have to get something that has bounds on both sides.

Stanton

Fortunately, that is not one of the problems I have to deal with every day. We have tried to work with some people on repairing designs, and it is indeed a tremendously challenging problem.

SYMPOSIUM SUMMARY AND CONCLUDING REMARKS

L. A. SCHMIT
University of California–Los Angeles
Los Angeles, California

Because of my very keen interest in design optimization generally, and structural optimization in particular, it gives me special pleasure to see what used to be a hardy little band of optimists growing into a real force such as we see represented here. I want to commend General Motors for organizing this outstanding symposium. I have been impressed by the vigor, candor and constructive character of the discussions evoked by the papers. While there is really very little that I can add to what has already been said here, I am going to share with you what I see as the lessons learned and issues raised by this symposium.

By and large the thoughts and observations that follow have been organized around the things that one integrates when seeking to put together an automated or semiautomated design procedure, namely, (1) design modeling, (2) analysis modeling, (3) behavior sensitivity analysis, (4) optimization and (5) creative control via interactive graphics.

DESIGN MODELING

When posing a design problem the first thing one is concerned with is design modeling, and we have heard a great deal about this subject during the symposium. One of the most important lessons learned here is that design modeling should generally be placed above analysis modeling in the formulation decision hierarchy. Particularly in the context of shape optimization, one wants to work with a small number of "natural design variables" in defining the design optimization task. Selection of these design variables should be guided by the way designers think about describing a part for manufacture. This will generally lead to the selection of a relatively small number of powerful design variables, well matched to product definition from a fabrication and manufacturing point of view. Posing design optimization problems in terms of natural design variables generally leads to a well-posed optimization problem. This is in contrast to the difficulties that have been encountered in using nodal coordinates as design variables after having chosen a finite element analysis model.

Several papers presented during the symposium clearly illustrate the power and importance of posing shape optimization in terms of natural design variables describing both the external and internal geometry of a part or component. In the chapter by Botkin et al., Figures 2 and 3 (pp. 237 and 238) use key nodes and boundary elements to define the shape of constant thickness portions of a part. Then the use of natural design variables is extended to nonplanar stamped parts where each segment has one completely closed exterior boundary, and provision is made for describing one or more interior cutouts. Segments are then joined along straight edges using rotation angles α_i (see Figure 7 in Botkin et al., p. 242) to describe the configuration of the entire part. Another illustration of natural design variables for shape optimization will be found in Figure 8 of Fleury's chapter (p. 316) where control nodes and design elements are used to form a shape design model. In this instance, the design variables are selected as the locations of moving control nodes along directions that are defined in advance by using fixed control nodes.

The idea of using natural design variables can be extended to internal geometry when designing with fiber composite materials. Stanton's Figure 3 (p. 369) illustrates the use of vector fields to define material orientation at every material point within a three-dimensional component. As Stanton shows in Chapter 15, it is possible to use a small number of global design variables for internal geometry optimization. Each of the foregoing illustrations of posing shape optimization problems in terms of natural design variables is independent of the analysis model which is subsequently adopted for predicting structural behavior. This is a key difference which distinguishes current state-of-the art shape optimization capabilities from early work on configuration optimization which often used finite element nodal coordinates as design variables.

ANALYSIS MODELING

Analysis modeling has always been an important part of the overall design optimization task because one has to be able to predict structural behavior for various trial designs in order to guide the design improvement process. It is worth noting that much of the early work in structural optimization focused on sizing of trusses where the design modeling is straightforward and does not need to be updated. Shape optimization of structural components involving two- and three-dimensional states of stress has made it necessary to carefully reexamine analysis modeling issues. Two points clearly made by this symposium are that for shape optimization (1) analysis modeling should take place after selection of the design model; and (2) consideration needs to be given to periodic updating of the analysis model.

Having selected a design model, attention turns to analysis modeling with the objective of predicting structural behavior with accuracy that is adequate to guide the design improvement process. In the context of the widely used displacement type finite element methods, this involves selecting finite element types and mesh generation. Given a design and a characteristic length, mesh generation techniques (such as those described in Chapter 9 by Botkin et al.) can be used to produce an analysis model automatically (see Figure 4, p. 240). This figure graphically illustrates (a) initial uniform mesh generation, (b) use of strain energy density contours to identify regions requiring mesh refinement and (c) a refined mesh based on first pass analysis results. This type of adaptive mesh generation adds displacement degrees of freedom to the initial analysis model, and it is usually used with finite elements that have low interpolation order.

The stress amplification factor idea (see Figure 5 in Botkin et al., p. 241) should also be noted because it appears to be an effective way of obtaining conservative estimates of critical stresses, based on an initial analysis and at least one refined mesh analysis. The judicious use of stress amplification factors can significantly reduce computational time for analysis by frequently making it possible to bypass the high number of degrees of freedom required by refined mesh analyses during the design optimization process. When higher-order finite elements (e.g., isoparametric elements with 16 degrees of freedom) are used to generate the analysis model (see Fleury, Figure 7(c), p. 242), it may not be necessary to refine the initial mesh in Figure 4(c) (p. 240 in Botkin et al.) by adding additional degrees of freedom. However, as the optimization process changes the shape, it is still necessary to update the mesh relocating the position of the nodes (see Fleury's Figure 7(d), p. 315) so that the analysis is well behaved.

Within the context of finite element analysis modeling for shape optimization, very little attention has been focused on problems involving multiple load conditions. Yet when multiple load conditions are combined with the idea of adaptive mesh refinement (see Figure 4 in Botkin et al., p. 240), the result is a different finite element analysis model for each load condition. One way of coping with this situation is to use an envelope approach in which the refinements that have occurred for each load condition are all combined. Adopting the envelope approach can easily lead to a refined analysis model with a large number of degrees of freedom. However, this difficulty can probably be alleviated by using iterative methods for the refined mesh analysis, so that interpolation from the cruder mesh displacement solution can be used as an initial estimate of the refined mesh solution.

While shape optimization based on finite element analysis has been successfully pursued, these efforts have required that careful attention be given to dynamic updating and/or refinement of the analysis model. The boundary element method of analysis is attractive in the context of shape optimization because it inherently eliminates the need for interior mesh refinement. As indicated by Mota Soares and Choi (Chapter 8), the boundary element method of analysis is well matched to shape optimization of shafts as well as some two- and three-dimensional elasticity problems involving only static constraints (i.e., no buckling or dynamic response constraints). While the boundary element method of analysis is less versatile than the finite element method, it may well emerge as the analysis tool of choice for a limited class of shape optimization problems.

Before leaving the subject of analysis modeling, it should be observed that during this symposium the accuracy of predicted static stresses on the boundary and in the interior of structural components has come up frequently, particularly in the context of shape optimization based on finite element analysis (displacement methods). This observation leads to the suggestion that if a great deal of attention is to be focused on the design of parts that are primarily stress critical under static load conditions, then a reexamination of force method finite element analysis may well be appropriate. After all, the force method is known to satisfy equilibrium requirements exactly, and prior experience has shown that it predicts stresses with good accuracy. It may even be useful to give some consideration to the so-called mixed method of finite element analysis based on the Hellinger-Reissner variational principle. In this approach compatibility is satisfied exactly while equilibrium and stress displacement relations are satisfied approximately, leading to more nearly uniform error in stress and displacement prediction. It is interesting that in Zienkiewicz et al. (Chapter 1), the idea of using the mixed method formulation as a basis for error estimation was set forth. Finally, it should be noted that predicting stresses and strains more accurately, both on the boundary and in the interior, can be of crucial importance in the context of variational design sensitivity analysis (VDSA) (See Haug and Choi, Chapter 2, and Yang and Botkin, Chapter 3).

SENSITIVITY ANALYSIS

At the outset it may be useful to clear up a semantic point having to do with the extensive use of the term *design sensitivity analysis* during this symposium. This term has frequently been used to refer to methods for determining the rates of change of response quantities (e.g., static displacements and static stresses) with respect to changes in individual design variables, all other design variables held constant. It is suggested here that the term *behavior sensitivity analysis* be used to refer to rates of change of response quantities with respect to individual design variables, thus reserving the term *optimum design sensitivity analysis* to refer to methods for determining the rates of change of optimum design variable values with respect to changes in design problem parameters (e.g., applied loads, allowable stresses, etc.). The important concept of optimum design sensitivity analysis was first brought to the attention of the structural optimization community by [1]. The distinction between *behavior sensitivity analysis* and *optimum design sensitivity analysis* should be well understood. To that end, Table 1 drawn from reference [2] is included here because it offers a helpful comparison of the two concepts. While the subject of optimum design sensitivity analysis has not been addressed during this symposium, it is a next logical step. After all, as soon as one learns how to do reliable analysis, people start to ask how the behavior response will change if the design is modified. Similarly, as soon as one learns how to routinely generate good near-optimum designs, people will start to ask how the optimum design will change if the loading, allowable stress or some other problem parameter is modified. It is suggested that the application of optimum design sensitivity analysis to shape optimization problems warrants further study.

Table 1

Sensitivity analysis analogy

Behavioral response sensitivity	Optimum design sensitivity
Rates of change of response quantitites with respect to design variables $\left[\frac{\partial \mathbf{u}}{\partial x_i}, \frac{\partial(\omega)^2}{\partial x_i}, \text{etc.}\right]$	Rates of change of optimal primal and dual variables with respect to problem parameters $\left[\frac{\partial \boldsymbol{\alpha}^*}{\partial p_k}, \frac{\partial \boldsymbol{\lambda}^*}{\partial p_k}\right]$
Implicit differentiation of pertinent analysis equations $\frac{\partial}{\partial x_i}\left\{\begin{array}{c}\mathbf{Ku} = \mathbf{P} \\ \mathbf{Kv} = \omega^2 M\mathbf{v}\end{array}\right\}$	Implicit differentiation of necessary conditions at optimum $\frac{\partial}{\partial p_k}\left\{\begin{array}{l}\frac{\partial M}{\partial x_i} - \sum_{q\in Q_{cr}} \lambda_q \frac{\partial g_q}{\partial x_i} = 0 \quad ; \quad i \in \tilde{I} \\ g_q = 0 \quad ; \quad q \in Q_{cr}\end{array}\right\}$
Interactive design modification	Trade-off studies
Selection of intermediate design variables	Selection of intermediate problem parameters
Construct explicit approximations for response quantitites	Construct explicit approximations for optimum design variable values
Bypass reanalysis	Bypass reoptimization

Turning attention back to behavior sensitivity analysis, it should be noted that this kind of first derivative information is essential to the optimization process. These derivatives are used to determine a search direction in the direct application of nonlinear programming algorithms, while in the approximation concepts approach they are used to construct explicit approximations for the critical and potentially critical behavior constraints. During the symposium there has been a great deal of discussion about the accuracy of various behavior sensitivity analysis methods. How accurate does behavior sensitivity information have to be? It is suggested here that this remains an open question which deserves further study. It may well be possible to use relatively crude estimates of behavior sensitivity derivatives early in the design optimization process, refining them gradually as the design is improved. The idea of gradual refinement could also be applied to the basic analysis.

The question of accuracy aside, this symposium has focused a great deal of

References p. 397

attention on alternative approaches to behavior sensitivity analysis (see chapters by Haug and Choi, Yang and Botkin and by Choi and Seong). There are two basic approaches, namely, direct sensitivity analysis (DSA) and variational design sensitivity analysis (VDSA). The DSA approach is carried out via implicit differentiation of the discretized analysis equations (e.g., $\mathbf{K}\bar{\mathbf{u}} = \bar{\mathbf{P}}$) with respect to natural design variables controlling shape. On the other hand, the VDSA approach involves implicit differentiation of the functionals of the variational formulation of the appropriate continuum mechanics analysis problem. The essential difference between DSA and VDSA is that in the direct approach you take the derivative with respect to the design variable after you have discretized the mechanics problem, while in the variational approach you take the derivative of the functionals of the variational statement before discretization of the mechanics problem. The central question has to do with whether or not these two operations are commutative. There is some evidence that analysis discretization and implicit differentiation with respect to shape design variables may be commutative, as indicated in Yang and Botkin (Chapter 3), at least for problems involving shape design changes on traction-free boundaries.

Direct behavior sensitivity analysis in the context of the commonly used displacement type finite element analysis method involves the solution of the following well-known matrix equation, for $\partial\bar{\mathbf{u}}/\partial b_i$;

$$\mathbf{K}\frac{\partial \mathbf{u}}{\partial b_i} + \frac{\partial \mathbf{K}}{\partial b_i}\mathbf{u} = \frac{\partial \mathbf{P}}{\partial b_i} \tag{1}$$

where b_i is understood to represent the ith independent design variable. It should be remembered here that the stiffness matrix $\mathbf{K}$ depends upon the strain distribution in, and the numerical integration over, the domain of each finite element making up the analysis model. This means that in order to obtain the stiffness matrix, you have to differentiate the assumed displacement functions inherently associated with the finite element formulation. Also, it should be noted the matrix $\partial\mathbf{K}/\partial b_i$ must, in many instances, be approximated by finite difference calculation, that is, $\partial\mathbf{K}/\partial b_i \approx \Delta\mathbf{K}/\Delta b_i$. Perhaps the primary current objection to DSA, in the context of shape optimization, is that this feature is simply not an available option in commercially available finite element analysis codes. It is noted in passing that even the existing DSA for sizing, currently available in MSC/NASTRAN, is semianalytic in that it depends on a finite difference approximation of the matrix $\partial\mathbf{K}/\partial b_i$.

In the VDSA approach used by Haug and Choi (Chapter 2), Yang and Botkin (Chapter 3) and Choi and Seong (Chapter 13), behavior sensitivity analysis is carried out by implicit differentiation of the functionals of the variational formulation (of the appropriate continuum analysis problem) with respect to shape parameter type design variables. By letting change in a shape defining parameter be analogous to a change in time, change in position of material particles (making up a body) per unit change in the shape defining parameter becomes analogous to a velocity field. This leads to the idea of using the *material derivative* concept of continuum mechanics to obtain sensitivity analysis formulations in terms of boundary integrals or domain integrals (via the Gauss theorem). In the

VDSA approach, discretization of the analysis and variation with respect to the independent displacement degrees of freedom are carried out after the implicit differentiation of the functionals of the variational formulation with respect to the shape design variables. It may well be that the main advantage of the VDSA methods is that shape sensitivity information can be obtained using stress (and strain) recovery output that is generally available from existing commercial finite element analysis codes. However, it should be noted that the VDSA boundary integral method depends on the accuracy with which boundary stresses can be predicted as well as the boundary numerical integration scheme employed. Also, it is observed that the VDSA domain integral method used by Haug and Choi (Chapter 2) depends upon the accuracy of interior stresses due to the actual loads, the accuracy of interior strains due to the adjoint loads, and the domain numerical integration scheme employed.

In Yang and Botkin (Chapter 3), the influence on displacement behavior sensitivity of refining the finite element analysis mesh is examined. The plane stress example problem employed is shown in Figure 1 (p. 69) along with the mesh refinement scheme (eight noded isoparametric plane stress elements are used). It should be noted that (1) the sensitivity of the vertical displacement at point A to change in the depth dimension b is examined, and (2) the external load of 100 pounds is parabolically applied at the right-hand end of the thin plate. The results, shown in Figure 2 (p. 70), indicate that refinement of the mesh, which improves the accuracy of boundary stress information, increases the value of the sensitivity derivative obtained and shows that the value of the sensitivity derivative converges rapidly when using the DSA approach with either finite difference or analytic determination of $\partial \mathbf{K}/\partial b_i$. All of the results shown in Figure 2 on p. 70 depend upon the discretized analysis modeling.

Alternative methods for behavior sensitivity analysis, in the context of shape optimization, received a great deal of attention during this symposium. Nevertheless, it would appear that at this time we do not know for sure which is the best way to obtain behavior sensitivity derivatives with respect to shape design variables. Therefore, it is recommended that some comparative studies of alternative methods be undertaken for reference problems where a closed form solution of the analysis exists. These comparative studies would assess relative efficiency; accuracy would be measured using a closed form reference solution that is independent of analysis model discretization. We need to know more about how much error the various discretizations introduce (e.g., those associated with the finite element model, finite differencing of the stiffness matrix, numerical integrations of boundary and domain integrals) in the alternative behavior sensitivity solution methods (e.g., DSA analytic or finite difference, and VDSA boundary integral or domain integral), using an exact reference solution as a standard for comparison. In Figure 1 an obvious example taken from a standard elasticity text [3] is suggested. We can assume that (1) the edge $x = 0$ is subject to a loading $\tau_{xy}(0,y) = -(P/2I)[(d^2/4) - y^2]$; (2) the edge $x = L$ is subject to a loading $\tau_{xy}(L,y) = -(P/2I)[(d^2/4) - y^2]$ and $\sigma_x(L,y) = -(PL/I)y$; (3) the point $x = L$, $y = 0$ is fixed (i.e., $u = v = dv/dx = 0$ at $x = L$, $y = 0$). The

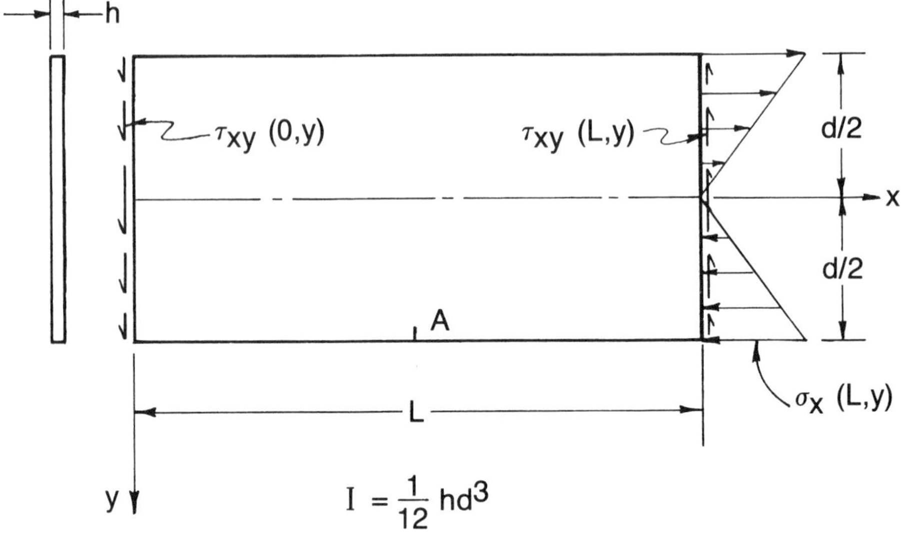

$$\sigma_x = -\frac{P\,x\,y}{I}; \quad \sigma_y = 0; \quad \tau_{xy} = -\frac{P}{2I}\left(\frac{d^2}{4} - y^2\right)$$

$$u = -\frac{P}{2EI}x^2y + \frac{P}{3EI}\left(1 + \frac{\nu}{2}\right)y^3 + \frac{P}{2EI}\left[L^2 - (1+\nu)\frac{d^2}{2}\right]y$$

$$v = \frac{\nu P}{2EI}xy^2 + \frac{P}{6EI}x^3 - \frac{PL^2}{2EI}x + \frac{PL^3}{3EI}.$$

Note: this assumes $u = v = \frac{dv}{dx} = 0$; at $(x = L,\ y = 0)$.

Figure 1. Proposed reference problem for behavior sensitivity analysis.

exact plane stress solution for the displacement and stress distribution is given by the equations in Figure 1, and behavior sensitivity derivatives for any pertinent response quantity can be obtained by closed form partial differentiation with respect to d or L. For example, the behavior sensitivity derivative for the vertical displacement at A (i.e., $x = L/2$, $y = +d/2$)

$$v(\frac{L}{2}, \frac{d}{2}) = \frac{3\nu PL}{4Edh} + \frac{5PL^3}{4Ehd^3} \tag{2}$$

with respect to the shape design variable d is

$$\frac{\partial v}{\partial d}(\frac{L}{2}, \frac{d}{2}) = -\frac{3\nu PL}{4Ehd^2} - \frac{15PL^3}{4Ehd^4}. \tag{3}$$

It is suggested that closed form solutions such as this be used to assess the accuracy of alternative behavior sensitivity analysis methods involving various discretizations and numerical integration schemes. When using VDSA methods to solve test problems of this kind, care must be taken to select an appropriate velocity field.

Before closing these remarks on behavior sensitivity analysis, there is one point I want to reemphasize. While at this time there remains some controversy over which method is the best for obtaining behavior sensitivity information, there is one sure conclusion. If you want to do shape optimization using an existing finite element code and you don't want to or can't break into that code, then the VDSA boundary integral and domain integral methods provide a ready tool for calculating behavior sensitivity derivatives.

OPTIMIZATION

During this symposium relatively little has been said about optimization algorithms per se, except for the presentation by Fleury (Chapter 12) which traces the emergence of CONLIN, a method of convex linearization, as an effective algorithmic tool for shape optimization. For the class of shape optimization problems dealt with during this symposium, the insightful use of natural design variables leads to design optimization problems with relatively small numbers of independent design variables, usually 10 to 30, and certainly less than 100 design variables. For these problems there already exists an adequate supply of robust optimization algorithms, such as those available in the ADS code.

Although there is an adequate supply of effective optimizers, they can be burdened and the number of analyses required for convergence can become prohibitively large, unless we remember to focus attention on solving a sequence of approximate problems rather than executing a complete analysis for every small change in the design. In this connection it is worth noting that for problems with 10 to 30 design variables, the most important part of the CONLIN approach presented by Fleury in Chapter 12 is the construction of robust approximations (via the use of hybrid or mixed variable approximations) that are convex, separable and

References p. 397

conservative. The subsequent generation of an explicit dual function, which is maximized subject to nonnegativity constraints on the dual variables, only contributes modest incremental gains in overall efficiency. This is because the computational burden, for most structural optimization problems, depends primarily on the number of times one must execute the complete analysis and the associated behavior sensitivity analysis.

Most of the papers presented during the symposium made effective use of temporary constraint deletion strategies and move limits. It was interesting to see that design variable side constraints and/or move limits have taken on new importance in the context of shape optimization, particularly in terms of trying to restrict optimization to the region of analysis validity in the design space. It may be useful to look more carefully at the selection of move limits. For example, more general but explicit relations might be developed to restrict the design search to the region of analysis validity. Also, for practical applications in an industrial setting, there are obvious advantages to be gained by using efficient algorithms that tend to generate a sequence of steadily improving feasible designs (i.e., funneling down the middle of the feasible region in design space). This does not necessarily mean the use of interior penalty functions; rather it urges that preference be given to any efficient algorithm exhibiting this desirable characteristic (e.g., CONLIN).

Under the heading of optimization, there is one last suggestion that I would like to offer. Rather than focusing all of our effort on weight minimization, some attention should be given to using optimization as a tool for finding balanced designs, i.e., a design that is satisfactory with respect to the weight budget, and which is also not critical with respect to any of the pertinent behavior constraints. Balanced designs that are uniformly buffered from the behavior constraints driving the design process can be expected to have a higher reliability. The notion here is to strive indirectly for designs with improved reliability, by using a deterministic formulation which yields a balanced design. This goal can be achieved by using the methods we already have available in a slightly different way. Specifically, instead of solving the familiar weight minimization problem

$$W(\mathbf{x}) \to \min \tag{4}$$

such that

$$G_j(\mathbf{x}) \geq 0; \quad j \in J \tag{5}$$

$$\mathbf{x}^\ell \leq \mathbf{x} \leq \mathbf{x}^u \tag{6}$$

attention is redirected toward solving an alternative problem

$$\beta \to \max \tag{7}$$

$$G_j(\mathbf{x}) \geq \beta; \quad j \in J \tag{8}$$

$$W(\mathbf{x}) \leq W_{\text{budget}} \tag{9}$$

$$\mathbf{x}^\ell \leq \mathbf{x} \leq \mathbf{x}^u \tag{10}$$

where a single additional variable β is maximized. In this problem all the constraints are buffered by at least an amount β, and the critical constraints are buffered uniformly by an amount β. By solving the optimization problem posed in equations (7)–(10), we are building in a uniform safety factor. The formulation given in equations (7)–(10) can also be modified so that buffering of individual constraints is tailored to reflect insights based on prior experience. This is accomplished by introducing preassigned weighting factors θ_j into equation (8) as follows:

$$G_j(\mathbf{x}) \geq \theta_j \beta; \qquad j \in J. \tag{8}$$

While the foregoing approach is certainly not a new idea, it is suggested that in practice more attention should be given to using modern structural optimization tools as a means of finding balanced designs.

CREATIVE CONTROL VIA INTERACTIVE GRAPHICS

It was clear from Botkin et al. (Chapter 9) and Fleury (Chapter 12) that interactive graphics will play an important role in helping the designer to make effective use of shape optimization tools. First-generation applications of interactive graphics for computer aided design provided an interface between analysis results and the designer. With the emergence of effective modern structural optimization tools, the use of interactive graphics is being shifted up to a higher level in the design process. In Chapter 9, Botkin and coauthors described a pilot preprocessor program for stating shape optimization problems. Such programs make it possible to use interactive graphics to describe a shape optimization design model in terms of few natural design variables. These preprocessors will also include design variable linking and preassigned parameter features. The design variable linking is not necessarily one-to-one; rather it will be capable of imposing some linear constraints between design variables so as to impose tangency requirements. The preassigned parameter feature can be used to fix any portion of the design. It is suggested that this feature may be particularly useful in separating out detail design problems around trouble spots. For example, one can imagine fixing the design around supports and load introduction areas and then using shape optimization tools to get a feel for what the overall system looks like. This could then be followed up with a very detailed treatment of the trouble spots. The foregoing use of the preassigned parameter feature points to the need for a multilevel approach in some types of design problems. The capability to describe applied loads and displacement boundary conditions in terms of design model parameters (rather than in terms of analysis model parameters) is another feature that one would like to have available in preprocessor programs.

It was apparent from Fleury (Chapter 12) that postprocessors for interactive graphic assessment of design optimization progress will play an increasingly important role. They will allow the designer to use shape optimization tools while maintaining creative control. For example, it should be relatively easy to implement postprocessors that will allow the designer to routinely look at a display of

References p. 397

the changing shape generated by the optimization program. This will allow the designer to keep track of how the design is progressing. In some instances it is easy to imagine that by observing the sequence of designs being generated, the designer may want to stop the program and redefine the design model by adding or deleting some design variables. It will also be useful to be able to display iteration histories online as they occur—not just objective function values, but the values of critical and near-critical constraints that are driving the design process.

Another very interesting use of interactive graphics was suggested by Fleury (Chapter 12), namely the display of two-dimensional design space slices. This idea is not only useful as a teaching tool, but it is also a potentially powerful means of dealing pragmatically with the relative minima problem. By graphically displaying the trace of constraint functions and the objective function on a particular two-dimensional subspace, one may be able to see relative minima that couldn't otherwise be discovered. This is important because convexity of the basic NLP design problem statement can only be proven for a few very special cases (see, for example, [4]). Therefore, almost all of the modern structural optimization methods developed in recent years must, strictly speaking, be regarded as tools for efficient automated design improvement.

Looking beyond shape optimization of structural components, it is apparent that there is a great deal more to be done. In Vanderplaats (Chapter 10), the applications of shape optimization in other disciplines (e.g., aerodynamics) were reviewed. It is important for structures people to encourage engineers in other disciplines to start doing more than just zero-order analysis. They should be urged to think about (1) first- and second-order behavior sensitivity analysis, (2) the selection of intermediate variables based on experience and physical insight, (3) the construction of high-quality explicit approximations for key response quantities and (4) the application of existing optimization algorithms to a sequence of explicit approximate problems. It will also be important for engineers working in other disciplines to keep in mind the distinction between design modeling and analysis modeling. They should be able to benefit from the experience of the structures community by (1) placing design modeling above analysis modeling in the formulative decision hierarchy, and (2) identifying natural design variables that are well matched to product definition from a fabrication and manufacturing point of view.

Looking to the future, we can expect to see many of the ideas discussed during this symposium extended to other disciplines and used for interdisciplinary optimization. Interdisciplinary design optimization will probably involve the use of multilevel methods, decomposition schemes and concurrent computing techniques such as those being actively explored by the IRO at the NASA Langley Research Center (see references [5] and [6]).

REFERENCES

1. J. Sobieszczanski-Sobieski, J. F. Barthelemy and K. M. Riley, Sensitivity of optimum solutions to problem parameters. *AIAA J.* **20** (9) 1291–1299 (Sept. 1982).
2. L. A. Schmit and K. J. Chang, Optimum design sensitivity based on approximation concepts and dual methods. *Int. J. Numer. Meth. Eng.* **20** 39–75 (1984).
3. C-T. Wang, *Applied Elasticity.* McGraw-Hill, 1953, pp. 46–49.
4. K. Svanberg, On local and global minima in structural optimization, *Proc. Int. Symposium on Optimum Structural Design*, pp. 6-3–6-10. Univ. of Arizona (Oct. 1981).
5. J. Sobieszczanski-Sobieski, A linear decomposition method for large optimization problems—Blueprint for development. NASA TM-83248. (Feb. 1982).
6. J. Sobieski, Multidisciplinary systems optimization by linear decomposition. *Proc. NASA Symposium on Recent Experiences in Multidisciplinary Analysis and Optimization-Part 1*, pp. 343–66. NASA CP-2327 (Apr. 1984).

L. A. Schmit

SYMPOSIUM PARTICIPANTS

Abel, J.
Cornell University
Ithaca, NY 14853

Anderson, W. J.
The University of Michigan
Ann Arbor, MI 48109

Arora, J. S.
University of Iowa
Iowa City, IA 52242

Atrek, E.
Engineering Mechanics Research Corp.
Troy, MI 48099

Babuška, I.
University of Maryland
College Park, MD 20742

Barone, M. R.
General Motors Research Laboratories
Warren, MI 48090-9055

Barthelemy, B.
Virginia Polytechnic Institute and
State University
Blacksburg, VA 24061

Beers, M.
University of California—Santa Barbara
Santa Barbara, CA 93106

Belegundu, A. D.
Pennsylvania State University
University Park, PA 16802

Benedict, R. L.
Goodyear Tire & Rubber Company
Akron, OH 44316

Bennett, J. A.
General Motors Research Laboratories
Warren, MI 48090-9055

Berke, L.
NASA Lewis Research Center
Cleveland, OH 44135

Bernhard, R. J.
Purdue University
West Lafayette, IN 47907

Blaser, D. A.
General Motors Research Laboratories
Warren, MI 48090-9055

Botkin, M. E.
General Motors Research Laboratories
Warren, MI 48090-9055

Boyse, J. W.
General Motors Research Laboratories
Warren, MI 48090-9055

Brown, K. W.
Pratt & Whitney Aircraft
East Hartford, CT 06108

Burns, S.
University of Illinois
Urbana, IL 61801

Buzan, L. R.
General Motors Research Laboratories
Warren, MI 48090-9055

Cannon, D.
Sverdrup Technology, Inc.
Tullahoma, TN 37388

Cavendish, J. C.
General Motors Research Laboratories
Warren, MI 48090-9055

Chang, D. C.
CPC Group
General Motors Corporation
Pontiac, MI 48058-1493

Cheng, F. Y.
University of Missouri - Rolla
Rolla, MO 65401

Chian, C-T.
Jet Propulsion Laboratory
Pasadena, CA 91109

Choi, K. K.
University of Iowa
Iowa City, IA 52242

Chon, C. T.
Ford Motor Company
Dearborn, MI 48121

Chun, Y. W.
Villanova University
Villanova, PA 19085

Chung, K.
The University of Michigan
Ann Arbor, MI 48109

Consentino, G. B.
NASA Ames Research Center
Moffet Field, CA 94035

Cox, J.
Garret Turbine Engine Company
Phoenix, AZ 85010

Davis, M. W.
United Technologies Research Center
East Hartford, CN 06108

DeVries, R.
Ford Motor Company
Dearborn, MI 48121

Dodd, G. G.
General Motors Research Laboratories
Warren, MI 48090-9055

Dopker, B.
University of Iowa
Iowa City, IA 52242

Eason, E. D.
Failure Analysis Associates
Palo Alto, CA 94303

Eisley, J. G.
The University of Michigan
Ann Arbor, MI 48109

Esselink, L.
General Motors Current Engineering and Manufacturing Services Staff
Warren, MI 48090-9010

Fenyes, P. A.
Cornell University
Ithaca, NY 14853

Field, D. A.
General Motors Research Laboratories
Warren, MI 48090-9055

Fleury, C.
University of California—Los Angeles
Los Angeles, CA 90024

Frey, W. H.
General Motors Research Laboratories
Warren, MI 48090-9055

Friedman, D.
Computervision Corporation
Bedford, MA 01730

Frosch, R. A.
General Motors Research Laboratories
Warren, MI 48090-9055

Gallagher, R. H.
Worcester Polytechnic Institute
Worcester, MA 01609

Grandhi, R. V.
Wright State University
Dayton, OH 45435

Grierson, D. E.
University of Waterloo
Waterloo, Ontario, Canada

Haber, R.
University of Illinois
Urbana, IL 61801

Haftka, R. T.
Virginia Polytechnic Institute and State University
Blacksburg, VA 24061

Haug, E. J.
University of Iowa
Iowa City, IA 52242

Holzwarth, J. C.
General Motors Research Laboratories
Warren, MI 48090-9055

Hou, J. W.
Old Dominion University
Norfolk, VA 23508

Howell, L. J.
General Motors Research Laboratories
Warren, MI 48090-9055

Hurst, T.
Brigham Young University
Provo, UT 84602

Izadpanah, K.
The MacNeal-Schwendler Corporation
Los Angeles, CA 90041

Johnston, L. E.
Electronic Data Systems
Warren, MI 48090-9010

Kamal, M. M.
General Motors Research Laboratories
Warren, MI 48090-9055

Kamel, H. A.
University of Arizona
Tucson, AZ 85721

Kane, J.
University of Bridgeport
Bridgeport, CT 06602

Katnik, R. B.
CPC Group
General Motors Corporation
Warren, MI 48090-9060

Katz, S.
General Motors Research Laboratories
Warren, MI 48090-9055

Kikuchi, N.
The University of Michigan
Ann Arbor, MI 48109

Kirsch, U.
Technion - Israel Institute of Technology
Haifa, 32000, Israel

Kline, K. A.
Wayne State University
Detroit, MI 48202

Klosterman, A. L.
Structural Dynamics Research Corp.
Milford, OH 45150

Kohn, R.
Courant Institute of Mathematics
New York, NY 10012

Kothawala, K. S.
Engineering Mechanics Research Corp.
Troy, MI 48099

Kumar, V.
General Electric Company
Schenectady, NY 12301

Landefeld, C. F.
General Motors Research Laboratories
Warren, MI 48090-9055

Leucht, P. M.
General Motors Truck & Bus Group
Pontiac, MI 48053

Levy, R.
Jet Propulsion Laboratory
Pasadena, CA 91103

Lust, R. V.
General Motors Research Laboratories
Warren, MI 48090-9055

Magee, C.
Ford Motor Company
Dearborn, MI 48121

Manning, R.
University of California—Los Angeles
Los Angeles, CA 90024

McDonald, R. J.
General Motors Research Laboratories
Warren, MI 48090-9055

Miura, H.
NASA Ames Research Center
Moffet Field, CA 94035

Mohanty, B. P.
Allison Gas Turbines Division
General Motors Corporation
Indianapolis, IN 46206-0420

Mota Soares, C. A.
Centro de Mecanica e Materiais
da Universidade Tecnica de Lisboa
1096 Lisboa, Portugal

Mróz, Z.
Polish Academy of Science
00-049 Warsaw, Poland

Muench, N. L.
General Motors Research Laboratories
Warren, MI 48090-9055

Nack, W.
General Motors Current Engineering
and Manufacturing Services Staff
Warren, MI 48090-9010

Narayanaswami, R.
Computerized Structural Analysis
and Research Corporation
Northridge, CA 91324

Nelson, M. F.
General Motors Advanced Concepts Center
Newbury Park, CA 91320

Oh, K. P.
General Motors Research Laboratories
Troy, MI 48084

Parkinson, A.
Brigham Young University
Provo, UT 84602

Prasad, B.
Electronic Data Systems
Warren, MI 48090

Pratt, T. K.
Pratt & Whitney Aircraft
East Harford, CT 06108

Rai, N. S.
CPC Group
General Motors Corporation
Warren, MI 48090-9060

Rajan, S. D.
Arizona State University
Tempe, AZ 85287

Rosalik, J. J.
BOC Group
General Motors Corporation
Detroit, MI 48232

Rouse, N.
Machine Design Magazine
Cleveland, OH 44114

Saigal, S.
Purdue University
West Lafayette, IN 47907

Sandgren, E.
University of Missouri
Columbia, MO 65201

Sarigul, N.
Ohio State University
Columbus, OH 43210

Schmit, L. A.
University of California—Los Angeles
Los Angeles, CA 90024

Shephard, M. S.
Rensselaer Polytechnic Institute
Troy, NY 12181

Sobieski, J.
NASA Langley Research Center
Hampton, VA 23665

Song, J. O.
General Motors Research Laboratories
Warren, MI 48090-9055

Spewock, N. A.
Electronic Data Systems
Warren, MI 48090

Stanton, E. L.
PDA Engineering
Santa Ana, CA 92705-5475

Steinetz, B. M.
NASA Lewis Research Center
Cleveland, OH 44135

Striz, F.
University of Oklahoma
Norman, OK 73019

Stroud, W. J.
NASA Langley Research Center
Hampton, VA 23665

Suh, C.
Advanced Engineering Staff
General Motors Corporation
Warren, MI 48090-9010

Swanson, J. A.
Swanson Analysis Systems, Inc.
Houston, PA 15342-0065

Szabo, B. A.
Washington University
St. Louis, MO 63130

Taylor, J.
The University of Michigan
Ann Arbor, MI 48109

Thompson, E. G.
Colorado State University
Fort Collins, CO 80523

Torigaki, T.
The University of Michigan
Ann Arbor, MI 48109

Trompette, Ph.
French National Institute of Science
69621 Villeurbanne, France

Tuesday, C. S.
General Motors Research Laboratories
Warren, MI 48090-9055

Vanderplaats, G. N.
University of California—Santa Barbara
Santa Barbara, CA 93107

Vasilopoulos, D.
General Motors Research Laboratories
Warren, MI 48090-9055

Vella, J.
CPC Group
General Motors Corporation
Pontiac, MI 48058-1493

Venkayya, V. B.
Wright-Patterson AFB
Dayton, OH 45433

Vogt, J. M.
CPC Group
General Motors Corporation
Pontiac, MI 48058-1493

Wang, B. P.
University of Texas
Arlington, TX 76019

Wilde, D. J.
Stanford University
Stanford, CA 94305

Yerry, M.
Aries Technology Incorporated
Lowell, MA 01854

Zienkiewicz, O. C.
University of Wales—Swansea
Swansea SA2 8PP, United Kingdom

Wilson, C.
The MacNeal-Schwendler Corporation
Los Angeles, CA 90041

Yang, R. J.
General Motors Research Laboratories
Warren, MI 48090-9055

AUTHOR AND CONTRIBUTOR INDEX

The page numbers for bibliographic citations appear in italics

SUBJECT INDEX